Lecture Notes in Applied and Computational Mechanics

Volume 60

This series aims to report new developments in applied and computational mechanics—quickly, informally and at a high level. This includes the fields of fluid, solid and structural mechanics, dynamics and control, and related disciplines. The applied methods can be of analytical, numerical and computational nature.

More information about this series at http://www.springer.com/series/4623

Tarek I. Zohdi

Modeling and Simulation of Functionalized Materials for Additive Manufacturing and 3D Printing: Continuous and Discrete Media

Continuum and Discrete Element Methods

Springer

Tarek I. Zohdi
University of California
Berkeley, CA
USA

ISSN 1613-7736 ISSN 1860-0816 (electronic)
Lecture Notes in Applied and Computational Mechanics
ISBN 978-3-319-70077-9 ISBN 978-3-319-70079-3 (eBook)
https://doi.org/10.1007/978-3-319-70079-3

Library of Congress Control Number: 2017957665

Printed on acid-free paper

This Springer imprint is published by Springer Nature
The registered company is Springer International Publishing AG
The registered company address is: Gewerbestrasse 11, 6330 Cham, Switzerland

To my patient and loving wife, Britta Schöenfelder-Zohdi and to the memory of my close friend, colleague and mentor, David Dornfeld.

Preface

Within the last decade, several industrialized countries have stressed the importance of advanced manufacturing to their economies. Many of these plans have highlighted the development of additive manufacturing techniques, such as 3D printing which, as of 2018, are still in their infancy. The objective is to develop superior products, produced at lower overall operational costs. For these goals to be realized, a deep understanding of the essential ingredients comprising the materials involved in additive manufacturing is needed. The combination of rigorous material modeling theories coupled with the dramatic increase of computational power can potentially play a significant role in the analysis, control, and design of many emerging additive manufacturing processes. Specialized materials and the precise design of their properties are key factors in these processes. Specifically, particle-functionalized materials play a central role in this field, in three main regimes:

- (1) To enhance overall filament-based material properties, by embedding particles within a binder, which is then passed through a heating element and deposited onto a surface,
- (2) To "functionalize" inks by adding particles to freely flowing solvents forming a mixture, which is then deposited onto a surface, and
- (3) To directly deposit particles, as dry powders, onto surfaces and then to heat them with a laser, e-beam, or other external sources, in order to fuse them into place.

The goal of these processes is primarily to build surface structures which are extremely difficult to construct using classical manufacturing methods. The objective of this monograph is to introduce the readers to basic techniques which can allow them to rapidly develop and analyze particulate-based materials needed in such additive manufacturing processes.

This monograph is broken into two main parts: "Continuum Method" (CM) approaches and "Discrete Element Method" (DEM) approaches. The materials associated with methods (1) and (2) are closely related types of continua (particles embedded in a continuous binder) and are treated using continuum approaches. The materials in method (3), which are of a discrete particulate character, are analyzed using discrete element methods. I am certain that, despite painstaking efforts, there remain errors of one sort or another in this monograph. Therefore, I would be grateful if readers who find such flaws could contact me at *zohdi@berkeley.edu*.

Berkeley, USA
September 2017

Tarek I. Zohdi

Contents

List of Figures

Chapter 1
Introduction: Additive/3D Printing Materials—Filaments, Functionalized Inks, and Powders

Additive manufacturing (AM) is usually defined as the process of joining materials to make objects from 3D model data, typically layer upon layer, as opposed to subtractive manufacturing methodologies, which remove material (American Society for Testing and Materials, ASTM). We refer the reader to the recent overview article by Huang et al. [1] on the wide array of activities in the manufacturing community in this area. One subclass of AM, so-called 3D printing (3DP), has received a great deal of attention over the last few years. Typically, such a process takes CAD drawings and slices them into layers, printing layer by layer. 3DP was pioneered by Hull [2] of the 3D Systems Corporation in 1984. 3DP was a 2.2 billion dollar industry in 2014, with applications ranging from motor vehicles, consumer products, medical devices, military hardware, and the arts.

A key ingredient of these processes is the specialized materials and the precise design of their properties, enabled by the use of fine-scale "functionalizing" particles. The rapid rise in the use of particle-based materials has been made possible by the large-scale production of consistent, high-quality particles, which are produced in a variety of ways, such as: (a) sublimation from a raw solid to a gas, which condenses into particles that are recaptured (harvested), (b) atomization of liquid streams into droplets by breaking jets of metal, (c) reduction of metal oxides, and (d) comminution/pulverizing of bulk material. As mentioned in the preface, particle-functionalized materials play a central role in this field, in three main ways:

(1) To enhance overall filament-based material properties, by embedding particles within a binder, which is then passed through a heating element and deposited onto a surface,
(2) To "functionalize" inks by adding particles to freely flowing solvents forming a mixture, which is then deposited onto a surface, and
(3) To directly deposit particles, as dry powders, onto surfaces and then to heat them with a laser, e-beam, or other external source, in order to fuse them into place.

T. I. Zohdi, *Modeling and Simulation of Functionalized Materials for Additive Manufacturing and 3D Printing: Continuous and Discrete Media*, Lecture Notes in Applied and Computational Mechanics 60, https://doi.org/10.1007/978-3-319-70079-3_1

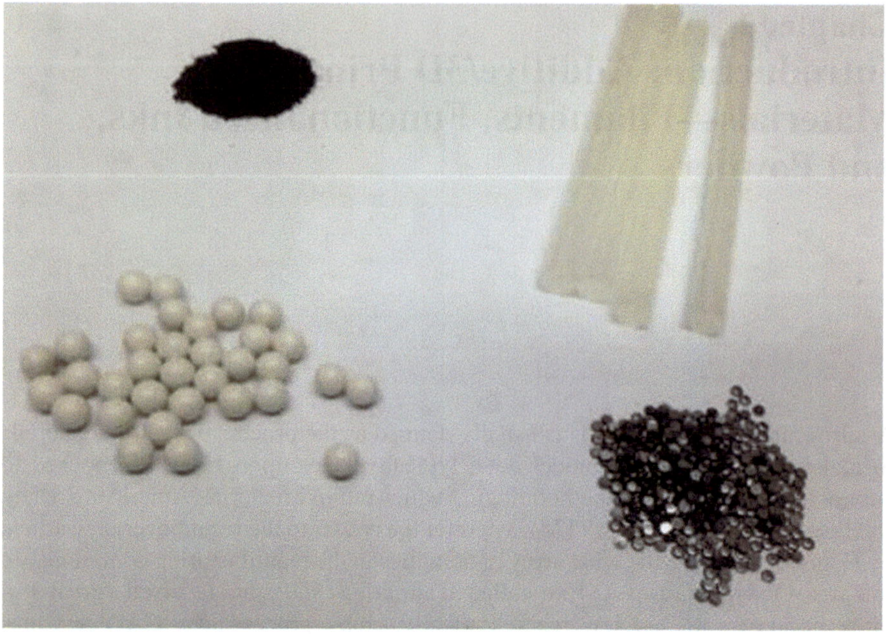

Fig. 1.1 Typical printing ingredients: top left: finely ground metallic powder (iron). Top right: extruded PLA. Bottom left: ABS pellets and bottom right: coarsely ground steel flakes

In more detail, we have (see Fig. 1.1):

- **Heated filament-based materials** (historically for prototyping) are comprised of thermoplastics. To extend the materials to applications beyond prototyping, second-phase particles are added to the heated mixture which solidify (cure) to form the overall material properties comprised of particles in a binding matrix when deposited onto a substrate. The particles are used to "tune" the binding matrix properties to the desired overall state. Specifically, much of the commercial additive manufacturing processes are polymer-based, with second-phase particles added to enhance the properties of the binder, which is typically either (1) polylactic acid or polylactide (PLA), which is a biodegradable thermoplastic aliphatic polyester or (2) acrylonitrile butadiene styrene (ABS) which is a common thermoplastic polymer. In 2015, PLA had the second highest consumption volume of any bioplastic of the world. PLA is derived from renewable resources, such as plants (corn starch, sugarcane, etc.). ABS is a terpolymer that is significantly stronger than PLA. It is made by polymerizing styrene and acrylonitrile in the presence of polybutadiene. The styrene gives the plastic a reflective surface, while the rubbery polybutadiene endows toughness. The overall properties are created by rubber toughening, where fine particles of elastomer are distributed throughout the rigid matrix. Typically, metal and ceramic particles are also added to endow specific mechanical, thermal, electrical, and magnetic effective overall properties.

- **Functionalized ink materials** (primarily for printed electronics) are comprised of particles in a solvent/lubricant which cure when deposited. Oftentimes, these inks are used to lay down electric circuit lines or to have some other specific electromagnetic function on a surface. One application where such functionalized inks are important is printed electronics on flexible foundational substrates, such as flexible solar cells and smart electronics. One important technological obstacle is to develop inexpensive, durable electronic material units that reside on flexible platforms or substrates which can be easily deployed onto large surface areas. Ink-based printing methods involving particles are, in theory, ideal for large-scale electronic applications and provide a framework for assembling electronic circuits by mounting printed electronic devices on flexible plastic substrates, such as polyimide and "PEEK" (polyether ether ketone, a flexible thermoplastic polymer) film. There are many variants of this type of technology, which is sometimes referred to as flexible electronics or flex circuits. Flex circuits can be, for example, screen-printed silver circuits on polyester. For an early history of the printed electronics field, see Gamota [3]. In order to develop flexible micro-/nanoelectronics for large area deployment, traditional methods of fabrication using silicon-based approaches have become limited for applications that involve large area coverage, due to high cost of materials and equipment (which frequently need a vacuum environment). For flexibility and lower cost, the ability to develop these electronics on plastics is necessary. To accomplish this task, print-based technologies are starting to become popular for these applications. In many cases, this requires the development of nanoparticle-functionalized "inks." These nanoparticles include germanium (which has higher mobility and better tailorable absorption spectrum for ambient light than silicon) and silver (which is being studied due to the possibility to sinter the particles without the need of directly applied intense heating). Other semiconductor nanoparticles, including zinc- and cadmium-based compounds and metals, such as gold and copper, can be considered. Precise patterning of (nanoparticle-functionalized) prints is critical for a number of different applications. For example, some recent applications include optical coatings and photonics (Nakanishi et al. [4]), MEMS applications (Fuller et al. [5], Samarasinghe et al. [6], and Gamota et al. [3]), and biomedical devices (Ahmad et al. [7]). In terms of processing techniques, we refer the reader to Sirringhaus et al. [8], Wang et al. [9], Huang et al. [10], Choi et al. [11–14], and Demko et al. [36, 37] for details.[1] We further mention that electromagnetically sensitive fluids are typically constructed ("functionalized") by embedding charged or electromagnetically sensitive particles in a neutral fluid. Such fluids date back, at least, to Winslow [19, 20] in 1947. While the most widely used class of such fluids are electrorheological fluids, which are comprised of extremely fine suspensions of charged particles (on the order of 50 microns) in an electrically neutral fluid, there has been a renewed interest in this class of materials because of so-called e-inks (electrically func-

[1] For reviews of optical coatings and photonics, see Nakanishi et al. [4] and Maier and Atwater [15], for biosensors, see Alivisatos [16], for catalysts, see Haruta [17], and for MEMS applications, see Fuller et al. [5] and Ho et al. [18].

tionalized inks) driven by printed electronics. Inkjet printing is attractive due to its simplicity, high throughput, and low material loss. However, patterning with inkjet printing is limited to a resolution of around 20–50 μm with current printers (Ridley et al. [21]) with higher resolution possible by adding complexity to the substrate prior to printing (Wang et al. [9]). Electrohydrodynamic printing has also been proposed to increase the resolution beyond the limits of inkjet printing, achieving a line resolution as small as 700 nm (Park et al. [22]).

- **Dry powder-based materials** (primarily for sintered load-bearing structures) are deposited onto a surface and then heated by a laser, e-beam, or other external source, in order to fuse them into place. These types of applications and associated technology are closely related to those in the area of spray coatings, and we refer the reader to the extensive works of Sevostianov and Kachanov [23–25], Nakamura and coworkers: Dwivedi et al. [26], Liu et al. [27, 28], Nakamura and Liu [29], Nakamura et al. [30] and Qian et al. [31] and to Martin [32, 33] for the state of the art in deposition technologies. In powder-based processes, after deposition, laser processing is applied to heat particles in a powder to desired temperatures to either subsequently soften, sinter, melt or ablate them. Selective laser sintering was pioneered by Householder [34] in 1979 and Deckard and Beaman [35] in the mid-1980s.[2] Laser-based heating is quite attractive because of the degree of targeted precision that it affords.[3] Because of the monochromatic and collimated nature of lasers, they are a highly controllable way to process powdered materials, in particular with pulsing, via continuous beam chopping or modulation of the voltage. Carbon dioxide (CO_2) and yttrium aluminum garnet (YAG) lasers are commonly used. The range of power of a typical industrial laser is relatively wide, ranging from approximately 100–10000 W. Typically, the initial beam produced is in the form of collimated (parallel) rays, which are then focused with a lens onto a small focal point as fine as 0.00001 m in diameter. *However,* a chief concern of manufacturers are residual stresses and the microstructural defects generated in additively manufactured products, created by imprecisely controlled heat-affected zones, brought on by miscalibration of the laser power needed for a specific goal. In particular, because many substrates can become thermally damaged, for example, from thermal stresses, ascertaining the appropriate amount of laser input is critical.

[2]A closely related method, electron beam melting, fully melts the material and produces dense solids that are void-free.

[3]There are a variety of other techniques that may be involved in an overall additive manufacturing processes, such as: (a) electron beam melting, which is a process by where powder is bonded together layer per layer with an electron beam in a high vacuum, (b) aerosol jetting, which consists of utilizing streams of atomized particles at high velocities toward a substrate, and (c) inkjet printing, which works by projecting small droplets of ink toward a substrate through a small orifice by pressure, heat, and vibration. The deposited material is then heated by UV light or other means to rapidly dry.

1.1 Objectives

In order for *emerging additive manufacturing approaches to succeed, such as the ones mentioned, one must draw upon rigorous theory and computation to guide and simultaneously develop design rules for the proper selection of particle, binder, and solvent combinations for upscaling to industrial manufacturing levels (Fig.* 1.2*). This motivates the content of this monograph.* This monograph is broken into two main methodologies: "Continuum Method" (CM) approaches and "Discrete Element Method" (DEM) approaches. The materials associated with heated filament and functionalized-ink methods are closely related types of continua (particles embedded in a continuous binder) and are analyzed using continuum approaches. The dry powder materials, which are of a discrete particulate character, are analyzed using discrete element methods for the deposition phase of the analysis, and continuum approaches are used for the curing (cooling) stress analysis. This monograph seeks to introduce the reader to some of the main approaches for modeling and simulation of particle-based materials used in additive manufacturing, namely:

- Basic continuum mechanics,
- Continuum characterization of particle-functionalized materials,
- Continuum properties of mixtures and optimization,
- CM approaches for ascertaining time-transient thermo-mechanical responses, residual stresses, and laser processing,
- DEM approaches for modeling the deposition of dry powders,
- DEM approaches for modeling laser–particle interaction, and
- DEM approaches for modeling of advanced processing and the associated multi-physical effects.

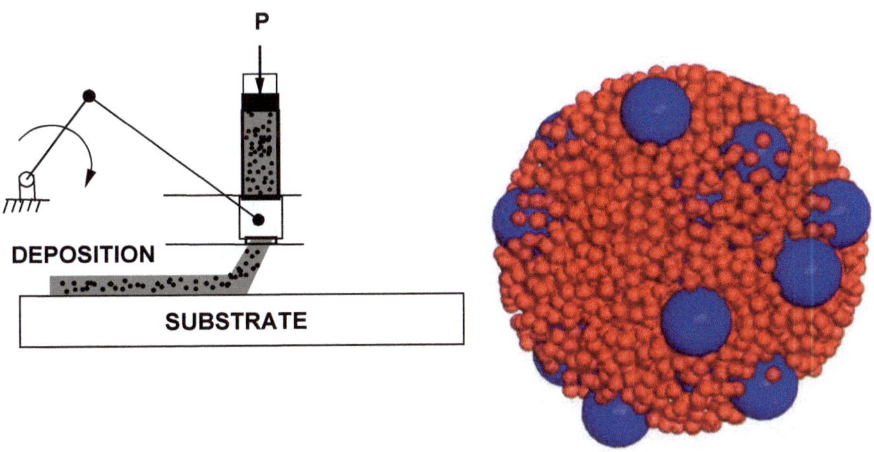

Fig. 1.2 Left: a linkage schematic of a 3D printer. Right: a multiphase droplet representation using the Discrete Element Method

In addition to appendices within the chapters themselves (labeled "Chapter Appendices"), background material is also included in the "Monograph Appendices" on the following related topics:

- Monograph Appendix 1: A review of essential mathematics,
- Monograph Appendix 2: Continuum electrical properties of mixtures,
- Monograph Appendix 3: Continuum properties of multiphase mixtures,
- Monograph Appendix 4: Continuum fluid properties of mixtures, and
- Monograph Appendix 5: Combining DEM and continuum approaches.

References

1. Huang, Y., Leu, M.C., Mazumdar, J., Donmez, A.: Additive manufacturing: current state, future potential, gaps and needs, and recommendation. J. Manufact. Sci. Eng **137**, 014001–1 (2015)
2. Hull, C.: Apparatus for production of three-dimensional objects by stereolithography. U.S. Patent 4,575,330, (1984)
3. Gamota, D., Brazis, P., Kalyanasundaram, K., Zhang, J.: Printed Organic and Molecular Electronics. Kluwer Academic Publishers, New York (2004)
4. Nakanishi, H., Bishop, K.J.M., Kowalczyk, B., Nitzan, A., Weiss, E.A., Tretiakov, K.V., Apodaca, M.M., Klajn, R., Stoddart, J.F., Grzybowski, B.A.: Photoconductance and inverse photoconductance in thin films of functionalized metal nanoparticles. Nature **460**, 371–375 (2009)
5. Fuller, S.B., Wilhelm, E.J., Jacobson, J.M.: Ink-jet printed nanoparticle microelectromechanical systems. J. Microelectromech. Syst. **11**, 54–60 (2002)
6. Samarasinghe, S.R., Pastoriza-Santos, I., Edirisinghe, M.J., Reece, M.J., Liz-Marzan, L.M.: Printing Gold Nanoparticles with an Electrohydrodynamic Direct Write Device. Gold Bulletin. **39**, 48–53 (2006)
7. Ahmad, Z., Rasekh, M., Edirisinghe, M.: Electrohydrodynamic direct writing of biomedical polymers and composites. Macromol. Mater. Eng. **295**, 315–319 (2010)
8. Sirringhaus, H., Kawase, T., Friend, R.H., Shimoda, T., Inbasekaran, M., Wu, W., Woo, E.P.: High-resolution inkjet printing of all-polymer transistor circuits. Science **290**, 2123–2126 (2000)
9. Wang, J.Z., Zheng, Z.H., Li, H.W., Huck, W.T.S., Sirringhaus, H.: Dewetting of conducting polymer inkjet droplets on patterned surfaces. Nat. Mater. **3**, 171–176 (2004)
10. Huang, D., Liao, F., Molesa, S., Redinger, D., Subramanian, V.: Plastic-compatible low-resistance printable gold nanoparticle conductors for flexible electronics. J. Electrochem. Soc. **150**(7), G412–417 (2003)
11. Choi, S., Park, I., Hao, Z., Holman, H.Y., Pisano, A.P., Zohdi, T.I.: Ultra-fast self-assembly of micro-scale particles by open channel flow. Langmuir **26**(7), 4661–4667 (2010)
12. Choi, S., Stassi, S., Pisano, A.P., Zohdi, T.I.: Coffee-ring effect-based three dimensional patterning of micro, nanoparticle assembly with a single droplet. Langmuir **26**(14), 11690–11698 (2010)
13. Choi, S., Jamshidi, A., Seok, T.J., Zohdi, T.I., Wu., M.C., Pisano, A.P.: Fast, High-throughput creation of size-tunable micro, nanoparticle clusters via evaporative self-assembly in picoliter-scale droplets of particle suspension. Langmuir **28**(6), 3102–11 (2012)
14. Choi, S., Pisano, A.P., Zohdi, T.I.: An Analysis of Evaporative Self-Assembly of Micro Particles in Printed Picoliter Suspension Droplets. J. Thin Solid Films **537**(30), 180–189 (2013)
15. Maier, S.A., Atwater, H.A.: Plasmonics: localization and guiding of electromagnetic energy in metal/dielectric structures. J. Appl. Phys. **2005**(98), 011101 (2005)
16. Alivisatos, P.: The use of nanocrystals in biological detection. Nat. Biotechnol. **22**(1), 47–52 (2004)

17. Haruta, M.: Catalysis of gold nanoparticles deposited on metal oxides. Cattech **6**(3), 102–115 (2002)
18. Ho, C., Steingart, D., Salminent, J., Sin, W., Rantala, T., Evans, J., Wright, P.: Dispenser printed electrochemical capacitors for power management of millimeter scale lithium ion polymer microbatteries for wireless sensors. In: 6th International Workshop on Micro and Nanotechnology for Power Generation and Energy Conversion Applications (PowerMEMS 2006), Berkeley, CA (2006)
19. Winslow, W.M.: Method and means for translating electrical impulses into mechanical force. U.S. Patent 2,417,850, (1947)
20. Winslow, W.M.: Induced fibration of suspensions. J. Appl. Phys. **20**(12), 1137–1140 (1949)
21. Ridley, B.A., Nivi, B., Jacobson, J.M.: All-inorganic field effect transistors fabricated by printing. Science **286**, 746–749 (1999)
22. Park, J.-U., Hardy, M., Kang, S.J., Barton, K., Adair, K., Mukhopadhyay, D.K., Lee, C.Y., Strano, M.S., Alleyne, A.G., Georgiadis, J.G., Ferreira, P.M., Rogers, J.A.: High-resolution electrohydrodynamic jet printing. Nat. Mater. **6**, 782–789 (2007)
23. Sevostianov, I., Kachanov, M.: Modeling of the anisotropic elastic properties of plasma-sprayed coatings in relation to their microstructure. Acta Mater. **48**(6), 1361–1370 (2000)
24. Sevostianov, I., Kachanov, M.: Thermal conductivity of plasma sprayed coatings in relation to their microstructure. J. Therm. Spray Technol. **9**(4), 478–482 (2001)
25. Sevostianov, I., Kachanov, M.: Plasma-sprayed ceramic coatings: anisotropic elastic and conductive properties in relation to the microstructure; cross-property correlations. Mater. Sci. Eng.-A **297**, 235–243 (2001)
26. Dwivedi, G., Wentz, T., Sampath, S., Nakamura, T.: Assessing process and coating reliability through monitoring of process and design relevant coating properties. J. Therm. Spray Technol. **19**, 695–712 (2010)
27. Liu, Y., Nakamura, T., Dwivedi, G., Valarezo, A., Sampath, S.: Anelastic behavior of plasma sprayed zirconia coatings. J. Am. Ceram. Soc. **91**, 4036–4043 (2008)
28. Liu, Y., Nakamura, T., Srinivasan, V., Vaidya, A., Gouldstone, A., Sampath, S.: Nonlinear elastic properties of plasma sprayed zirconia coatings and associated relationships to processing conditions. Acta mater. **55**, 4667–4678 (2007)
29. Nakamura, T., Liu, Y.: Determination of nonlinear properties of thermal sprayed ceramic coatings via inverse analysis. Int. J. Solids Struct. **44**, 1990–2009 (2007)
30. Nakamura, T., Qian, G., Berndt, C.C.: Effects of pores on mechanical properties of plasma sprayed ceramic coatings. J. Am. Ceram. Soc. **83**, 578–584 (2000)
31. Qian, G., Nakamura, T., Berndt, C.C.: Effects of thermal gradient and residual stresses on thermal barrier coating fracture. Mech. Mater. **27**, 91–110 (1998)
32. Martin, P.: Handbook of deposition technologies for films and coatings. 3rd (Ed.) Elsevier (2009)
33. Martin, P.: Introduction to surface engineering and functionally engineered materials. Scrivener and Elsevier (2011)
34. Householder, R.: Molding Process. U.S. Patent 4,247,508, (1979)
35. Deckard, C.: Method and apparatus for producing parts by selective sinterin. U.S. Patent 4,863,538, (1986)
36. Demko, M., Choi, S., Zohdi, T.I., Pisano, A.P.: High resolution patterning of nanoparticles by evaporative self-assembly enabled by in-situ creation and mechanical lift-off of a polymer template. Appl. Phys. Lett. **99**(25), 253102-1–253102-3 (2012)
37. Demko, M.T., Cheng, J.C., Pisano, A.P.: High-resolution direct patterning of gold nanoparticles by the microfluidic molding process. Langmuir 412–417 (2010)

Chapter 2
Continuum Methods (CM): Basic Continuum Mechanics

2.1 Notation

Throughout this work, boldface symbols denote vectors or tensors. Furthermore, we exclusively employ a Cartesian basis. For the inner product of two vectors (first-order tensors), $\boldsymbol{u}$ and $\boldsymbol{v}$, we have $\boldsymbol{u} \cdot \boldsymbol{v} = u_i v_i = u_1 v_1 + u_2 v_2 + u_3 v_3$ in three dimensions, where a Cartesian basis and Einstein index summation notation are used. In this introduction, for clarity of presentation, *we will ignore the difference between second-order tensors and matrices*. Accordingly, if we consider the second-order tensor $\boldsymbol{A} = A_{ik}\, \boldsymbol{e}_i \otimes \boldsymbol{e}_k$, then a first-order contraction (inner product) of two second-order tensors $\boldsymbol{A} \cdot \boldsymbol{B}$ is defined by the matrix product $[\boldsymbol{A}][\boldsymbol{B}]$, with components of $A_{ij} B_{jk} = C_{ik}$. It is clear that the range of the inner index j must be the same for $[\boldsymbol{A}]$ and $[\boldsymbol{B}]$. For three dimensions, we have $i, j = 1, 2, 3$. The inner product of a tensor (matrix) with a vector is defined as $\boldsymbol{A} \cdot \boldsymbol{v} = A_{ij} v_j$. The second-order inner (scalar) product of two tensors (matrices) is defined as $\boldsymbol{A} : \boldsymbol{B} = A_{ij} B_{ij} = tr([\boldsymbol{A}]^T [\boldsymbol{B}])$. Monograph Appendix 1 provides a basic mathematical review.

2.2 Kinematics of Deformations

In this chapter, we synopsize a more detailed discussion found in Zohdi and Wriggers [1]. The term deformation refers to a change in the shape of a continuum between a reference configuration and current configuration. In the reference configuration, a representative particle of a continuum occupies a point P in space and has the position vector (Fig. 2.1)

$$\boldsymbol{X} = X_1 \boldsymbol{e}_1 + X_2 \boldsymbol{e}_2 + X_3 \boldsymbol{e}_3, \tag{2.1}$$

where $\boldsymbol{e}_1, \boldsymbol{e}_2, \boldsymbol{e}_3$ is a Cartesian reference triad, and X_1, X_2, X_3 (with center O) can be thought of as labels for a material point. Sometimes the coordinates or labels

T. I. Zohdi, *Modeling and Simulation of Functionalized Materials for Additive Manufacturing and 3D Printing: Continuous and Discrete Media*, Lecture Notes in Applied and Computational Mechanics 60, https://doi.org/10.1007/978-3-319-70079-3_2

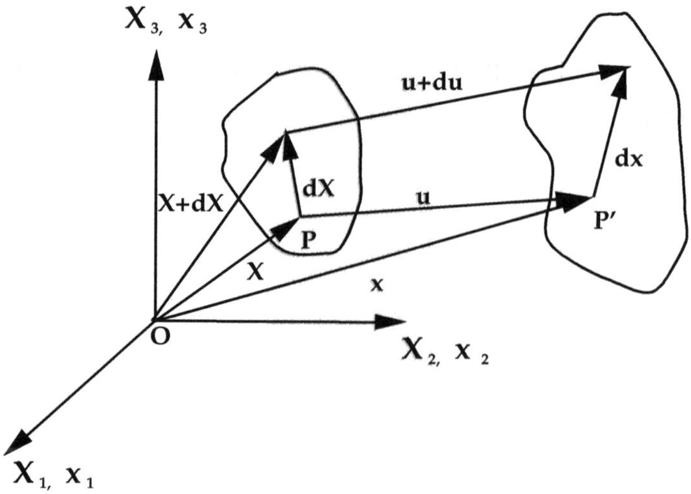

Fig. 2.1 Different descriptions of a deforming body. Ω_o is the reference configuration, and Ω is the current configuration

(X_1, X_2, X_3) are called the referential or material coordinates. In the current configuration, the particle originally located at point P (at time $t = 0$) is located at point P' and can be also expressed in terms of another position vector $\boldsymbol{x}$, with coordinates (x_1, x_2, x_3). These are called the current coordinates. In this framework, the displacement is $\boldsymbol{u} = \boldsymbol{x} - \boldsymbol{X}$ for a point originally at $\boldsymbol{X}$ and with final coordinates $\boldsymbol{x}$.

When a continuum undergoes deformation (or flow), its points move along various paths in space. This motion may be expressed as a function of $\boldsymbol{X}$ and t as

$$\boldsymbol{x}(\boldsymbol{X}, t) = \boldsymbol{u}(\boldsymbol{X}, t) + \boldsymbol{X}(t), \tag{2.2}$$

which gives the present location of a point at time t, written in terms of the referential coordinates X_1, X_2, X_3. The previous position vector may be interpreted as a mapping of the initial configuration onto the current configuration. In classical approaches, it is assumed that such a mapping is one-to-one and continuous, with continuous partial derivatives to whatever order is required. The description of motion or deformation expressed previously is known as the Lagrangian formulation. Alternatively, if the independent variables are the coordinates $\boldsymbol{x}$ and time t, then $\boldsymbol{x}(x_1, x_2, x_3, t) = \boldsymbol{u}(x_1, x_2, x_3, t) + \boldsymbol{X}(x_1, x_2, x_3, t)$, and the formulation is denoted as Eulerian (Fig. 2.1).[1]

[1]Frequently, analysts consider the referential configuration to be fixed in time; thus, in that case it is not a function of time, $\boldsymbol{X} \neq \boldsymbol{X}(t)$. We shall consider $\boldsymbol{X} \neq \boldsymbol{X}(t)$ for the remainder of the monograph.

2.2.1 Deformation of Line Elements

Partial differentiation of the displacement vector $\boldsymbol{u} = \boldsymbol{x} - \boldsymbol{X}$, with respect to $\boldsymbol{X}$, produces the following displacement gradient:

$$\nabla_X \boldsymbol{u} = \boldsymbol{F} - \boldsymbol{1}, \tag{2.3}$$

where

$$\boldsymbol{F} \stackrel{\text{def}}{=} \nabla_X \boldsymbol{x} \stackrel{\text{def}}{=} \frac{\partial \boldsymbol{x}}{\partial \boldsymbol{X}} = \begin{bmatrix} \frac{\partial x_1}{\partial X_1} & \frac{\partial x_1}{\partial X_2} & \frac{\partial x_1}{\partial X_3} \\ \frac{\partial x_2}{\partial X_1} & \frac{\partial x_2}{\partial X_2} & \frac{\partial x_2}{\partial X_3} \\ \frac{\partial x_3}{\partial X_1} & \frac{\partial x_3}{\partial X_2} & \frac{\partial x_3}{\partial X_3} \end{bmatrix}. \tag{2.4}$$

$\boldsymbol{F}$ is known as the material deformation gradient.

Now, consider the length of a differential element in the reference configuration $d\boldsymbol{X}$ and $d\boldsymbol{x}$ in the current configuration, $d\boldsymbol{x} = \nabla_X \boldsymbol{x} \cdot d\boldsymbol{X} = \boldsymbol{F} \cdot d\boldsymbol{X}$. Taking the difference in the squared magnitudes of these elements yields

$$\begin{aligned} d\boldsymbol{x} \cdot d\boldsymbol{x} - d\boldsymbol{X} \cdot d\boldsymbol{X} &= (\nabla_X \boldsymbol{x} \cdot d\boldsymbol{X}) \cdot (\nabla_X \boldsymbol{x} \cdot d\boldsymbol{X}) - d\boldsymbol{X} \cdot d\boldsymbol{X} \\ &= d\boldsymbol{X} \cdot (\boldsymbol{F}^T \cdot \boldsymbol{F} - \boldsymbol{1}) \cdot d\boldsymbol{X} \stackrel{\text{def}}{=} 2\, d\boldsymbol{X} \cdot \boldsymbol{E} \cdot d\boldsymbol{X}. \end{aligned} \tag{2.5}$$

Equation (2.5) defines the so-called *Lagrangian* strain tensor

$$\boldsymbol{E} \stackrel{\text{def}}{=} \tfrac{1}{2}(\boldsymbol{F}^T \cdot \boldsymbol{F} - \boldsymbol{1}) = \tfrac{1}{2}[\nabla_X \boldsymbol{u} + (\nabla_X \boldsymbol{u})^T + (\nabla_X \boldsymbol{u})^T \cdot \nabla_X \boldsymbol{u}]. \tag{2.6}$$

Remark 1 It should be clear that $d\boldsymbol{x}$ can be reinterpreted as the result of a mapping $\boldsymbol{F} \cdot d\boldsymbol{X} \rightarrow d\boldsymbol{x}$ or a change in configuration (reference to current). An important quantity is the Jacobian of the deformation gradient, $J \stackrel{\text{def}}{=} det \boldsymbol{F}$, which relates differential volumes in the reference configuration ($d\Omega_0$) to differential volumetric domains in the current configuration ($d\Omega$) via $d\Omega = J\, d\Omega_0$. The Jacobian of the deformation gradient must remain positive, otherwise we obtain physically impossible "negative" volumes. One way is compare the differential volume of mutually orthogonal triad of differential vectors in the reference configuration $d\boldsymbol{X}^{(1)}, d\boldsymbol{X}^{(2)}, d\boldsymbol{X}^{(3)}$ (forming the edges of a cube), given by the triple product $d\boldsymbol{X}^{(1)} \cdot (d\boldsymbol{X}^{(2)} \times d\boldsymbol{X}^{(3)})$ to the volume of the mapping of the triad $d\boldsymbol{x}^{(1)} = \boldsymbol{F} \cdot d\boldsymbol{X}^{(1)}, d\boldsymbol{x}^{(2)} = \boldsymbol{F} \cdot d\boldsymbol{X}^{(2)}, d\boldsymbol{x}^{(3)} = \boldsymbol{F} \cdot d\boldsymbol{X}^{(3)}$, given by $d\boldsymbol{x}^{(1)} \cdot (d\boldsymbol{x}^{(2)} \times d\boldsymbol{x}^{(3)})$. Another way to prove this is by formulating a conservation of mass over an arbitrary volume within the domain

$$\int_{\omega_o} \rho_o\, d\omega_o = \int_{\omega} \rho\, d\omega = \int_{\omega_o} \rho J\, d\omega_o, \tag{2.7}$$

which immediately leads to the conclusion that $\rho_o = \rho J$, since ω_o is arbitrary. For more details, we refer the reader to the texts of Malvern [2], Gurtin [3], Chandrasekharaiah and Debnath [4], and Zohdi and Wriggers [1].

Remark 2 One may develop so-called Eulerian formulations, employing the current configuration coordinates to generate Eulerian strain tensor measures (see Zohdi and Wriggers [1]).

2.3 Equilibrium/Kinetics of Continua

The balance of linear momentum in the deformed (current) configuration is

$$\underbrace{\int_{\partial\omega} \boldsymbol{t}\, da}_{\text{surface forces}} + \underbrace{\int_{\omega} \rho \boldsymbol{b}\, d\omega}_{\text{body forces}} = \underbrace{\frac{d}{dt}\int_{\omega} \rho \dot{\boldsymbol{u}}\, d\omega}_{\text{inertial forces}}, \tag{2.8}$$

where $\omega \subset \Omega$ is an arbitrary portion of the body (Fig. 2.1), with boundary $\partial\omega$, ρ is the material density, $\boldsymbol{b}$ is the body force per unit mass, and $\dot{\boldsymbol{u}}$ is the time derivative of the displacement. The surface force densities, $\boldsymbol{t}$, are commonly referred to as "tractions."

2.3.1 *Postulates on Volume and Surface Quantities*

Now, consider a tetrahedron (commonly referred to as a Cauchy tetrahedron) in equilibrium, as shown in Fig. 2.2, where a balance of forces yields

$$\boldsymbol{t}^{(n)}\Delta A^{(n)} + \boldsymbol{t}^{(-1)}\Delta A^{(1)} + \boldsymbol{t}^{(-2)}\Delta A^{(2)} + \boldsymbol{t}^{(-3)}\Delta A^{(3)} + \rho \boldsymbol{b}\Delta V = \rho \ddot{\boldsymbol{u}}\Delta V, \tag{2.9}$$

where $\Delta A^{(n)}$ is the surface area of the face of the tetrahedron with normal $\boldsymbol{n}$ and ΔV is the tetrahedron volume. As the distance (h) between the tetrahedron

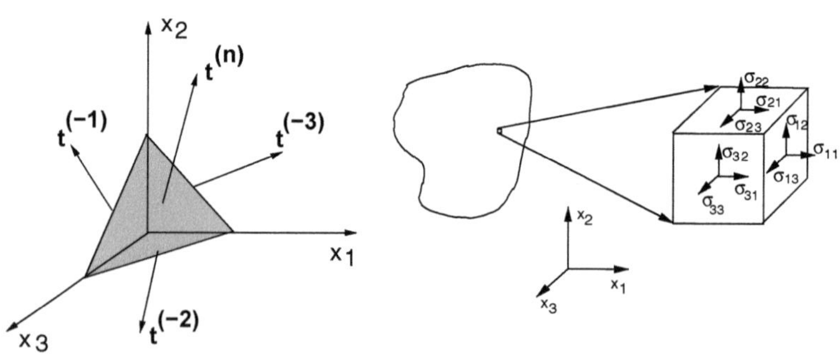

Fig. 2.2 Left: Cauchy tetrahedron: a "sectioned point" and right: stress at a point

base (located at (0, 0, 0)) and the surface center goes to zero ($h \rightarrow 0$), we have $\Delta A^{(n)} \rightarrow 0 \Rightarrow \frac{\Delta V}{\Delta A^{(n)}} \rightarrow 0$. Geometrically, we have $\frac{\Delta A^{(i)}}{\Delta A^{(n)}} = cos(x_i, x_n) \stackrel{\text{def}}{=} n_i$, and therefore $\boldsymbol{t}^{(n)} + \boldsymbol{t}^{(-1)} cos(x_1, x_n) + \boldsymbol{t}^{(-2)} cos(x_2, x_n) + \boldsymbol{t}^{(-3)} cos(x_3, x_n) = \boldsymbol{0}$, where (x_i, x_n) indicates the angle between the x_i and x_n directions. It is clear that forces on the surface areas could be decomposed into three linearly independent, mutually orthogonal, components. It is convenient to introduce the concept of stress at a point, representing the surface forces (tractions) there, pictorially represented by a cube surrounding a point. The fundamental issue that must be resolved is the characterization of these surface forces. We can represent the traction on a surface by the component representation:

$$\boldsymbol{t}^{(i)} \stackrel{\text{def}}{=} \left\{ \begin{array}{c} \sigma_{i1} \\ \sigma_{i2} \\ \sigma_{i3} \end{array} \right\}, \tag{2.10}$$

where the second index represents the direction of the component and the first index represents components of the normal to corresponding coordinate plane. We have $\boldsymbol{t}^{(n)} = \boldsymbol{\sigma}^T \cdot \boldsymbol{n}$, where

$$\boldsymbol{\sigma} \stackrel{\text{def}}{=} \left[\begin{array}{ccc} \sigma_{11} & \sigma_{12} & \sigma_{13} \\ \sigma_{21} & \sigma_{22} & \sigma_{23} \\ \sigma_{31} & \sigma_{32} & \sigma_{33} \end{array} \right], \tag{2.11}$$

or explicitly ($\boldsymbol{t}^{(1)} = -\boldsymbol{t}^{(-1)}, \boldsymbol{t}^{(2)} = -\boldsymbol{t}^{(-2)}, \boldsymbol{t}^{(3)} = -\boldsymbol{t}^{(-3)}$)

$$\boldsymbol{t}^{(n)} = \boldsymbol{t}^{(1)} n_1 + \boldsymbol{t}^{(2)} n_2 + \boldsymbol{t}^{(3)} n_3 = \boldsymbol{\sigma}^T \cdot \boldsymbol{n} = \left[\begin{array}{ccc} \sigma_{11} & \sigma_{12} & \sigma_{13} \\ \sigma_{21} & \sigma_{22} & \sigma_{23} \\ \sigma_{31} & \sigma_{32} & \sigma_{33} \end{array} \right]^T \left\{ \begin{array}{c} n_1 \\ n_2 \\ n_3 \end{array} \right\}, \tag{2.12}$$

where $\boldsymbol{\sigma}$ is the so-called Cauchy stress tensor. Henceforth, we will drop the superscript notation of $\boldsymbol{t}^{(n)}$, where it is implicitly assumed that $\boldsymbol{t} \stackrel{\text{def}}{=} \boldsymbol{t}^{(n)}$.

Remark In the absence of micromoment stresses, a balance of angular momentum implies a symmetry of stress, $\boldsymbol{\sigma} = \boldsymbol{\sigma}^T$, and thus, the difference in notations becomes immaterial. Explicitly, starting with an angular momentum balance, under the assumptions that no infinitesimal "micromoments" or so-called couple-stresses exist, then it can be shown that the stress tensor must be symmetric,[2] i.e., $\int_{\partial\omega} \boldsymbol{x} \times \boldsymbol{t}\, da + \int_{\omega} \boldsymbol{x} \times \rho \boldsymbol{b}\, d\omega = \frac{d}{dt} \int_{\omega} \boldsymbol{x} \times \rho \dot{\boldsymbol{u}}\, d\omega$; that is, $\boldsymbol{\sigma}^T = \boldsymbol{\sigma}$.

[2] It is somewhat easier to simply consider a differential element, such as in Fig. 2.2, and to simply sum moments about the center. Doing this, one immediately obtains $\sigma_{12} = \sigma_{21}, \sigma_{23} = \sigma_{32}$ and $\sigma_{13} = \sigma_{31}$. Consequently, $\boldsymbol{t} = \boldsymbol{\sigma} \cdot \boldsymbol{n} = \boldsymbol{\sigma}^T \cdot \boldsymbol{n}$.

2.3.2 Balance Law Formulations

Substitution of Eq. 2.12 into Eq. 2.8 yields ($\omega \subset \Omega$)

$$\underbrace{\int_{\partial\omega} \boldsymbol{\sigma} \cdot \boldsymbol{n}\, da}_{\text{surface forces}} + \underbrace{\int_{\omega} \rho \boldsymbol{b}\, d\omega}_{\text{body forces}} = \underbrace{\frac{d}{dt} \int_{\omega} \rho \dot{\boldsymbol{u}}\, d\omega}_{\text{inertial forces}} . \tag{2.13}$$

A relationship can be determined between the densities in the current and reference configurations, $\int_{\omega} \rho d\omega = \int_{\omega_0} \rho J d\omega_0 = \int_{\omega_0} \rho_0 d\omega_0$. Therefore, the Jacobian can also be interpreted as the ratio of material densities at a point. Since the volume is arbitrary, we can assume that $\rho J = \rho_0$ holds at every point in the body. Therefore, we may write $\frac{d}{dt}(\rho_0) = \frac{d}{dt}(\rho J) = 0$, when the system is mass conservative over time. This leads to writing the last term in Eq. 2.13 as $\frac{d}{dt}\int_{\omega} \rho \dot{\boldsymbol{u}}\, d\omega = \int_{\omega_0} \frac{d(\rho J)}{dt} \dot{\boldsymbol{u}}\, d\omega_0 + \int_{\omega_0} \rho \ddot{\boldsymbol{u}} J\, d\omega_0 = \int_{\omega} \rho \ddot{\boldsymbol{u}}\, d\omega$. From Gauss's divergence theorem, and an implicit assumption that $\boldsymbol{\sigma}$ is differentiable, we have $\int_{\omega} (\nabla_x \cdot \boldsymbol{\sigma} + \rho \boldsymbol{b} - \rho \ddot{\boldsymbol{u}})\, d\omega = \boldsymbol{0}$. If the volume is selected as being arbitrary, then the integrand must be equal to zero at every point, yielding

$$\nabla_x \cdot \boldsymbol{\sigma} + \rho \boldsymbol{b} = \rho \ddot{\boldsymbol{u}}. \tag{2.14}$$

2.4 The First Law of Thermodynamics/An Energy Balance

The interconversions of mechanical, thermal, and chemical energy in a system are governed by the first law of thermodynamics, which states that the time rate of change of the total energy, $\mathcal{K} + \mathcal{I}$, is equal to the rate of input of energy $\dot{W}$. Specifically, we can relate the kinetic and potential energy states at two instances of time by

$$\mathcal{K}(t) + \mathcal{I}(t) + \Delta W = \mathcal{K}(t + \Delta t) + \mathcal{I}(t + \Delta t) \tag{2.15}$$

or, as $\Delta t \to 0$,

$$\frac{d}{dt}(\mathcal{K} + \mathcal{I}) = \dot{W} = \mathcal{P} + \mathcal{H} + \mathcal{Q}, \tag{2.16}$$

where the mechanical power is $\mathcal{P}$ and the net heat supplied from sources and conduction is $\mathcal{Q} + \mathcal{H}$. Here, the kinetic energy of a subvolume of material contained in Ω, denoted ω, is

$$\mathcal{K} \stackrel{\text{def}}{=} \int_{\omega} \frac{1}{2} \rho \dot{\boldsymbol{u}} \cdot \dot{\boldsymbol{u}}\, d\omega, \tag{2.17}$$

the power (rate of work) of the external forces acting on ω is given by

$$\mathcal{P} \stackrel{\text{def}}{=} \int_{\omega} \rho \boldsymbol{b} \cdot \dot{\boldsymbol{u}} \, d\omega + \int_{\partial\omega} \boldsymbol{\sigma} \cdot \boldsymbol{n} \cdot \dot{\boldsymbol{u}} \, da, \tag{2.18}$$

the heat flow into the volume by conduction is

$$\mathcal{Q} \stackrel{\text{def}}{=} - \int_{\partial\omega} \boldsymbol{q} \cdot \boldsymbol{n} \, da = - \int_{\omega} \nabla_x \cdot \boldsymbol{q} \, d\omega, \tag{2.19}$$

$\boldsymbol{q}$ being the heat flux, the heat generated due to sources, *such as chemical reactions*, is

$$\mathcal{H} \stackrel{\text{def}}{=} \int_{\omega} \rho z \, d\omega, \tag{2.20}$$

where z is the reaction source rate per unit mass, and the internal energy is

$$\mathcal{I} \stackrel{\text{def}}{=} \int_{\omega} \rho w \, d\omega, \tag{2.21}$$

where w being the internal energy per unit mass. Differentiating the kinetic energy yields

$$\begin{aligned} \frac{d\mathcal{K}}{dt} = \frac{d}{dt} \int_{\omega} \frac{1}{2} \rho \dot{\boldsymbol{u}} \cdot \dot{\boldsymbol{u}} \, d\omega &= \int_{\omega_0} \frac{d}{dt} \frac{1}{2} (\rho J \dot{\boldsymbol{u}} \cdot \dot{\boldsymbol{u}}) \, d\omega_0 \\ &= \int_{\omega_0} (\frac{d}{dt} \rho_0) \frac{1}{2} \dot{\boldsymbol{u}} \cdot \dot{\boldsymbol{u}} \, d\omega_0 + \int_{\omega} \rho \frac{d}{dt} \frac{1}{2} (\dot{\boldsymbol{u}} \cdot \dot{\boldsymbol{u}}) \, d\omega \\ &= \int_{\omega} \rho \dot{\boldsymbol{u}} \cdot \ddot{\boldsymbol{u}} \, d\omega, \end{aligned} \tag{2.22}$$

where we have assumed that the mass in the system is constant. We also have

$$\frac{d\mathcal{I}}{dt} = \frac{d}{dt} \int_{\omega} \rho w \, d\omega = \frac{d}{dt} \int_{\omega_0} \rho J w \, d\omega_0 = \int_{\omega_0} \underbrace{\frac{d}{dt}(\rho_0)}_{=0} w \, d\omega_0 + \int_{\omega} \rho \dot{w} \, d\omega = \int_{\omega} \rho \dot{w} \, d\omega. \tag{2.23}$$

By using the divergence theorem, we obtain

$$\int_{\partial\omega} \boldsymbol{\sigma} \cdot \boldsymbol{n} \cdot \dot{\boldsymbol{u}} \, da = \int_{\omega} \nabla_x \cdot (\boldsymbol{\sigma} \cdot \dot{\boldsymbol{u}}) \, d\omega = \int_{\omega} (\nabla_x \cdot \boldsymbol{\sigma}) \cdot \dot{\boldsymbol{u}} \, d\omega + \int_{\omega} \boldsymbol{\sigma} : \nabla_x \dot{\boldsymbol{u}} \, d\omega. \tag{2.24}$$

Combining the results, and enforcing a balance of linear momentum, leads to

$$\begin{aligned} &\int_{\omega} (\rho \dot{w} + \dot{\boldsymbol{u}} \cdot (\rho \ddot{\boldsymbol{u}} - \nabla_x \cdot \boldsymbol{\sigma} - \rho \boldsymbol{b}) - \boldsymbol{\sigma} : \nabla_x \dot{\boldsymbol{u}} + \nabla_x \cdot \boldsymbol{q} - \rho z) \, d\omega = \\ &\int_{\omega} (\rho \dot{w} - \boldsymbol{\sigma} : \nabla_x \dot{\boldsymbol{u}} + \nabla_x \cdot \boldsymbol{q} - \rho z) \, d\omega = 0. \end{aligned} \tag{2.25}$$

Since the volume ω is arbitrary, the integrand must hold locally and we have

$$\rho\dot{w} - \boldsymbol{\sigma} : \nabla_x \dot{\boldsymbol{u}} + \nabla_x \cdot \boldsymbol{q} - \rho z = 0. \tag{2.26}$$

When dealing with multifield problems, this equation is used extensively.

2.5 Linearly Elastic Constitutive Equations

We now discuss relationships between the stress and the strain, so-called *material laws* or *constitutive relations* for linearly elastic cases (infinitesimal deformations).

2.5.1 The Infinitesimal Strain Case

In infinitesimal deformation theory, the displacement gradient components are considered small enough that higher-order terms like $(\nabla_X \boldsymbol{u})^T \cdot \nabla_X \boldsymbol{u}$ and $(\nabla_x \boldsymbol{u})^T \cdot \nabla_x \boldsymbol{u}$ can be neglected in the strain measure $\boldsymbol{E} = \frac{1}{2}(\nabla_X \boldsymbol{u} + (\nabla_X \boldsymbol{u})^T + (\nabla_X \boldsymbol{u})^T \cdot \nabla_X \boldsymbol{u})$, leading to $\boldsymbol{E} \approx \boldsymbol{\epsilon} \stackrel{\text{def}}{=} \frac{1}{2}[\nabla_X \boldsymbol{u} + (\nabla_X \boldsymbol{u})^T]$. If the displacement gradients are small compared with unity, $\boldsymbol{\epsilon}$ coincides closely to $\boldsymbol{E}$. If we assume $\frac{\partial}{\partial X} \approx \frac{\partial}{\partial x}$, we may use $\boldsymbol{E}$ or $\boldsymbol{\epsilon}$ interchangeably. Usually $\boldsymbol{\epsilon}$ is the symbol used for infinitesimal strains. Furthermore, to avoid confusion, when using models employing the geometrically linear infinitesimal strain assumption, we use the symbol of ∇ with no $\boldsymbol{X}$ or $\boldsymbol{x}$ subscript. Hence, the infinitesimal strains are defined by

$$\boldsymbol{\epsilon} = \frac{1}{2}(\nabla \boldsymbol{u} + (\nabla \boldsymbol{u})^T). \tag{2.27}$$

2.5.2 Material Response

If we neglect thermal effects, Eq. 2.26 implies $\rho\dot{w} = \boldsymbol{\sigma} : \nabla_x \dot{\boldsymbol{u}}$ which, in the infinitesimal strain linearly elastic case, is $\rho\dot{w} = \boldsymbol{\sigma} : \dot{\boldsymbol{\epsilon}}$. From the chain rule of differentiation, we have

$$\rho\dot{w} = \rho\frac{\partial w}{\partial \boldsymbol{\epsilon}} : \frac{d\boldsymbol{\epsilon}}{dt} = \boldsymbol{\sigma} : \dot{\boldsymbol{\epsilon}} \Rightarrow \boldsymbol{\sigma} = \rho\frac{\partial w}{\partial \boldsymbol{\epsilon}}. \tag{2.28}$$

The starting point to develop a constitutive theory is to assume a stored elastic energy function exists, a function denoted $W \stackrel{\text{def}}{=} \rho w$, which depends only on the mechanical deformation. The simplest function that fulfills $\boldsymbol{\sigma} = \rho\frac{\partial w}{\partial \boldsymbol{\epsilon}}$ is $W = \frac{1}{2}\boldsymbol{\epsilon} : \mathbb{E} : \boldsymbol{\epsilon}$, where $\mathbb{E}$ is the fourth rank elasticity tensor. Such a function satisfies the intuitive physical requirement that, for any small strain from an undeformed state, energy must be stored in the material. Subsequently, a small strain material law can be derived from $\boldsymbol{\sigma} = \frac{\partial W}{\partial \boldsymbol{\epsilon}}$ and $W \approx c_0 + \mathbf{c}_1 : \boldsymbol{\epsilon} + \frac{1}{2}\boldsymbol{\epsilon} : \mathbb{E} : \boldsymbol{\epsilon} + \ldots$ which implies $\boldsymbol{\sigma} \approx \mathbf{c}_1 + \mathbb{E} : \boldsymbol{\epsilon} + \ldots$. We are free to set $c_0 = 0$ (it is arbitrary) in order to have zero strain energy at zero strain, and furthermore, we assume that no stresses exist in the reference state ($\mathbf{c}_1 = \mathbf{0}$). With these assumptions, we obtain the familiar relation

$$\boldsymbol{\sigma} = \mathbb{E} : \boldsymbol{\epsilon}. \tag{2.29}$$

This is a linear relation between stresses and strains. The existence of a strictly positive stored energy function in the reference configuration implies that the linear elasticity tensor must have positive eigenvalues at every point in the body. Typically, different materials are classified according to the number of independent components in $\mathbb{E}$. In theory, $\mathbb{E}$ has 81 components, since it is a fourth-order tensor relating nine components of stress to strain. However, the number of components can be reduced to 36 since the stress and strain tensors are symmetric. This is observed from the matrix representation[3] of $\mathbb{E}$:

$$\underbrace{\begin{Bmatrix} \sigma_{11} \\ \sigma_{22} \\ \sigma_{33} \\ \sigma_{12} \\ \sigma_{23} \\ \sigma_{31} \end{Bmatrix}}_{\stackrel{\text{def}}{=}\{\boldsymbol{\sigma}\}} = \underbrace{\begin{bmatrix} E_{1111} & E_{1122} & E_{1133} & E_{1112} & E_{1123} & E_{1113} \\ E_{2211} & E_{2222} & E_{2233} & E_{2212} & E_{2223} & E_{2213} \\ E_{3311} & E_{3322} & E_{3333} & E_{3312} & E_{3323} & E_{3313} \\ E_{1211} & E_{1222} & E_{1233} & E_{1212} & E_{1223} & E_{1213} \\ E_{2311} & E_{2322} & E_{2333} & E_{2312} & E_{2323} & E_{2313} \\ E_{1311} & E_{1322} & E_{1333} & E_{1312} & E_{1323} & E_{1313} \end{bmatrix}}_{\stackrel{\text{def}}{=}[\mathbb{E}]} \underbrace{\begin{Bmatrix} \epsilon_{11} \\ \epsilon_{22} \\ \epsilon_{33} \\ 2\epsilon_{12} \\ 2\epsilon_{23} \\ 2\epsilon_{31} \end{Bmatrix}}_{\stackrel{\text{def}}{=}\{\boldsymbol{\epsilon}\}}. \tag{2.30}$$

The existence of a scalar energy function forces $\mathbb{E}$ to be symmetric since the strains are symmetric; in other words, $W = \frac{1}{2}\boldsymbol{\epsilon} : \mathbb{E} : \boldsymbol{\epsilon} = \frac{1}{2}(\boldsymbol{\epsilon} : \mathbb{E} : \boldsymbol{\epsilon})^T = \frac{1}{2}\boldsymbol{\epsilon}^T : \mathbb{E}^T : \boldsymbol{\epsilon}^T = \frac{1}{2}\boldsymbol{\epsilon} : \mathbb{E}^T : \boldsymbol{\epsilon}$ which implies $\mathbb{E}^T = \mathbb{E}$. Consequently, $\mathbb{E}$ has only 21 independent components. The nonnegativity of W imposes the restriction that $\mathbb{E}$ remains positive definite. At this point, based on many factors that depend on the material microstructure, it can be shown that the components of $\mathbb{E}$ may be written in terms of anywhere between 21 and 2 independent parameters. Accordingly, for isotropic materials, we have two planes of symmetry and an infinite number of planes of directional independence (two free components), yielding

$$\mathbb{E} \stackrel{\text{def}}{=} \begin{bmatrix} \kappa + \frac{4}{3}\mu & \kappa - \frac{2}{3}\mu & \kappa - \frac{2}{3}\mu & 0 & 0 & 0 \\ \kappa - \frac{2}{3}\mu & \kappa + \frac{4}{3}\mu & \kappa - \frac{2}{3}\mu & 0 & 0 & 0 \\ \kappa - \frac{2}{3}\mu & \kappa - \frac{2}{3}\mu & \kappa + \frac{4}{3}\mu & 0 & 0 & 0 \\ 0 & 0 & 0 & \mu & 0 & 0 \\ 0 & 0 & 0 & 0 & \mu & 0 \\ 0 & 0 & 0 & 0 & 0 & \mu \end{bmatrix}. \tag{2.31}$$

[3]The symbol $[\cdot]$ is used to indicate the matrix notation equivalent to a tensor form, while $\{\cdot\}$ is used to indicate the vector representation.

In this case, we have

$$\sigma = I\!E : \epsilon = 3\kappa \frac{tr\epsilon}{3}\mathbf{1} + 2\mu\epsilon' \Rightarrow W = \frac{1}{2}\epsilon : I\!E : \epsilon = \frac{9}{2}\kappa(\frac{tr\epsilon}{3})^2 + \mu\epsilon' : \epsilon', \tag{2.32}$$

where $tr\epsilon = \epsilon_{ii}$ and $\epsilon' = \epsilon - \frac{1}{3}(tr\epsilon)\mathbf{1}$ is the deviatoric strain. The eigenvalues of an isotropic elasticity tensor are $(3\kappa, 2\mu, 2\mu, \mu, \mu, \mu)$. Therefore, we must have $\kappa > 0$ and $\mu > 0$ to retain positive definiteness of $I\!E$. All of the material components of $I\!E$ may be spatially variable, as in the case of composite particulate-functionalized media.

2.5.3 *Material Component Interpretation*

There are a variety of ways to write isotropic constitutive laws, each time with a physically meaningful pair of material values.

Splitting the Strain

It is sometimes important to split infinitesimal strains into two physically meaningful parts

$$\epsilon = \frac{tr\epsilon}{3}\mathbf{1} + (\epsilon - \frac{tr\epsilon}{3}\mathbf{1}). \tag{2.33}$$

An expansion of the Jacobian of the deformation gradient yields $J = det(\mathbf{1} + \nabla_X \boldsymbol{u}) \approx 1 + tr\nabla_X \boldsymbol{u} + \mathcal{O}(\nabla_X \boldsymbol{u}) = 1 + tr\epsilon + \ldots$. Therefore, with infinitesimal strains, $(1 + tr\epsilon)d\omega_0 = d\omega$ and we can write $tr\epsilon = \frac{d\omega - d\omega_0}{d\omega_0}$. Hence, $tr\epsilon$ is associated with the *volumetric part of the deformation*. Furthermore, since $\epsilon' \stackrel{\text{def}}{=} \epsilon - \frac{tr\epsilon}{3}\mathbf{1}$, the so-called strain deviator describes distortion in the material.

Infinitesimal Strain Material Laws

The stress $\boldsymbol{\sigma}$ can be split into two parts (dilatational and a deviatoric):

$$\sigma = \frac{tr\sigma}{3}\mathbf{1} + (\sigma - \frac{tr\sigma}{3}\mathbf{1}) \stackrel{\text{def}}{=} -p\mathbf{1} + \sigma', \tag{2.34}$$

where we call the symbol p the hydrostatic pressure and $\boldsymbol{\sigma}'$ the stress deviator. With (2.32), we write

$$p = -3\,\kappa\left(\frac{tr\epsilon}{3}\right) \quad \text{and} \quad \sigma' = 2\,\mu\,\epsilon'. \tag{2.35}$$

This is one form of Hooke's law. The resistance to change in the volume is measured by κ. We note that $(\frac{tr\boldsymbol{\sigma}}{3}\mathbf{1})' = \mathbf{0}$, which indicates that this part of the stress produces no distortion.

Another fundamental form of Hooke's law is

$$\boldsymbol{\sigma} = \frac{E^y}{1+\nu}\left(\boldsymbol{\epsilon} + \frac{\nu}{1-2\nu}(tr\boldsymbol{\epsilon})\mathbf{1}\right), \tag{2.36}$$

and the inverse form is

$$\boldsymbol{\epsilon} = \frac{1+\nu}{E^y}\boldsymbol{\sigma} - \frac{\nu}{E^y}(tr\boldsymbol{\sigma})\mathbf{1}, \tag{2.37}$$

where E^y, the Young's modulus, is the ratio of the uniaxial stress to the corresponding strain component and the Poisson ratio, ν, is the ratio of the transverse strains to the uniaxial strain. To interpret the material values, consider an idealized uniaxial tension test (pulled in the x_1-direction inducing a uniform stress state) where $\sigma_{12} = \sigma_{13} = \sigma_{23} = 0$, which implies $\epsilon_{12} = \epsilon_{13} = \epsilon_{23} = 0$. Also, we have $\sigma_{22} = \sigma_{33} = 0$. Under these conditions, we have $\sigma_{11} = E^y\epsilon_{11}$ (axial stiffness) and $\epsilon_{22} = \epsilon_{33} = -\nu\epsilon_{11}$ (the ratio of transverse to axial strain).

Another commonly used set of stress–strain forms are the Lamé relations,

$$\boldsymbol{\sigma} = \lambda(tr\boldsymbol{\epsilon})\mathbf{1} + 2\mu\boldsymbol{\epsilon} \quad \text{or} \quad \boldsymbol{\epsilon} = -\frac{\lambda}{2\mu(3\lambda+2\mu)}(tr\boldsymbol{\sigma})\mathbf{1} + \frac{\boldsymbol{\sigma}}{2\mu}, \tag{2.38}$$

where λ is referred to a the Lame parameter. To interpret the material values, consider a homogeneous pressure test (uniform stress) where $\sigma_{12} = \sigma_{13} = \sigma_{23} = 0$ and where $\sigma_{11} = \sigma_{22} = \sigma_{33}$. Under these conditions, we have

$$\kappa = \lambda + \frac{2}{3}\mu = \frac{E^y}{3(1-2\nu)} \quad \text{and} \quad \mu = \frac{E^y}{2(1+\nu)}, \tag{2.39}$$

and consequently

$$\frac{\kappa}{\mu} = \frac{2(1+\nu)}{3(1-2\nu)}. \tag{2.40}$$

We observe that $\frac{\kappa}{\mu} \to \infty$ implies $\nu \to \frac{1}{2}$, and $\frac{\kappa}{\mu} \to 0$ implies $\Rightarrow \nu \to -1$. Therefore, since both κ and μ must be positive and finite, this implies $-1 < \nu < 1/2$ and $0 < E^y < \infty$. For example, some polymeric foams exhibit $\nu < 0$, steels $\nu \approx 0.3$, and some forms of rubber have $\nu \to 1/2$. We note that λ *can be positive or negative*. For more details, see Malvern [2], Gurtin [3], Chandrasekharaiah and Debnath [4].

Remark See Zohdi and Wriggers [1] for a variety of different finite-deformation constitutive laws.

References

1. Zohdi, T.I., Wriggers, P.: Introduction to Computational Micromechanics. Springer (2008). Second Reprinting (Peer Reviewed) (2005)
2. Malvern, L.: Introduction to the Mechanics of a Continuous Medium. Prentice Hall (1968)
3. Gurtin, M.: An Introduction to Continuum Mechanics. Academic Press (1981)
4. Chandrasekharaiah, D.S., Debnath, L.: Continuum Mechanics. Academic Press (1994)

Chapter 3
CM Approaches: Characterization of Particle-Functionalized Materials

3.1 Introduction

During the development of new particulate-functionalized materials, experiments to determine the appropriate combinations of particulate and matrix phases are time-consuming and expensive. Therefore, elementary "microstructure-macroproperty" methods have been generated over the last century in order to analyze and guide new material development. The overall properties of such materials are the aggregate response of the collection of interacting components (Fig. 3.1). The macroscopic properties can be tailored to the specific application, for example, in structural engineering applications, by choosing a harder particulate phase that serves as a stiffening agent for a ductile, easy to form, base matrix material. "Microstructure-macroproperty" (micro–macro) methods are referred to by many different terms, such as "homogenization," "regularization," "mean field theory," "upscaling" in various scientific communities to compute effective properties of heterogeneous materials. We will use these terms interchangeably in this chapter. The usual approach is to compute a constitutive "relation between averages," relating volume-averaged field variables, resulting in effective properties. Thereafter, the effective properties can be used in a macroscopic analysis. The volume averaging takes place over a statistically representative sample of material, referred to in the literature as a representative volume element (RVE). The internal fields, which are to be volumetrically averaged, must be computed by solving a series of boundary value problems with test loadings. There is a vast literature on such methods, dating back to Maxwell [1, 2] and Lord Rayleigh [3], for estimating the overall macroscopic properties of heterogeneous materials. For an authoritative review of the general theory of random heterogeneous media, see Torquato [4]; for more mathematical homogenization aspects, see Jikov et. al. [5]; for solid mechanics inclined accounts of the subject, see Hashin [6], Mura [7], Nemat-Nasser and Hori [8], and Huet [9, 10]; for analyses of cracked media, see Sevostianov et. al. [11]; and for computational aspects, see Zohdi and Wriggers [12–42], Ghosh [43], and Ghosh and Dimiduk [44].

T. I. Zohdi, *Modeling and Simulation of Functionalized Materials for Additive Manufacturing and 3D Printing: Continuous and Discrete Media*, Lecture Notes in Applied and Computational Mechanics 60, https://doi.org/10.1007/978-3-319-70079-3_3

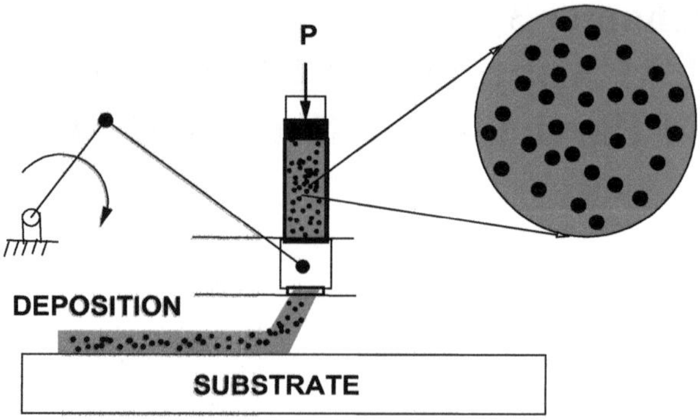

Fig. 3.1 A matrix binder and particulate additives

Our objective in this chapter is to provide some very basic concepts in this area, illustrated by a linear elasticity framework, where the mechanical properties of microheterogeneous materials are characterized by a spatially variable elasticity tensor $\mathbb{E}$. In order to characterize the effective (homogenized) macroscopic response of such materials, a relation between averages,

$$\langle \boldsymbol{\sigma} \rangle_\Omega = \mathbb{E}^* : \langle \boldsymbol{\epsilon} \rangle_\Omega , \tag{3.1}$$

is sought, where

$$\langle \cdot \rangle_\Omega \stackrel{\text{def}}{=} \frac{1}{|\Omega|} \int_\Omega \cdot \, d\Omega \, , \tag{3.2}$$

and where $\boldsymbol{\sigma}$ and $\boldsymbol{\epsilon}$ are the stress and strain tensor fields within a statistically representative volume element (RVE) of volume $|\Omega|$. The quantity $\mathbb{E}^*$ is known as the effective property. It is the elasticity tensor used in usual structural analyses. Similarly, one can describe other effective quantities such as conductivity or diffusivity, in virtually the same manner, relating other volumetrically averaged field variables. However, for the sake of brevity, we restrict ourselves to linear elastostatics problems.

3.2 Basic Micro–Macro Concepts

For a relation between averages to be useful, it must be computed over a sample containing a statistically representative amount of material. This is a requirement that can be formulated in a concise mathematical form. A commonly accepted macro-/microcriterion used in effective property calculations is the so-called Hill's condition, $\langle \boldsymbol{\sigma} : \boldsymbol{\epsilon} \rangle_\Omega = \langle \boldsymbol{\sigma} \rangle_\Omega : \langle \boldsymbol{\epsilon} \rangle_\Omega$. Hill's condition [45] dictates the size requirements on the RVE. The classical argument is as follows. For any perfectly bonded heterogeneous body, in the absence of body forces, two physically important loading states satisfy Hill's condition: (1) linear displacements of the form $\boldsymbol{u}|_{\partial\Omega} = \mathcal{E} \cdot \boldsymbol{x} \Rightarrow \langle \boldsymbol{\epsilon} \rangle_\Omega = \mathcal{E}$ and (2) pure tractions in the form $\boldsymbol{t}|_{\partial\Omega} = \mathcal{L} \cdot \boldsymbol{n} \Rightarrow \langle \boldsymbol{\sigma} \rangle_\Omega = \mathcal{L}$, where $\mathcal{E}$ and $\mathcal{L}$

are constant strain and stress tensors, respectively. Applying (1)- or (2)-type boundary conditions to a large sample is a way of reproducing approximately what may be occurring in a statistically representative microscopic sample of material in a macroscopic body. *The requirement is that the sample must be large enough to have relatively small boundary field fluctuations relative to its size and small enough relative to the macroscopic engineering structure. These restrictions force us to choose boundary conditions that are uniform.*

3.2.1 Testing Procedures

To determine $\mathbb{E}^*$, one specifies six linearly independent loadings of the form,

(1) $\boldsymbol{u}|_{\partial\Omega} = \mathcal{E}^{(1\to 6)} \cdot \boldsymbol{x}$ or
(2) $\boldsymbol{t}|_{\partial\Omega} = \mathcal{L}^{(1\to 6)} \cdot \boldsymbol{n}$,

where $\mathcal{E}^{(1\to 6)}$ and $\mathcal{L}^{(1\to 6)}$ are symmetric second-order strain and stress tensors, with spatially constant (nonzero) components. This loading is applied to a sample of microheterogeneous material. Each independent loading yields six different averaged stress components and hence provides six equations to determine the constitutive constants in $\mathbb{E}^*$. In order for such an analysis to be valid, i.e., to make the material data reliable, the sample of material must be small enough that it can be considered as a material point with respect to the size of the domain under analysis, but large enough to be a statistically representative sample of the microstructure.

If the effective response is assumed to be isotropic, then only one test loading (instead of usually six), containing nonzero dilatational ($\frac{tr\boldsymbol{\sigma}}{3}$ and $\frac{tr\boldsymbol{\epsilon}}{3}$) and deviatoric components ($\boldsymbol{\sigma}' \stackrel{\text{def}}{=} \boldsymbol{\sigma} - \frac{tr\boldsymbol{\sigma}}{3}\mathbf{1}$ and $\boldsymbol{\epsilon}' \stackrel{\text{def}}{=} \boldsymbol{\epsilon} - \frac{tr\boldsymbol{\epsilon}}{3}\mathbf{1}$), is necessary to determine the effective bulk (κ) and shear (μ) moduli:

$$3\kappa^* \stackrel{\text{def}}{=} \frac{\langle \frac{tr\boldsymbol{\sigma}}{3} \rangle_\Omega}{\langle \frac{tr\boldsymbol{\epsilon}}{3} \rangle_\Omega} \quad \text{and} \quad 2\mu^* \stackrel{\text{def}}{=} \sqrt{\frac{\langle \boldsymbol{\sigma}' \rangle_\Omega : \langle \boldsymbol{\sigma}' \rangle_\Omega}{\langle \boldsymbol{\epsilon}' \rangle_\Omega : \langle \boldsymbol{\epsilon}' \rangle_\Omega}}. \tag{3.3}$$

In general, in order to determine the material properties of a microheterogeneous material, one computes 36 constitutive constants[1] E^*_{ijkl} in the following relation between averages,

$$\begin{Bmatrix} \langle\sigma_{11}\rangle_\Omega \\ \langle\sigma_{22}\rangle_\Omega \\ \langle\sigma_{33}\rangle_\Omega \\ \langle\sigma_{12}\rangle_\Omega \\ \langle\sigma_{23}\rangle_\Omega \\ \langle\sigma_{13}\rangle_\Omega \end{Bmatrix} = \begin{bmatrix} E^*_{1111} & E^*_{1122} & E^*_{1133} & E^*_{1112} & E^*_{1123} & E^*_{1113} \\ E^*_{2211} & E^*_{2222} & E^*_{2233} & E^*_{2212} & E^*_{2223} & E^*_{2213} \\ E^*_{3311} & E^*_{3322} & E^*_{3333} & E^*_{3312} & E^*_{3323} & E^*_{3313} \\ E^*_{1211} & E^*_{1222} & E^*_{1233} & E^*_{1212} & E^*_{1223} & E^*_{1213} \\ E^*_{2311} & E^*_{2322} & E^*_{2333} & E^*_{2312} & E^*_{2323} & E^*_{2313} \\ E^*_{1311} & E^*_{1322} & E^*_{1333} & E^*_{1312} & E^*_{1323} & E^*_{1313} \end{bmatrix} \begin{Bmatrix} \langle\epsilon_{11}\rangle_\Omega \\ \langle\epsilon_{22}\rangle_\Omega \\ \langle\epsilon_{33}\rangle_\Omega \\ 2\langle\epsilon_{12}\rangle_\Omega \\ 2\langle\epsilon_{23}\rangle_\Omega \\ 2\langle\epsilon_{13}\rangle_\Omega \end{Bmatrix}. \tag{3.4}$$

[1] There are, of course, only 21 constants, since $\mathbb{E}^*$ is symmetric.

As mentioned before, each independent loading leads to six equations, and hence, in total 36 equations are generated by the independent loadings, which are used to determine the tensor relation between average stress and strain, $\mathbb{E}^*$. $\mathbb{E}^*$ *is exactly what appears in engineering literature as the "property" of a material.* The usual choices for the six independent load cases are

$$\mathcal{E} \ or \ \mathcal{L} = \begin{bmatrix} \beta\,0\,0 \\ 0\,0\,0 \\ 0\,0\,0 \end{bmatrix}, \begin{bmatrix} 0\,0\,0 \\ 0\,\beta\,0 \\ 0\,0\,0 \end{bmatrix}, \begin{bmatrix} 0\,0\,0 \\ 0\,0\,0 \\ 0\,0\,\beta \end{bmatrix}, \begin{bmatrix} 0\,\beta\,0 \\ \beta\,0\,0 \\ 0\,0\,0 \end{bmatrix}, \begin{bmatrix} 0\,0\,0 \\ 0\,0\,\beta \\ 0\,\beta\,0 \end{bmatrix}, \begin{bmatrix} 0\,0\,\beta \\ 0\,0\,0 \\ \beta\,0\,0 \end{bmatrix}, \tag{3.5}$$

where β is a load parameter. For completeness, we record a few related fundamental results, which are useful in micro–macro mechanical analysis.

3.2.2 The Average Strain Theorem

If a heterogeneous body, see Fig. 3.2, has the following uniform loading on its surface: $\boldsymbol{u}|_{\partial\Omega} = \mathcal{E} \cdot \boldsymbol{x}$, then

$$\begin{aligned} \langle\epsilon\rangle_\Omega &= \frac{1}{2|\Omega|}\int_\Omega (\nabla\boldsymbol{u} + (\nabla\boldsymbol{u})^T)\,d\Omega \\ &= \frac{1}{2|\Omega|}\left(\int_{\Omega_1} (\nabla\boldsymbol{u} + (\nabla\boldsymbol{u})^T)\,d\Omega + \int_{\Omega_2} (\nabla\boldsymbol{u} + (\nabla\boldsymbol{u})^T)\,d\Omega\right) \\ &= \frac{1}{2|\Omega|}\left(\int_{\partial\Omega_1} (\boldsymbol{u}\otimes\boldsymbol{n} + \boldsymbol{n}\otimes\boldsymbol{u})\,dA + \int_{\partial\Omega_2} (\boldsymbol{u}\otimes\boldsymbol{n} + \boldsymbol{n}\otimes\boldsymbol{u})\,dA\right) \\ &= \frac{1}{2|\Omega|}\left(\int_{\partial\Omega} ((\mathcal{E}\cdot\boldsymbol{x})\otimes\boldsymbol{n} + \boldsymbol{n}\otimes(\mathcal{E}\cdot\boldsymbol{x}))\,dA + \int_{\partial\Omega_1\cap\partial\Omega_2} ([\![\boldsymbol{u}]\!]\otimes\boldsymbol{n} + \boldsymbol{n}\otimes[\![\boldsymbol{u}]\!])\,dA\right) \\ &= \frac{1}{2|\Omega|}\left(\int_{\Omega} (\nabla(\mathcal{E}\cdot\boldsymbol{x}) + \nabla(\mathcal{E}\cdot\boldsymbol{x})^T)\,d\Omega + \int_{\partial\Omega_1\cap\partial\Omega_2} ([\![\boldsymbol{u}]\!]\otimes\boldsymbol{n} + \boldsymbol{n}\otimes[\![\boldsymbol{u}]\!])\,dA\right) \\ &= \mathcal{E} + \frac{1}{2|\Omega|}\int_{\partial\Omega_1\cap\partial\Omega_2} ([\![\boldsymbol{u}]\!]\otimes\boldsymbol{n} + \boldsymbol{n}\otimes[\![\boldsymbol{u}]\!])\,dA, \end{aligned} \tag{3.6}$$

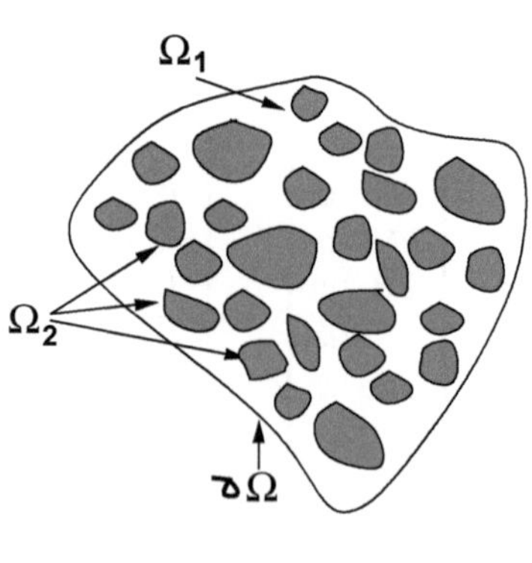

Fig. 3.2 Nomenclature for the averaging theorems (for a general body)

where $(\boldsymbol{u} \otimes \boldsymbol{n} \stackrel{\text{def}}{=} u_i n_j)$ is a tensor product of the vector $\boldsymbol{u}$ and vector $\boldsymbol{n}$. $[\![\boldsymbol{u}]\!]$ describes the displacement jumps at the interfaces between Ω_1 and Ω_2. *Therefore, only if the material is perfectly bonded, then* $\langle \epsilon \rangle_\Omega = \mathcal{E}$. Note that the presence of finite body forces does not affect this result. Also note that the third line in Eq. 3.6 is not an outcome of the divergence theorem, but of a generalization that can be found in a variety of books, for example, Chandrasekharaiah and Debnath [46].

3.2.3 The Average Stress Theorem

Again, we consider a body (in static equilibrium) with $\boldsymbol{t}|_{\partial\Omega} = \mathcal{L} \cdot \boldsymbol{n}$, where $\mathcal{L}$ is a constant tensor. We make use of the identity $\nabla \cdot (\boldsymbol{\sigma} \otimes \boldsymbol{x}) = (\nabla \cdot \boldsymbol{\sigma}) \otimes \boldsymbol{x} + \boldsymbol{\sigma} \cdot \nabla \boldsymbol{x} = -\boldsymbol{f} \otimes \boldsymbol{x} + \boldsymbol{\sigma}$, where $\boldsymbol{f}$ represents the body forces. Substituting this into the definition of the average stress yields

$$
\begin{aligned}
\langle \boldsymbol{\sigma} \rangle_\Omega &= \frac{1}{|\Omega|} \int_\Omega \nabla \cdot (\boldsymbol{\sigma} \otimes \boldsymbol{x})\, d\Omega + \frac{1}{|\Omega|} \int_\Omega (\boldsymbol{f} \otimes \boldsymbol{x})\, d\Omega \\
&= \frac{1}{|\Omega|} \int_{\partial\Omega} (\boldsymbol{\sigma} \otimes \boldsymbol{x}) \cdot \boldsymbol{n}\, dA + \frac{1}{|\Omega|} \int_\Omega (\boldsymbol{f} \otimes \boldsymbol{x})\, d\Omega \\
&= \frac{1}{|\Omega|} \int_{\partial\Omega} (\mathcal{L} \otimes \boldsymbol{x}) \cdot \boldsymbol{n}\, dA + \frac{1}{|\Omega|} \int_\Omega (\boldsymbol{f} \otimes \boldsymbol{x})\, d\Omega \\
&= \mathcal{L} + \frac{1}{|\Omega|} \int_\Omega (\boldsymbol{f} \otimes \boldsymbol{x})\, d\Omega.
\end{aligned}
\tag{3.7}
$$

If there are no body forces, $\boldsymbol{f} = \boldsymbol{0}$, then $\langle \boldsymbol{\sigma} \rangle_\Omega = \mathcal{L}$. *Note that debonding (interface separation) does not change this result.*

3.2.4 Satisfaction of Hill's Energy Condition

Consider a body (in static equilibrium) with a perfectly bonded microstructure and $\boldsymbol{f} = \boldsymbol{0}$. This condition yields

$$
\int_{\partial\Omega} \boldsymbol{u} \cdot \boldsymbol{t}\, dA = \int_{\partial\Omega} \boldsymbol{u} \cdot \boldsymbol{\sigma} \cdot \boldsymbol{n}\, dA = \int_\Omega \nabla \cdot (\boldsymbol{u} \cdot \boldsymbol{\sigma})\, d\Omega. \tag{3.8}
$$

With $\nabla \cdot \boldsymbol{\sigma} = \boldsymbol{0}$, it follows that $\int_\Omega \nabla \cdot (\boldsymbol{u} \cdot \boldsymbol{\sigma})\, d\Omega = \int_\Omega \nabla \boldsymbol{u} : \boldsymbol{\sigma}\, d\Omega = \int_\Omega \epsilon : \boldsymbol{\sigma}\, d\Omega$.
If $\boldsymbol{u}|_{\partial\Omega} = \mathcal{E} \cdot \boldsymbol{x}$ and $\boldsymbol{f} = \boldsymbol{0}$, then

$$\int_{\partial\Omega} \boldsymbol{u} \cdot \boldsymbol{t}\, dA = \int_{\partial\Omega} \mathcal{E} \cdot \boldsymbol{x} \cdot \boldsymbol{\sigma} \cdot \boldsymbol{n}\, dA = \int_{\Omega} \nabla \cdot (\mathcal{E} \cdot \boldsymbol{x} \cdot \boldsymbol{\sigma})\, d\Omega \tag{3.9}$$
$$= \int_{\Omega} \nabla(\mathcal{E} \cdot \boldsymbol{x}) : \boldsymbol{\sigma}\, d\Omega = \mathcal{E} : \langle\boldsymbol{\sigma}\rangle_\Omega |\Omega|.$$

Noting that $\langle\boldsymbol{\epsilon}\rangle_\Omega = \mathcal{E}$, we have $\langle\boldsymbol{\epsilon}\rangle_\Omega : \langle\boldsymbol{\sigma}\rangle_\Omega = \langle\boldsymbol{\epsilon} : \boldsymbol{\sigma}\rangle_\Omega$. If $\boldsymbol{t}|_{\partial\Omega} = \mathcal{L} \cdot \boldsymbol{n}$ and $\boldsymbol{f} = \boldsymbol{0}$, then $\int_{\partial\Omega} \boldsymbol{u} \cdot \boldsymbol{t}\, dA = \int_{\partial\Omega} \boldsymbol{u} \cdot \mathcal{L} \cdot \boldsymbol{n}\, dA = \int_\Omega \nabla \cdot (\boldsymbol{u} \cdot \mathcal{L})\, d\Omega = \int_\Omega \nabla \boldsymbol{u} : \mathcal{L}\, d\Omega = \mathcal{L} : \int_\Omega \boldsymbol{\epsilon}\, d\Omega$. Therefore, since $\langle\boldsymbol{\sigma}\rangle_\Omega = \mathcal{L}$, as before we have $\langle\boldsymbol{\epsilon}\rangle_\Omega : \langle\boldsymbol{\sigma}\rangle_\Omega = \langle\boldsymbol{\epsilon} : \boldsymbol{\sigma}\rangle_\Omega$. Satisfaction of Hill's condition guarantees that the microscopic and macroscopic energies will be the same, and it implies the use of the two mentioned test boundary conditions on sufficiently large samples of material.

3.2.5 *The Hill–Reuss–Voigt Bounds*

Until recently, the direct computation of micromaterial responses was very difficult. Classical approaches have sought to approximate or bound the effective material responses. Many classical approaches start by splitting the stress field within a sample into a volume average and a purely fluctuating part, $\boldsymbol{\epsilon} = \langle\boldsymbol{\epsilon}\rangle_\Omega + \tilde{\boldsymbol{\epsilon}}$, and we directly obtain

$$\begin{aligned}
0 &\le \int_\Omega \tilde{\boldsymbol{\epsilon}} : \boldsymbol{I\!E} : \tilde{\boldsymbol{\epsilon}}\, d\Omega = \int_\Omega (\boldsymbol{\epsilon} : \boldsymbol{I\!E} : \boldsymbol{\epsilon} - 2\langle\boldsymbol{\epsilon}\rangle_\Omega : \boldsymbol{I\!E} : \boldsymbol{\epsilon} + \langle\boldsymbol{\epsilon}\rangle_\Omega : \boldsymbol{I\!E} : \langle\boldsymbol{\epsilon}\rangle_\Omega)\, d\Omega \\
&= (\langle\boldsymbol{\epsilon}\rangle_\Omega : \boldsymbol{I\!E}^* : \langle\boldsymbol{\epsilon}\rangle_\Omega - 2\langle\boldsymbol{\epsilon}\rangle_\Omega : \langle\boldsymbol{\sigma}\rangle_\Omega + \langle\boldsymbol{\epsilon}\rangle_\Omega : \langle\boldsymbol{I\!E}\rangle_\Omega : \langle\boldsymbol{\epsilon}\rangle_\Omega)|\Omega| \\
&= \langle\boldsymbol{\epsilon}\rangle_\Omega : (\langle\boldsymbol{I\!E}\rangle_\Omega - \boldsymbol{I\!E}^*) : \langle\boldsymbol{\epsilon}\rangle_\Omega |\Omega|.
\end{aligned} \tag{3.10}$$

Similarly, for the complementary case, with $\boldsymbol{\sigma} = \langle\boldsymbol{\sigma}\rangle_\Omega + \tilde{\boldsymbol{\sigma}}$, and the following assumption (microscopic energy equals the macroscopic energy)

$$\underbrace{\langle\boldsymbol{\sigma} : \boldsymbol{I\!E}^{-1} : \boldsymbol{\sigma}\rangle_\Omega}_{\text{micro energy}} = \underbrace{\langle\boldsymbol{\sigma}\rangle_\Omega : \boldsymbol{I\!E}^{*-1} : \langle\boldsymbol{\sigma}\rangle_\Omega}_{\text{macro energy}}, \quad \text{where} \quad \langle\boldsymbol{\epsilon}\rangle_\Omega = \boldsymbol{I\!E}^{*-1} : \langle\boldsymbol{\sigma}\rangle_\Omega, \tag{3.11}$$

we have

$$\begin{aligned}
0 &\le \int_\Omega \tilde{\boldsymbol{\sigma}} : \boldsymbol{I\!E}^{-1} : \tilde{\boldsymbol{\sigma}}\, d\Omega \\
&= \int_\Omega (\boldsymbol{\sigma} : \boldsymbol{I\!E}^{-1} : \boldsymbol{\sigma} - 2\langle\boldsymbol{\sigma}\rangle_\Omega : \boldsymbol{I\!E}^{-1} : \boldsymbol{\sigma} + \langle\boldsymbol{\sigma}\rangle_\Omega : \boldsymbol{I\!E}^{-1} : \langle\boldsymbol{\sigma}\rangle_\Omega)\, d\Omega \\
&= (\langle\boldsymbol{\sigma}\rangle_\Omega : \boldsymbol{I\!E}^{*-1} : \langle\boldsymbol{\sigma}\rangle_\Omega - 2\langle\boldsymbol{\epsilon}\rangle_\Omega : \langle\boldsymbol{\sigma}\rangle_\Omega + \langle\boldsymbol{\sigma}\rangle_\Omega : \langle\boldsymbol{I\!E}^{-1}\rangle_\Omega : \langle\boldsymbol{\sigma}\rangle_\Omega)|\Omega| \\
&= \langle\boldsymbol{\sigma}\rangle_\Omega : (\langle\boldsymbol{I\!E}^{-1}\rangle_\Omega - \boldsymbol{I\!E}^{*-1}) : \langle\boldsymbol{\sigma}\rangle_\Omega |\Omega|.
\end{aligned} \tag{3.12}$$

Invoking Hill's condition, which is loading-independent in this form, we have

$$\underbrace{\langle \mathbb{E}^{-1}\rangle_{\Omega}^{-1}}_{\text{Reuss}} \le \mathbb{E}^{*} \le \underbrace{\langle \mathbb{E}\rangle_{\Omega}}_{\text{Voigt}}. \tag{3.13}$$

This inequality means that the eigenvalues of the tensors $\mathbb{E}^{*} - \langle \mathbb{E}^{-1}\rangle_{\Omega}^{-1}$ and $\langle \mathbb{E}\rangle_{\Omega} - \mathbb{E}^{*}$ are nonnegative. The practical outcome of the analysis is that bounds on effective properties are obtained. These bounds are commonly known as the Hill–Reuss–Voigt bounds, for historical reasons. Voigt [47], in 1889, assumed that the strain field within a sample of aggregate of polycrystalline material was uniform (constant), under uniform strain exterior loading. If the constant strain Voigt field is assumed within the RVE, $\epsilon = \epsilon^{0}$, then $\langle \sigma\rangle_{\Omega} = \langle \mathbb{E} : \epsilon\rangle_{\Omega} = \langle \mathbb{E}\rangle_{\Omega} : \epsilon^{0}$, which implies $\mathbb{E}^{*} = \langle \mathbb{E}\rangle_{\Omega}$. The dual assumption was made by Reuss [48], in 1929, who approximated the stress fields within the aggregate of polycrystalline material as uniform (constant), $\sigma = \sigma^{0}$, leading to $\langle \epsilon\rangle_{\Omega} = \langle \mathbb{E}^{-1} : \sigma\rangle_{\Omega} = \langle \mathbb{E}^{-1}\rangle_{\Omega} : \sigma^{0}$, and thus to $\mathbb{E}^{*} = \langle \mathbb{E}^{-1}\rangle_{\Omega}^{-1}$.

Remark Different boundary conditions (compared to the standard ones specified earlier) are often used in computational homogenization analysis. For example, periodic boundary conditions are sometimes employed. Although periodic conditions are really only appropriate for perfectly periodic media for many cases, it has been shown that, in some cases, their use can provide better effective responses than either linear displacement or uniform traction boundary conditions (e.g., see Terada et. al. [49] or Segurado and Llorca [50]). Periodic boundary conditions also satisfy Hill's condition a priori. Another related type of boundary condition is the so-called uniform-mixed type, whereby tractions are applied on some parts of the boundary and displacements on other parts, generating, in some cases, effective properties that match those produced with uniform boundary conditions, but with smaller sample sizes (e.g., see Hazanov and Huet [51]). Another approach is "framing," whereby the traction or displacement boundary conditions are applied to a large sample of material, with the averaging computed on an interior subsample to avoid possible boundary layer effects. This method is similar to exploiting a St. Venant type of effect, commonly used in solid mechanics, to avoid boundary layers. The approach provides a way of determining what the microstructure really experiences, without "bias" from the boundary loading. However, generally, the advantages of one boundary condition over another diminish as the sample increases in size.

3.2.6 Improved Estimates

Over the last half-century, improved estimates have been pursued, with a notable contribution being the Hashin–Shtrikman bounds [6, 52, 53]. The Hashin–Shtrikman bounds are the tightest possible bounds on isotropic effective responses, with isotropic microstructures, when the volume fractions and phase contrasts of the constituents are the only data known. For isotropic materials with isotropic effective (mechanical) responses, the Hashin–Shtrikman bounds (for a two-phase material) are as follows for the bulk modulus

$$\kappa^{*,-} \stackrel{\text{def}}{=} \kappa_1 + \frac{v_2}{\frac{1}{\kappa_2-\kappa_1} + \frac{3(1-v_2)}{3\kappa_1+4\mu_1}} \le \kappa^* \le \kappa_2 + \frac{1-v_2}{\frac{1}{\kappa_1-\kappa_2} + \frac{3v_2}{3\kappa_2+4\mu_2}} \stackrel{\text{def}}{=} \kappa^{*,+} \tag{3.14}$$

and for the shear modulus

$$\mu^{*,-} \stackrel{\text{def}}{=} \mu_1 + \frac{v_2}{\frac{1}{\mu_2-\mu_1} + \frac{6(1-v_2)(\kappa_1+2\mu_1)}{5\mu_1(3\kappa_1+4\mu_1)}} \le \mu^* \le \mu_2 + \frac{(1-v_2)}{\frac{1}{\mu_1-\mu_2} + \frac{6v_2(\kappa_2+2\mu_2)}{5\mu_2(3\kappa_2+4\mu_2)}} \stackrel{\text{def}}{=} \mu^{*,+}, \tag{3.15}$$

where κ_2 and κ_1 are the bulk moduli, μ_2 and μ_1 are the shear moduli of the respective phases ($\kappa_2 \ge \kappa_1$ and $\mu_2 \ge \mu_1$), and v_2 is the second-phase volume fraction. Note that no geometric or other microstructural information is required for the bounds.

Remark 1 There exist a multitude of other approaches which seek to estimate or bound the aggregate responses of microheterogeneous materials. A complete survey is outside the scope of the present work. We refer the reader to the works of Hashin [6], Mura [7], Aboudi [54], Nemat-Nasser and Hori [8], Torquato [4] and Zohdi and Wriggers [55] for such reviews.

Remark 2 Numerical methods have become a valuable tool in determining micro–macro relations, with the caveat being that local fields in the microstructure are resolved, which is important in being able to quantify the intensity of the loads experienced by the microstructure. This is important for ascertaining failure of the material. In particular, finite element-based methods are extremely popular for micro–macro calculations. Applying such methods entails generating a sample of material microstructure, meshing it to sufficient resolution for tolerable numerical accuracy, and solving a series of boundary value problems with different test loadings. The effective properties can be determined by post-processing (averaging over the RVE). For an extensive review of this topic, see Zohdi [12–42]. We also refer the reader to that work for more extensive mathematical details and background information.

Remark 3 If needed, one can post-process the effective bulk and shear modulus to obtain the effective Poisson ratio $\nu^* = \frac{3\kappa^*-2\mu^*}{2(3\kappa^*+\mu^*)}$ and the effective Young's modulus $E^* = 2\mu^*(1+\nu^*) = 3\kappa^*(1-2\nu^*)$.

References

1. Maxwell, J.C.: On the dynamical theory of gases. Philos. Trans. Soc. London. **157**, 49 (1867)
2. Maxwell, J.C.: A treatise on electricity and magnetism, 3rd edn. Clarendon Press, Oxford (1873)

3. Rayleigh, J.W.: On the influence of obstacles arranged in rectangular order upon properties of a medium. Phil. Mag. **32**, 481–491 (1892)
4. Torquato, S.: Random Heterogeneous Materials: Microstructure and Macroscopic Properties. Springer, New York (2002)
5. Jikov, V.V., Kozlov, S.M., Olenik, O.A.: Homogenization of Differential Operators and Integral Functionals. Springer (1994)
6. Hashin, Z.: Analysis of composite materials: a survey. ASME J. Appl. Mech. **50**, 481–505 (1983)
7. Mura, T.: Micromechanics of Defects in Solids, 2nd edn. Kluwer Academic Publishers (1993)
8. Nemat-Nasser, S., Hori, M.: Micromechanics: Overall Properties of Heterogeneous Solids, 2nd edn. Elsevier, Amsterdam (1999)
9. Huet, C.: Universal conditions for assimilation of a heterogeneous material to an effective medium. Mecha. Res. Commun. **9**(3), 165–170 (1982)
10. Huet, C.: On the definition and experimental determination of effective constitutive equations for heterogeneous materials. Mech. Res. Commun. **11**(3), 195–200 (1984)
11. Sevostianov, I., Kachanov, M.: Effective properties of heterogeneous materials: proper application of the non-interaction and the "dilute limit" approximations. Int. J. Eng. Sci. **58**, 124–128 (2012)
12. Zohdi, T.I.: Genetic design of solids possessing a random-particulate microstructure. Philos. Trans. R. Soc. Math. Phys. Eng. Sci. **361**(1806), 1021–1043 (2003)
13. Zohdi, T.I.: On the compaction of cohesive hyperelastic granules at finite strains. Proc. R. Soc. **454**(2034), 1395–1401 (2003)
14. Zohdi, T.I.: Constrained inverse formulations in random material design. Comput. Methods Appl. Mech. Eng. 1–20 **192**(28–30), 18, 3179–3194 (2003)
15. Zohdi, T.I.: Statistical ensemble error bounds for homogenized microheterogeneous solids. J. Appl. Math. Phys. (Zeitschrift für Angewandte Mathematik und Physik) **56**(3), 497–515 (2005)
16. Zohdi, T.I., Kachanov, M.: A note on the micromechanics of plastic yield of porous solids. Int. J. Fract./Lett. Micromechanics **133**, L31–L35 (2005)
17. Zohdi, T.I.: Particle collision and adhesion under the influence of near-fields. J. Mech. Mater. Struct. **2**(6), 1011–1018 (2007)
18. Zohdi, T.I.: On the computation of the coupled thermo-electromagnetic response of continua with particulate microstructure. Int. J. Numer. Methods Eng. **76**, 1250–1279 (2008)
19. Zohdi, T.I.: Mechanistic modeling of swarms. Comput. Methods Appl. Mech. Eng. **198**(21–26), 2039–2051 (2009)
20. Zohdi, T.I.: Charged wall-growth in channel-flow. Int. J. Eng. Sci. **48**, 1520 (2010)
21. Zohdi, T.I.: On the dynamics of charged electromagnetic particulate jets. Arch. Comput. Methods Eng. **17**(2), 109–135 (2010)
22. Zohdi, T.I., Kuypers, F.A., Lee, W.C.: Estimation of Red Blood Cell volume fraction from overall permittivity measurement. Int. J. Eng. Sci. **48**, 1681–1691 (2010)
23. Zohdi, T.I.: Simulation of coupled microscale multiphysical-fields in particulate-doped dielectrics with staggered adaptive FDTD. Comput. Methods Appl. Mech. Eng. **199**, 79–101 (2010)
24. Zohdi, T.I.: Dynamics of clusters of charged particulates in electromagnetic fields. Int. J. Numer. Methods Eng. **85**, 1140–1159 (2011)
25. Zohdi, T.I.: Joule-heating field phase-amplification in particulate-doped dielectrics. Int. J. Eng. Sci. **49**, 30–40 (2011)
26. Zohdi, T.I.: Electromagnetically-induced deformation of functionalized fabric. J. Elast. **105**(1–2), 381–398 (2011)
27. Zohdi, T.I.: Estimation of electrical-heating load-shares for sintering of powder mixtures. Proc. R. Soc. **468**, 2174–2190 (2012)
28. Zohdi, T.I.: Modeling and simulation of the optical response rod-functionalized reflective surfaces. Comput. Mech. **50**(2), 257–268 (2012)
29. Zohdi, T.I.: On the reduction of heat generation in lubricants using microscale additives. Int. J. Eng. Sci. **62**, 84–89 (2013)

30. Zohdi, T.I.: Electromagnetically-induced vibration in particulate-doped materials. ASME J. Vib. Acoust. **135**(3) (2013). https://doi.org/10.1115/1.4023251
31. Zohdi, T.I.: Numerical simulation of charged particulate cluster-droplet impact on electrified surfaces. J. Comput. Phys. **233**, 509–526 (2013)
32. Zohdi, T.I.: On inducing compressive residual stress in microscale print-lines for flexible electronics. Int. J. Eng. Sci. **62**, 157–164 (2013)
33. Zohdi, T.I.: Rapid simulation of laser processing of discrete particulate materials. Arch. Comput. Methods Eng. **20**, 309–325 (2013)
34. Zohdi, T.I.: A direct particle-based computational framework for electrically-enhanced thermo-mechanical sintering of powdered materials. Math. Mech. Solids. **19**(1), 93–113 (2014)
35. Zohdi, T.I.: On cross-correlation between thermal gradients and electric fields. Int. J. Eng. Sci. **74**, 143–150 (2014)
36. Zohdi, T.I.: Mechanically-driven accumulation of microscale material at coupled solid-fluid interfaces in biological channels. Proc. R. Soc. Interface **11**, 20130922 (2014)
37. Zohdi, T.I.: A computational modeling framework for heat transfer processes in laser-induced dermal tissue removal. Comput. Mech. Eng. Sci. **98**(3), 261–277 (2014)
38. Zohdi, T.I.: Additive particle deposition and selective laser processing-a computational manufacturing framework. Comput. Mech. **54**, 171–191 (2014)
39. Zohdi, T.I.: Embedded electromagnetically sensitive particle motion in functionalized fluids. Computat. Part. Mech. **1**, 27–45 (2014)
40. Zohdi, T.I.: Impact and penetration resistance of network models of coated lightweight fabric shielding. GAMM-Mitteilungen, vol. 37, Issue 1, p. 124150 (2014)
41. Zohdi, T.I.: Rapid computation of statistically-stable particle/feature ratios for consistent substrate stresses in printed flexible electronics. J. Manuf. Sci. Eng. ASME MANU-14-1476 (2015). https://doi.org/10.1115/1.4029327
42. Zohdi, T.I.: A computational modelling framework for high-frequency particulate obscurant cloud performance. Int. J. Eng. Sci. **89**, 75–85 (2015)
43. Ghosh, S.: Micromechanical Analysis and Multi-Scale Modeling Using the Voronoi Cell Finite Element Method. CRC Press/Taylor & Francis (2011)
44. Ghosh, S., Dimiduk, D.: Computational Methods for Microstructure-Property Relations. Springer, NY (2011)
45. Hill, R.: The elastic behaviour of a crystalline aggregate. Proc. Phys. Soc. (Lond.) **A65**, 349–354 (1952)
46. Chandrasekharaiah, D.S., Debnath, L.: Continuum Mechanics. Academic press (1994)
47. Voigt, W.: Über die Beziehung zwischen den beiden Elastizitätskonstanten isotroper Körper. Wied. Ann. **38**, 573–587 (1889)
48. Reuss, A.: Berechnung der Fliessgrenze von Mischkristallen auf Grund der Plastizitätsbedingung für Einkristalle. Z. Angew. Math. Mech. **9**, 49–58 (1929)
49. Terada, K., Hori, M., Kyoya, T., Kikuchi, N.: Simulation of the multi-scale convergence in computational homogenization approaches. Int. J. Solids Struct. **37**, 2229–2361 (2000)
50. Segurado, J., Llorca, J.: A numerical approximation to the elastic properties of sphere-reinforced composites. J. Mech. Phys. Solids **50** (2002)
51. Hazanov, S., Huet, C.: Order relationships for boundary conditions effect in heterogeneous bodies smaller than the representative volume. J. Mech. Phys. Solids **42**, 1995–2011 (1994)
52. Hashin, Z., Shtrikman, S.: On some variational principles in anisotropic and nonhomogeneous elasticity. J. Mech. Phys. Solids **10**, 335–342 (1962)
53. Hashin, Z., Shtrikman, S.: A variational approach to the theory of the elastic behaviour of multiphase materials. J. Mech. Phys. Solids **11**, 127–140 (1963)
54. Aboudi, J.: Mechanics of Composite Materials-a Unified Micromechanical Approach, p. 29. Elsevier (1992)
55. Zohdi, T.I., Wriggers, P.: Introduction to computational micromechanics, Springer Verlag (2008)

Chapter 4
CM Approaches: Estimation and Optimization of the Effective Properties of Mixtures

4.1 Combining Bounds

The typical use of the bounds from the previous chapter is to make an estimate of the effective properties by forming a convex combination of them in the following manner:

$$\kappa^* \approx \phi\kappa^{*,+} + (1-\phi)\kappa^{*,-} \tag{4.1}$$

and

$$\mu^* \approx \phi\mu^{*,+} + (1-\phi)\mu^{*,-}, \tag{4.2}$$

where $0 \le \phi \le 1$ is a parameter such that:

- If $\phi = 0$, we have the lower bound.
- If $\phi = 1$, we have the upper bound.
- If $\phi = 1/2$, we have the average of the bounds.

ϕ is a function of the microstructure and must be calibrated.

A critical observation is that the lower bound is more accurate when the material is composed of stiff particles that are surrounded by a soft matrix (denoted case 1), and the upper bound is more accurate for a stiff matrix surrounding soft particles (denoted case 2). This can be explained by considering two cases of material combinations, one with 50% soft material and the other with 50% stiff material. A material with a continuous soft binder (50%) will isolate the stiff particles (50%), and the overall system will not be stiff (this is case 1, and the lower bound is more accurate), while a material formed by a continuous stiff binder (50%) surrounding soft particles (50%, case 2) will, in an overall sense, be stiffer than case 1. Thus, case 2 is more closely approximated by the upper bound and case 1 is closer to the lower bound (Fig. 4.1).

T. I. Zohdi, *Modeling and Simulation of Functionalized Materials for Additive Manufacturing and 3D Printing: Continuous and Discrete Media*, Lecture Notes in Applied and Computational Mechanics 60, https://doi.org/10.1007/978-3-319-70079-3_4

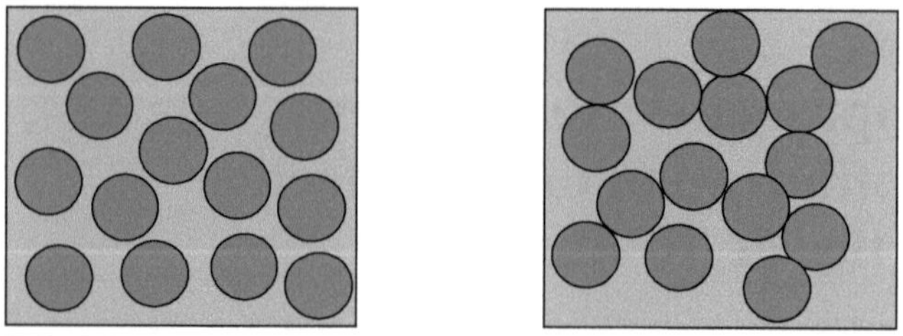

Fig. 4.1 Comparing microstructures with the same volume fractions. Cases where particles touch need a higher value of ϕ

As mentioned, for stiff spherical particles, at low volume fractions, for example under 15%, where the particles are not making contact, the lower bound is more accurate. Thus, one would pick $\phi = \phi^s \leq 1/2$ to bias the estimate to the lower bound. However, if we take the same volume fraction of particles, but make the flat flakes, they will certainly touch and produce stiff pathways. Their overall stiffness will be higher than those of spheres at the same volume fraction. Thus, one would pick $\phi = \phi^f > \phi^s$. One can calibrate ϕ by comparing it to different experiments. For example, for mechanical properties, see Zohdi et al. [1]. Essentially, the more the particles interact, for example, physically touch, the more the upper bound becomes relevant. The general trends are (a) for cases where the upper bound is more accurate, $\phi > \frac{1}{2}$ and (b) for cases when the lower bound is more accurate, $\phi < \frac{1}{2}$. ϕ indicates the degree of interaction of the particulate constituents.

Remark This same trend holds for electrical and thermal properties (see Monograph Appendices 2 and 3).

4.2 Local Fields: Stresses and Strains

The determination of the average load sharing between phases at the microstructural scale can be obtained from the overall effective mechanical properties of the micro-heterogeneous material, for example, comprised of particles suspended in a binding matrix.

The load carried by each phase in the microstructure is characterized via stress and strain concentration tensors, which we now discuss. These provide a measure of the deviation away from the mean fields throughout the material. One can decompose averages of an arbitrary quantity over Ω into averages over each of the phases in the following manner: $\langle \boldsymbol{A} \rangle_\Omega = (1/|\Omega|) \left(\int_{\Omega_1} \boldsymbol{A}\, d\Omega + \int_{\Omega_2} \boldsymbol{A}\, d\Omega \right) = v_1 \langle \boldsymbol{A} \rangle_{\Omega_1} + v_2 \langle \boldsymbol{A} \rangle_{\Omega_2}$. If we make use of this decomposition, we have

$$\begin{aligned}
\langle\boldsymbol{\sigma}\rangle_{\Omega} &= v_1\langle\boldsymbol{\sigma}\rangle_{\Omega_1} + v_2\langle\boldsymbol{\sigma}\rangle_{\Omega_2} \\
&= v_1\mathbb{E}_1 : \langle\boldsymbol{\epsilon}\rangle_{\Omega_1} + v_2\mathbb{E}_2 : \langle\boldsymbol{\epsilon}\rangle_{\Omega_2} \\
&= \mathbb{E}_1 : (\langle\boldsymbol{\epsilon}\rangle_{\Omega} - v_2\langle\boldsymbol{\epsilon}\rangle_{\Omega_2}) + v_2\mathbb{E}_2 : \langle\boldsymbol{\epsilon}\rangle_{\Omega_2} \\
&= \left(\mathbb{E}_1 + v_2(\mathbb{E}_2 - \mathbb{E}_1) : \boldsymbol{C}^{\epsilon,2}\right) : \langle\boldsymbol{\epsilon}\rangle_{\Omega},
\end{aligned} \tag{4.3}$$

where $\boldsymbol{C}^{\epsilon,2} \stackrel{\text{def}}{=} \left(\frac{1}{v_2}(\mathbb{E}_2 - \mathbb{E}_1)^{-1} : (\mathbb{E}^* - \mathbb{E}_1)\right)$ with $\boldsymbol{C}^{\epsilon,2} : \langle\boldsymbol{\epsilon}\rangle_{\Omega} = \langle\boldsymbol{\epsilon}\rangle_{\Omega_2}$. The strain concentration tensor $\boldsymbol{C}^{\epsilon,2}$ relates the average strain over the particle phase (2) to the average strain over all phases. Similarly, for the variation in the stress we have $\boldsymbol{C}^{\epsilon,2} : \mathbb{E}^{*-1} : \langle\boldsymbol{\sigma}\rangle_{\Omega} = \mathbb{E}_2^{-1} : \langle\boldsymbol{\sigma}\rangle_{\Omega_2}$, which reduces to $\mathbb{E}_2 : \boldsymbol{C}^{\epsilon,2} : \mathbb{E}^{*-1} : \langle\boldsymbol{\sigma}\rangle_{\Omega} \stackrel{\text{def}}{=} \boldsymbol{C}^{\sigma,2} : \langle\boldsymbol{\sigma}\rangle_{\Omega} = \langle\boldsymbol{\sigma}\rangle_{\Omega_2}$. $\boldsymbol{C}^{\sigma,2}$ is known as the stress concentration tensor; it relates the average stress in the particle phase to that in the whole RVE. Note that once either $\boldsymbol{C}^{\epsilon,2}$ or $\mathbb{E}^*$ are known, the other can be determined. In the case of isotropy, we may write

$$C_{\kappa}^{\sigma,2} \stackrel{\text{def}}{=} \frac{1}{v_2}\frac{\kappa_2}{\kappa^*}\frac{\kappa^* - \kappa_1}{\kappa_2 - \kappa_1} \quad \text{and} \quad C_{\mu}^{\sigma,2} \stackrel{\text{def}}{=} \frac{1}{v_2}\frac{\mu_2}{\mu^*}\frac{\mu^* - \mu_1}{\mu_2 - \mu_1} \tag{4.4}$$

where $C_{\kappa}^{\sigma,2}\langle\frac{tr\boldsymbol{\sigma}}{3}\rangle_{\Omega} = \langle\frac{tr\boldsymbol{\sigma}}{3}\rangle_{\Omega_2}$ and $C_{\mu}^{\sigma,2}\langle\boldsymbol{\sigma}'\rangle_{\Omega} = \langle\boldsymbol{\sigma}'\rangle_{\Omega_2}$. Clearly, the microstress fields are minimally distorted when $C_{\kappa}^{\sigma,2} = C_{\mu}^{\sigma,2} = 1$; there are no stress concentrations in a homogeneous material. For the matrix,

$$\begin{aligned}
\langle\boldsymbol{\sigma}\rangle_{\Omega_1} &= \frac{\langle\boldsymbol{\sigma}\rangle_{\Omega} - v_2\langle\boldsymbol{\sigma}\rangle_{\Omega_2}}{v_1} = \frac{\langle\boldsymbol{\sigma}\rangle_{\Omega} - v_2\boldsymbol{C}^{\sigma,2} : \langle\boldsymbol{\sigma}\rangle_{\Omega}}{v_1} \\
&= \frac{(\mathbf{1} - v_2\boldsymbol{C}^{\sigma,2}) : \langle\boldsymbol{\sigma}\rangle_{\Omega}}{v_1} \stackrel{\text{def}}{=} \boldsymbol{C}^{\sigma,1} : \langle\boldsymbol{\sigma}\rangle_{\Omega}.
\end{aligned} \tag{4.5}$$

Therefore, in the case of isotropy,

$$C_{\kappa}^{\sigma,1} \stackrel{\text{def}}{=} \frac{1}{v_1}(1 - v_2C_{\kappa}^{\sigma,2}) \quad \text{and} \quad C_{\mu}^{\sigma,1} \stackrel{\text{def}}{=} \frac{1}{v_1}(1 - v_2C_{\mu}^{\sigma,2}). \tag{4.6}$$

The fraction of the total stress carried by each phase can be determined by multiplying the concentration factors by the corresponding volume fractions

$$\begin{aligned}
\langle\boldsymbol{\sigma}\rangle_{\Omega} &= v_1\langle\boldsymbol{\sigma}\rangle_{\Omega_1} + v_2\langle\boldsymbol{\sigma}\rangle_{\Omega_2} \\
&= v_1\boldsymbol{C}^{\sigma,1} : \langle\boldsymbol{\sigma}\rangle_{\Omega} + v_2\boldsymbol{C}^{\sigma,2} : \langle\boldsymbol{\sigma}\rangle_{\Omega}.
\end{aligned} \tag{4.7}$$

Remark Similar to the stress, for the strain, we have for the matrix,

$$\begin{aligned}
\langle\boldsymbol{\epsilon}\rangle_{\Omega_1} &= \frac{\langle\boldsymbol{\epsilon}\rangle_{\Omega} - v_2\langle\boldsymbol{\epsilon}\rangle_{\Omega_2}}{v_1} = \frac{\langle\boldsymbol{\epsilon}\rangle_{\Omega} - v_2\boldsymbol{C}^{\epsilon,2} : \langle\boldsymbol{\epsilon}\rangle_{\Omega}}{v_1} \\
&= \frac{(\mathbf{1} - v_2\boldsymbol{C}^{\epsilon,2}) : \langle\boldsymbol{\epsilon}\rangle_{\Omega}}{v_1} \stackrel{\text{def}}{=} \boldsymbol{C}^{\epsilon,1} : \langle\boldsymbol{\epsilon}\rangle_{\Omega}.
\end{aligned} \tag{4.8}$$

Therefore, in the case of isotropy,

$$C_{\kappa}^{\epsilon,1} \stackrel{\text{def}}{=} \frac{1}{v_1}(1 - v_2 C_{\kappa}^{\epsilon,2}) \quad \text{and} \quad C_{\mu}^{\epsilon,1} \stackrel{\text{def}}{=} \frac{1}{v_1}(1 - v_2 C_{\mu}^{\epsilon,2}). \tag{4.9}$$

The fraction of the total strain carried by each phase can be determined by multiplying the concentration factors by the corresponding volume fractions

$$\begin{aligned} \langle \boldsymbol{\epsilon} \rangle_{\Omega} &= v_1 \langle \boldsymbol{\epsilon} \rangle_{\Omega_1} + v_2 \langle \boldsymbol{\epsilon} \rangle_{\Omega_2} \\ &= v_1 \boldsymbol{C}^{\epsilon,1} : \langle \boldsymbol{\epsilon} \rangle_{\Omega} + v_2 \boldsymbol{C}^{\epsilon,2} : \langle \boldsymbol{\epsilon} \rangle_{\Omega}. \end{aligned} \tag{4.10}$$

4.3 Optimization: Formulation of a Cost Function

The deviation in the particulate stress fields from the mean value is

$$\left|\frac{\langle tr\boldsymbol{\sigma} \rangle_{\Omega_2} - \langle tr\boldsymbol{\sigma} \rangle_{\Omega}}{\langle tr\boldsymbol{\sigma} \rangle_{\Omega}}\right| = |C_{\kappa}^{\sigma,2} - 1| \tag{4.11}$$

and

$$\sqrt{\frac{(\langle \boldsymbol{\sigma}' \rangle_{\Omega_2} - \langle \boldsymbol{\sigma}' \rangle_{\Omega}) : (\langle \boldsymbol{\sigma}' \rangle_{\Omega_2} - \langle \boldsymbol{\sigma}' \rangle_{\Omega})}{\langle \boldsymbol{\sigma}' \rangle_{\Omega} : \langle \boldsymbol{\sigma}' \rangle_{\Omega}}} = |C_{\mu}^{\sigma,2} - 1|, \tag{4.12}$$

and for the matrix material

$$\left|\frac{\langle tr\boldsymbol{\sigma} \rangle_{\Omega_1} - \langle tr\boldsymbol{\sigma} \rangle_{\Omega}}{\langle tr\boldsymbol{\sigma} \rangle_{\Omega}}\right| = |C_{\kappa}^{\sigma,1} - 1| \tag{4.13}$$

and

$$\sqrt{\frac{(\langle \boldsymbol{\sigma}' \rangle_{\Omega_1} - \langle \boldsymbol{\sigma}' \rangle_{\Omega}) : (\langle \boldsymbol{\sigma}' \rangle_{\Omega_1} - \langle \boldsymbol{\sigma}' \rangle_{\Omega})}{\langle \boldsymbol{\sigma}' \rangle_{\Omega} : \langle \boldsymbol{\sigma}' \rangle_{\Omega}}} = |C_{\mu}^{\sigma,1} - 1|. \tag{4.14}$$

In order to incorporate the deviation into a cost function, we introduce a tolerance where, ideally,

$$|C_{\kappa}^{\sigma,2} - 1| \leq TOL_{\kappa} \quad \text{and} \quad |C_{\mu}^{\sigma,2} - 1| \leq TOL_{\mu} \tag{4.15}$$

and

$$|C_{\kappa}^{\sigma,1} - 1| \leq TOL_{\kappa} \quad \text{and} \quad |C_{\mu}^{\sigma,1} - 1| \leq TOL_{\mu}. \tag{4.16}$$

If the normalized deviation exceeds the corresponding TOL, then the level of violation is incorporated as a multilateral constraint to the macroscopic objectives. As an example, our immediate goal is to formulate the design of the macroscale effective bulk and shear moduli κ^* and μ^*, using convex combinations of the Hashin–Shtrikman bounds as approximations for the effective moduli $\kappa^* \approx \phi\kappa^{*,+} + (1-\phi)\kappa^{*,-}$ and $\mu^* \approx \phi\mu^{*,+} + (1-\phi)\mu^{*,-}$, where $0 \leq \phi \leq 1$. The micro–macro objective function is

$$
\begin{aligned}
\Pi = w_1 \left|\frac{\kappa^*}{\kappa^{*,D}} - 1\right|^2 + w_2 \left|\frac{\mu^*}{\mu^{*,D}} - 1\right|^2 \\
+\hat{w}_3 \left(|C_\kappa^{\sigma,2} - 1| - TOL_\kappa\right)^2 + \hat{w}_4 \left(|C_\mu^{\sigma,2} - 1| - TOL_\mu\right)^2 \\
+\hat{w}_5 \left(|C_\kappa^{\sigma,1} - 1| - TOL_\kappa\right)^2 + \hat{w}_6 \left(|C_\mu^{\sigma,1} - 1| - TOL_\mu\right)^2,
\end{aligned}
$$

where (I) if $|C_\kappa^{\sigma,2} - 1| \leq TOL_\kappa$, then $\hat{w}_3 = 0$, (II) if $|C_\kappa^{\sigma,2} - 1| > TOL_\kappa$, then $\hat{w}_3 = w_3$, (III) if $|C_\mu^{\sigma,2} - 1| \leq TOL_\mu$, then $\hat{w}_4 = 0$, (IV) if $|C_\mu^{\sigma,2} - 1| > TOL_\mu$, then $\hat{w}_4 = w_4$, (V) if $|C_\kappa^{\sigma,1} - 1| \leq TOL_\kappa$, then $\hat{w}_5 = 0$, (VI) if $|C_\kappa^{\sigma,1} - 1| > TOL_\kappa$, then $\hat{w}_5 = w_5$, (VII) if $|C_\mu^{\sigma,1} - 1| \leq TOL_\mu$, then $\hat{w}_6 = 0$, (VIII) if $|C_\mu^{\sigma,1} - 1| > TOL_\mu$, then $\hat{w}_6 = w_6$. Here, the design variables are $\Lambda = \{\kappa_2, \mu_2\, v_2\}$, and their constrained ranges are $\kappa_2^{(-)} \leq \kappa_2 \leq \kappa_2^{(+)}$, $\mu_2^{(-)} \leq \mu_2 \leq \mu_2^{(+)}$, and $v_2^{(-)} \leq v_2 \leq v_2^{(+)}$. There are two characteristics of such a formulation which make the application of standard gradient-type minimization schemes, such as Newton's method, difficult:

- (I) The incorporation of limits on the microfield behavior, as well as design search space restrictions, renders the objective function not continuously differentiable in design space and
- (II) The objective function is nonconvex; i.e., the system Hessian is not positive definite (invertible) throughout design space.

One way to minimize such objective functions is by following a two-stage approach whereby one (1) determines promising optimal regions in parameter space using (nonderivative) algorithms (such as evolutionary "genetic" algorithms, simulated annealing) and then (2) applies classical gradient-based schemes in locally convex regions, if the objective functions are smooth, since they are generally extremely efficient for the minimization of smooth convex functions. As indicated, the search for convex "pockets" of Π can be achieved by using "genetic" algorithms (GA), before applying classical gradient-based schemes.[1]

Genetic algorithms are search methods based on the principles of natural selection, employing concepts of species evolution, such as reproduction, mutation, and crossover. Implementation typically involves a randomly generated population of fixed-length elemental strings, "genetic" information, each of which represents a specific choice of system parameters. The population of individuals undergoes "mating sequences" and other biologically inspired events in order to find promising regions

[1] An exhaustive review of these methods can be found in the texts of Luenberger [2] and Gill, Murray and Wright [3], while a state of the art can be found in Papadrakakis et al. [4].

of the search space. Such methods can be traced back, at least, to the work of John Holland [5]. For reviews of such methods, see, for example, Goldberg [6], Davis [7], Onwubiko [8], Kennedy and Eberhart [9], Lagaros et al. [10], Papadrakakis et al. [11–14], and Goldberg and Deb [15]. In Zohdi [16–33], a genetic algorithm has been developed to treat a wide variety of nonconvex inverse problems involving various aspects of multiparticle mechanics, and we refer the interested reader to that work. Specifically, the central idea is that the system parameters form a genetic string and a survival of the fittest algorithm is applied to a population of such strings.

The overall process is: (a) a population (S) of different parameter sets are generated at random within the parameter space, each represented by a ("genetic") string of the system (N) parameters, (b) the performance of each parameter set is tested, (c) the parameter sets are ranked from top to bottom according to their performance, (d) the best parameter sets (parents) are mated pair-wise producing two offspring (children); i.e., each best pair exchanges information by taking random convex combinations of the parameter set components of the parents' genetic strings, and (e) the worst performing genetic strings are eliminated, then new replacement parameter sets (genetic strings) are introduced into the remaining population of best-performing genetic strings, and the process (a–e) is then repeated.

The term "fitness" of a genetic string is used to indicate the value of the objective function. The most fit genetic string is the one with the smallest objective function. The retention of the top fit genetic strings from a previous generation (parents) is critical, since if the objective functions are highly nonconvex (the present case), there exists a clear possibility that the inferior offspring will replace superior parents. When the top parents are retained, the minimization of the cost function is guaranteed to be monotone (guaranteed improvement) with increasing generations. There is no guarantee of successive improvement if the top parents are not retained, even though nonretention of parents allows more new genetic strings to be evaluated in the next generation. In the scientific literature, numerical studies imply that, for sufficiently large populations, the benefits of parent retention outweigh this advantage and any disadvantages of "inbreeding," i.e., a stagnant population. For more details on this so-called inheritance property, see Davis [7] or Kennedy and Eberhart [9]. In the upcoming algorithm, inbreeding is mitigated since, with each new generation, new parameter sets, selected at random within the parameter space, are added to the population. Previous numerical studies of the author [16–33] have indicated that not retaining the parents is suboptimal due to the possibility that inferior offspring will replace superior parents. Additionally, parent retention is computationally less expensive, since these parameter sets do not have to be re-evaluated (or ranked) in the next generation. An implementation of such ideas is as follows [16–33]:

- **STEP 1:** Randomly generate a population of S starting genetic strings, $\boldsymbol{\Lambda}^i$, $(i = 1, ..., S)$:
  $\boldsymbol{\Lambda}^i \stackrel{\mathrm{def}}{=} \{\Lambda^i_1, \Lambda^i_2, \Lambda^i_3, \Lambda^i_4, ...\Lambda^i_N\} \stackrel{\mathrm{def}}{=} \{\kappa^i_2, \mu^i_2, v^i_2, ...\}$.
- **STEP 2:** Compute fitness of each string $\Pi(\boldsymbol{\Lambda}^i)$, (i = 1, ..., S).

- **STEP 3:** Rank genetic strings: $\boldsymbol{\Lambda}^i$, (i = 1, ..., S).
- **STEP 4:** Mate nearest pairs and produce two offspring, (i = 1, ..., S)
  $\boldsymbol{\lambda}^i \stackrel{\text{def}}{=} \Phi^{(I)}\boldsymbol{\Lambda}^i + (1-\Phi^{(I)})\boldsymbol{\Lambda}^{i+1}, \quad \boldsymbol{\lambda}^{i+1} \stackrel{\text{def}}{=} \Phi^{(II)}\boldsymbol{\Lambda}^i + (1-\Phi^{(II)})\boldsymbol{\Lambda}^{i+1}.$
- **NOTE:** $\Phi^{(I)}$ and $\Phi^{(II)}$ are random numbers, such that $0 \leq \Phi^{(I)}, \Phi^{(II)} \leq 1$, which are different for each component of each genetic string.
- **STEP 5:** Kill off bottom $M < S$ strings and keep top $K < N$ parents and top K offspring (K offspring+K parents+M=S).
- **STEP 6:** Repeat STEPS 1–6 with top gene pool (K offspring and K parents), plus M new, randomly generated, strings.
- **Option:** Rescale and restart search around best-performing parameter set every few generations.

Remark 1 STEPS 1–6, which are associated with the genetic part of the overall algorithm, attempt to collect multiple local minima.[2] At first glance, it seems somewhat superfluous to retain even the top parents in such an algorithm. However, many studies have shown that the retention of the top old fit genetic strings is critical for proper convergence. As alluded to earlier, by observing Fig. 4.2 one sees that if the objective functions are highly nonconvex, there exists a strong possibility that the inferior offspring will replace superior parents. Therefore, retaining the top parents is not only less computationally expensive, since these designs do not have to be re-evaluated, it is theoretically superior. With parent retention, the minimization of the cost function is guaranteed to be monotone with increasing generations, i.e., $\Pi(\boldsymbol{\Lambda}^{opt,I}) \geq \Pi(\boldsymbol{\Lambda}^{opt,I+1})$, where $\boldsymbol{\Lambda}^{opt,I+1}$ and $\boldsymbol{\Lambda}^{opt,I}$ are the best genetic strings from generations $I+1$ and I, respectively. There is no such guarantee if the top parents are not retained. While the nonretention of parents allows more newer genetic strings to be evaluated in the next generation, numerical studies conducted thus far imply, for sufficiently large populations, that the benefits of parent retention outweigh this advantage, as well as any disadvantages of "inbreeding," i.e., a stagnant population. The case of inbreeding is circumvented in the current algorithm due to the fact that, with each new generation, new material designs, selected at random within the design space, are introduced into the population. Not retaining the parents is suboptimal due to the possibility that inferior offspring will replace superior parents.

Remark 2 After application of such a global search algorithm, one can apply a gradient-based method if the objective function is sufficiently smooth in that region of the parameter space. In other words, if one has located a convex portion of the parameter space with a global genetic search, one can employ gradient-based procedures locally to minimize the objective function further, since they are generally much more efficient for convex optimization of smooth functions. In such procedures, in order to obtain a new directional step for $\boldsymbol{\Lambda}$, one must solve the following system

---

[2]It is remarked that if the function Φ is allowed to be greater than unity, one can consider the resulting convex combination (offspring) as a "mutation."

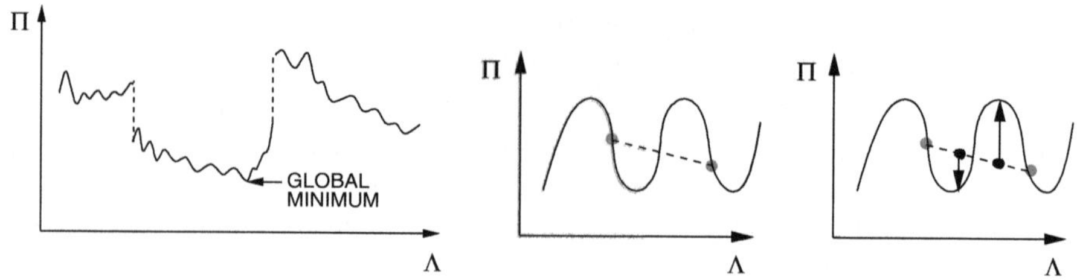

Fig. 4.2 Left: a characterization of the class of objective functions of interest. Right: a loss of superior older genetic strings if the top parents are not retained

$$[\boldsymbol{H}]\{\Delta\boldsymbol{\Lambda}\} = -\{g\}, \tag{4.17}$$

where $[\boldsymbol{H}]$ is the Hessian matrix ($N \times N$), $\{\Delta\boldsymbol{\Lambda}\}$ is the parameter increment ($N \times 1$), and $\{g\}$ is the gradient ($N \times 1$). We shall not employ this second (post-genetic) stage in this work. Specifically, this is determined by forcing the gradient of $\nabla_{\boldsymbol{\Lambda}}\Pi(\boldsymbol{\Lambda}) = \boldsymbol{0}$. Expanding (linearizing) around a first guess $\boldsymbol{\Lambda}^i$ yields

$$\nabla_{\boldsymbol{\Lambda}}\Pi(\boldsymbol{\Lambda}^{i+1}) \approx \nabla_{\boldsymbol{\Lambda}}\Pi(\boldsymbol{\Lambda}^i) + \nabla\left(\nabla_{\boldsymbol{\Lambda}}\Pi(\boldsymbol{\Lambda}^i)\right)\cdot(\boldsymbol{\Lambda}^{i+1} - \boldsymbol{\Lambda}^i) + \textit{higher order terms} \approx \boldsymbol{0} \tag{4.18}$$

or, in more streamlined matrix notation, defining the Hessian, $[\boldsymbol{H}] = \nabla\left(\nabla_{\boldsymbol{\Lambda}}\Pi(\boldsymbol{\Lambda})\right)$ and $\{g\} = \nabla_{\boldsymbol{\Lambda}}\Pi(\boldsymbol{\Lambda})$,

$$[\boldsymbol{H}]\{\Delta\boldsymbol{\Lambda}\} + \{g\} = \boldsymbol{0}. \tag{4.19}$$

Following a standard Newton-type multivariate search, a new design increment is computed,

$$\Delta\boldsymbol{\Lambda} = (\Delta\Lambda_1, \Delta\Lambda_2, \ldots\Delta\Lambda_N), \tag{4.20}$$

for a design vector, $\boldsymbol{\Lambda}$, by solving the following system, $[\boldsymbol{H}]\{\Delta\boldsymbol{\Lambda}\} = -\{g\}$, where $[\boldsymbol{H}]$ is the Hessian matrix ($N \times N$), with components

$$H_{ij} = \frac{\partial^2\Pi(\boldsymbol{\Lambda})}{\partial\Lambda_i\partial\Lambda_j}, \tag{4.21}$$

$\{g\}$ is the gradient ($N \times 1$), with components

$$g_i = \frac{\partial\Pi(\boldsymbol{\Lambda})}{\partial\Lambda_i} \tag{4.22}$$

and $\{\Delta \boldsymbol{\Lambda}\}$ is the design increment ($N \times 1$), with components $\Delta\Lambda_i$. After the design increment has been solved for, one then forms an updated design vector, $\boldsymbol{\Lambda}^{new} = \boldsymbol{\Lambda}^{old} + \Delta\boldsymbol{\Lambda}$, and the process is repeated until $||\Pi|| \leq TOL$. Explicitly, the incremental system is

$$\begin{bmatrix} \frac{\partial^2\Pi(\Lambda)}{\partial\Lambda_1\partial\Lambda_1} & \frac{\partial^2\Pi(\Lambda)}{\partial\Lambda_1\partial\Lambda_2} & \frac{\partial^2\Pi(\Lambda)}{\partial\Lambda_1\partial\Lambda_3} & \frac{\partial^2\Pi(\Lambda)}{\partial\Lambda_1\partial\Lambda_4} & \cdots\cdots \\ \frac{\partial^2\Pi(\Lambda)}{\partial\Lambda_2\partial\Lambda_1} & \frac{\partial^2\Pi(\Lambda)}{\partial\Lambda_2\partial\Lambda_2} & \frac{\partial^2\Pi(\Lambda)}{\partial\Lambda_2\partial\Lambda_3} & \frac{\partial^2\Pi(\Lambda)}{\partial\Lambda_2\partial\Lambda_4} & \cdots\cdots \\ \frac{\partial^2\Pi(\Lambda)}{\partial\Lambda_3\partial\Lambda_1} & \frac{\partial^2\Pi(\Lambda)}{\partial\Lambda_3\partial\Lambda_2} & \frac{\partial^2\Pi(\Lambda)}{\partial\Lambda_3\partial\Lambda_3} & \frac{\partial^2\Pi(\Lambda)}{\partial\Lambda_3\partial\Lambda_4} & \cdots\cdots \\ \frac{\partial^2\Pi(\Lambda)}{\partial\Lambda_4\partial\Lambda_1} & \frac{\partial^2\Pi(\Lambda)}{\partial\Lambda_4\partial\Lambda_2} & \frac{\partial^2\Pi(\Lambda)}{\partial\Lambda_4\partial\Lambda_3} & \frac{\partial^2\Pi(\Lambda)}{\partial\Lambda_4\partial\Lambda_4} & \cdots\cdots \\ \cdots\cdots & \cdots\cdots & \cdots\cdots & \cdots\cdots & \cdots\cdots \\ \cdots\cdots & \cdots\cdots & \cdots\cdots & \cdots\cdots & \cdots\cdots \\ \frac{\partial^2\Pi(\Lambda)}{\partial\Lambda_N\partial\Lambda_1} & \frac{\partial^2\Pi(\Lambda)}{\partial\Lambda_N\partial\Lambda_2} & \frac{\partial^2\Pi(\Lambda)}{\partial\Lambda_N\partial\Lambda_3} & \frac{\partial^2\Pi(\Lambda)}{\partial\Lambda_N\partial\Lambda_4} & \cdots\cdots \end{bmatrix} \begin{Bmatrix} \Delta\Lambda_1 \\ \Delta\Lambda_2 \\ \Delta\Lambda_3 \\ \Delta\Lambda_4 \\ \cdots\cdots \\ \cdots\cdots \\ \Delta\Lambda_N \end{Bmatrix} = - \begin{Bmatrix} \frac{\partial\Pi(\Lambda)}{\partial\Lambda_1} \\ \frac{\partial\Pi(\Lambda)}{\partial\Lambda_2} \\ \frac{\partial\Pi(\Lambda)}{\partial\Lambda_3} \\ \frac{\partial\Pi(\Lambda)}{\partial\Lambda_4} \\ \cdots\cdots \\ \cdots\cdots \\ \frac{\partial\Pi(\Lambda)}{\partial\Lambda_N} \end{Bmatrix}. \tag{4.23}$$

The derivatives must often be computed numerically:

- For the first derivative of Π at $(\Lambda_1, \Lambda_2, \Lambda_3)$:

$$\frac{\partial\Pi}{\partial\Lambda_1} \approx \frac{\Pi(\Lambda_1 + \Delta\Lambda_1, \Lambda_2, \Lambda_3) - \Pi(\Lambda_1 - \Delta\Lambda_1, \Lambda_2, \Lambda_3)}{2\Delta\Lambda_1}. \tag{4.24}$$

- For the second derivative at $(\Lambda_1, \Lambda_2, \Lambda_3)$:

$$\begin{aligned} \frac{\partial}{\partial\Lambda_1}\left(\frac{\partial\Pi}{\partial\Lambda_1}\right) &\approx \frac{\left(\frac{\partial\Pi}{\partial\Lambda_1}\right)|_{\Lambda_1+\frac{\Delta\Lambda_1}{2},\Lambda_2,\Lambda_3} - \left(\frac{\partial\Pi}{\partial\Lambda_1}\right)|_{\Lambda_1-\frac{\Delta\Lambda_1}{2},\Lambda_2,\Lambda_3}}{\Delta\Lambda_1} \\ &= \frac{1}{\Delta\Lambda_1}\left(\left(\frac{\Pi(\Lambda_1 + \Delta\Lambda_1), \Lambda_2, \Lambda_3 - \Pi(\Lambda_1, \Lambda_2, \Lambda_3)}{\Delta\Lambda_1}\right)\right. \\ &\quad \left. - \left(\frac{\Pi(\Lambda_1, \Lambda_2, \Lambda_3) - \Pi(\Lambda_1 - \Delta\Lambda_1, \Lambda_2, \Lambda_3)}{\Delta\Lambda_1}\right)\right). \end{aligned} \tag{4.25}$$

- For the cross-derivative at (Λ_1, Λ_2):

$$\begin{aligned} \frac{\partial}{\partial\Lambda_2}\left(\frac{\partial\Pi}{\partial\Lambda_1}\right) &\approx \frac{\partial}{\partial\Lambda_2}\left(\frac{\Pi(\Lambda_1 + \Delta\Lambda_1, \Lambda_2, \Lambda_3) - \Pi(\Lambda_1 - \Delta\Lambda_1, \Lambda_2, \Lambda_3)}{2\Delta\Lambda_1}\right) \\ &\approx \frac{1}{4\Delta\Lambda_1\Delta\Lambda_2}\left(\Pi(\Lambda_1 + \Delta\Lambda_1, \Lambda_2 + \Delta\Lambda_2, \Lambda_3) - \Pi(\Lambda_1 - \Delta\Lambda_1, \Lambda_2 + \Delta\Lambda_2, \Lambda_3)\right) \\ &- \left(\Pi(\Lambda_1 + \Delta\Lambda_1, \Lambda_2 - \Delta\Lambda_2, \Lambda_3) - \Pi(\Lambda_1 - \Delta\Lambda_1, \Lambda_2 - \Delta\Lambda_2, \Lambda_3)\right). \end{aligned} \tag{4.26}$$

An exhaustive review of these methods can be found in the texts of Luenberger [2] and Gill, Murray and Wright [3], while a state of the art can be found in Papadrakakis et al. [4].

4.4 Suboptimal Properties Due to Defects—Effects of Pores/voids

Often, during material processing, voids are present. Using the previous framework, one can estimate the reduction of the material quality as a function of the porous material by assuming that it is comprised of an isotropic elastic matrix, with a bulk modulus κ_m and shear modulus μ_m, while the porous void space is modeled by an elastic material with very low bulk and shear moduli $\kappa_v = \delta\kappa_m$, $\mu_v = \delta\mu_m$, with $0 \leq \delta << 1$. The exact case of voids corresponds to $\delta \to 0$. To estimate the properties of the material with voids, we employ the Hashin–Shtrikman bounds and assign the following (the harder material is the matrix, and the softer is the voids): $\kappa_v = \kappa_1$, $\mu_v = \mu_1$ and $\kappa_m = \kappa_2$, $\mu_m = \mu_2$, $v_v = v_1$ and $v_m = v_2$, and force $\mu_v \to 0$ and $\kappa_v \to 0$. Specifically, this yields:

$$0 \leq \kappa^{*,voids} \leq \kappa_m(1 - v_v\mathcal{F}(v_v)), \tag{4.27}$$

where

$$\mathcal{F}(v_v) = \frac{3\kappa_m + 4\mu_m}{3v_v\kappa_m + 4\mu_m} \tag{4.28}$$

and

$$0 \leq \mu^{*,voids} \leq \mu_m(1 - v_v C(v_v)), \tag{4.29}$$

where

$$C(v_v) = \frac{5(3\kappa_m + 4\mu_m)}{\kappa_m(9 + 6v_v) + \mu_m(8 + 12v_v)}. \tag{4.30}$$

One can then assign the effective properties of the void-free part of the particle-laden mixture to the matrix material, $\kappa^{*,no-voids} = \kappa_m$ and $\mu^{*,no-voids} = \mu_m$, leading to

$$0 \leq \kappa^{*,voids} \leq \kappa^{*,no-voids}(1 - v_v\mathcal{F}(v_v)) \tag{4.31}$$

and

$$0 \leq \mu^{*,voids} \leq \mu^{*,no-voids}(1 - v_v C(v_v)) \tag{4.32}$$

It is important to note that

- As $v_v \to 1$, $v_v\mathcal{F}(v_v) \to 1$ and $v_v C(v_v) \to 1$, thus $\mu^{*,voids} \to 0$ and
- As $v_v \to 0$, $v_v\mathcal{F}(v_v) \to 0$ and $v_v C(v_v) \to 0$, thus $\mu^{*,voids} \to \mu^{*,no-voids}$.

These expressions show the resulting effective property loss as a function of the voids. We remark that in some applications, such as biomedical devices, controlled porosity with pre-specified pore shapes, sizes, and distributions are sought after using, for example, Porogen Templating Processes. We refer the reader to Hong et al. [34] for a detailed overview of the state of the art of porogen patterning. Other emerging, cutting-edge approaches for controlled generation of desired porosity involve laser processing (Kongsuwan et al. [35]). This is particularly useful for precisely functionalized layered substrates.

Remark Monograph Appendices 2 and 3 discuss electrical and other properties of materials, as well as materials made of multiple phases (more than two).

References

1. Zohdi, T.I., Monteiro, P.J.M., Lamour, V.: Extraction of elastic moduli from granular compacts. Int. J. Fract. Lett. Micromech. **115**, L49–L54 (2002)
2. Luenberger, D.: Introduction to Linear and Nonlinear Programming. Addison-Wesley, Menlo Park (1974)
3. Gill, P., Murray, W., Wright, M.: Practical Optimization. Academic Press (1995)
4. Papadrakakis, M., Lagaros, N., Tsompanakis, Y., Plevris, V.: Large scale structural optimization: computational methods and optimization algorithms. Arch. Comput. Methods Eng. State art Rev. **8**(3), 239–301 (2001)
5. Holland, J.H.: Adaptation in Natural and Artificial Systems. University of Michigan Press, Ann Arbor, Mich (1975)
6. Goldberg, D. E.: Genetic Algorithms in Search, Optimization and Machine Learning. Addison-Wesley (1989)
7. Davis, S.H.: Theory of Solidification. Cambridge University Press (2001)
8. Onwubiko, C.: Introduction to Engineering Design Optimization. Prentice Hall (2000)
9. Kennedy, J., Eberhart, R.: Swarm Intelligence. Morgan Kaufmann Publishers (2001)
10. Lagaros, N., Papadrakakis, M., Kokossalakis, G.: Structural optimization using evolutionary algorithms. Comput. Struct. **80**, 571–589 (2002)
11. Papadrakakis, M., Lagaros, N., Thierauf, G., Cai, J.: Advanced solution methods in structural optimisation using evolution strategies. Eng. Comput. J. **15**(1), 12–34 (1998)
12. Papadrakakis, M., Lagaros, N., Tsompanakis, Y.: Structural optimization using evolution strategies and neutral networks. Comput. Methods Appl. Mech. Eng. **156**(1), 309–335 (1998)
13. Papadrakakis, M., Lagaros, N., Tsompanakis, Y.: Optimization of large-scale 3d trusses using evolution strategies and neural networks. Int. J. Space Struct. **14**(3), 211–223 (1999)
14. Papadrakakis, M., Tsompanakis, J., Lagaros, N.: Structural shape optimisation using evolution strategies. Eng. Optim. **31**, 515–540 (1999)
15. Goldberg, D.E., Deb, K.: Special issue on genetic algorithms. Comput. Methods Appl. Mech. Eng. **186**(2–4), 121–124 (2000)
16. Zohdi, T.I.: Genetic design of solids possessing a random-particulate microstructure. Phil. Trans. R. Soc. Math. Phys. Eng. Sci. **361**(1806) 1021–1043 (2003)
17. Zohdi, T.I.: On the compaction of cohesive hyperelastic granules at finite strains. In: Proceedings of the Royal Society. Vol. 454(2034) pp. 1395–1401 (2003)
18. Zohdi, T.I.: Constrained inverse formulations in random material design. Computer Methods in Applied Mechanics and Engineering. **192**(28–30), 18, 1–20, 3179–3194 (2003)
19. Zohdi, T.I.: Staggering error control for a class of inelastic processes in random microheterogeneous solids. Int. J. Nonlinear Mech. **39**, 281–297 (2004)
20. Zohdi, T.I.: Modeling and simulation of a class of coupled thermo-chemo-mechanical processes in multiphase solids. Comput. Methods Appl. Mech. Eng. **193**(6–8), 679–699 (2004)

21. Zohdi, T.I.: Statistical ensemble error bounds for homogenized microheterogeneous solids. J. Appl Math. Phys. (Zeitschrift für Angewandte Mathematik und Physik) **56**(3), 497–515 (2005)
22. Zohdi, T.I., Kachanov, M.: A note on the micromechanics of plastic yield of porous solids. Int. J. Fract. Lett. Micromech. **133**(3), L31–L35 (2005)
23. Zohdi, T.I.: On the optical thickness of disordered particulate media. Mech. Materials. **38**, 969–981 (2006)
24. Zohdi, T.I.: Computation of the coupled thermo-optical scattering properties of random particulate systems. Comput. Method Appl. Mech. Eng. **195**, 5813–5830 (2006)
25. Zohdi, T.I.: On the computation of the coupled thermo-electromagnetic response of continua with particulate microstructure. Int. J. Numer. Method Eng. **76**, 1250–1279 (2008)
26. Zohdi, T.I., Kuypers, F.A., Lee, W.C.: Estimation of red blood cell volume fraction from overall permittivity measurement. Int. J. Eng. Sci. **48**, 1681–1691 (2010)
27. Zohdi, T.I.: Simulation of coupled microscale multiphysical-fields in particulate-doped dielectrics with staggered adaptive FDTD. Comput. Method Appl. Mech. Eng. **199**, 79–101 (2010)
28. Zohdi, T.I.: Joule-heating field phase-amplification in particulate-doped dielectrics. Int. J. Eng. Sci. **49**, 30–40 (2011)
29. Zohdi, T.I.: Estimation of electrical-heating load-shares for sintering of powder mixtures. Proc. R. Soc. **468**, 2174–2190 (2012)
30. Zohdi, T.I.: Modeling and simulation of the optical response rod-functionalized reflective surfaces. Comput. Mech. **50**(2), 257–268 (2012)
31. Zohdi, T.I.: On the reduction of heat generation in lubricants using microscale additives. Int. J. Eng. Sci. **62**, 84–89 (2013)
32. Zohdi, T.I.: On cross-correlation between thermal gradients and electric fields. Int. J. Eng. Sci. **74**, 143–150 (2014)
33. Zohdi, T.I.: Modeling and simulation of cooling-induced residual stresses in heated particulate mixture depositions. Comput. Mech. **56**, 613–630 (2015)
34. Hong, Y., Zhou, J.G., Yao, D.: Porogen Templating Processes: an overview. J. Manuf. Sci. Eng. **136**(3), Paper MANU-13-1304. https://doi.org/10.1115/1.4026899 (2014)
35. Kongsuwan, P., Brandal, G., Yao, Y.L., Laser induced porosity and crystallinity modification of a bioactive glass coating on titanium substrates. J. Manuf. Sci. Eng. **137**(3), MANU-14-1245. https://doi.org/10.1115/1.4029566 (2015)

Chapter 5
CM Approaches: Numerical Thermo-Mechanical Formulations

The previous analytical expressions provide good way to estimate and optimize material combination for effective properties, while controlling local field fluctuations. However, in order to probe the response of a given material combination more deeply, in particular the time-dependent behavior when it is thermoformed, one must resort to numerical methods. Generally, the most practical strategy is to:

- Use estimates (e.g., based on bounds) to determine proposed optimal combinations of materials and
- Use numerical discretizations of the continuum to determine the detailed performance of the proposed optimal designs.

Accordingly, this chapter is concerned with the computational characterization of the evolution of the material response and residual stresses in materials with microstructures that arise from heated (or curing) deposited mixtures of particles. Residual stresses arise because the hot bonded materials cannot freely contract to their stress-free state, when cooled, due to their interaction with other components in the system and the surrounding environment to which they are joined. The objective of this work is to develop a straightforward computational framework that researchers in the field can easily implement and use as a computationally efficient design tool. Generally speaking, there is thermo-mechanical multifield coupling present, along with material changes associated with material hardening, elasto-plasticity, and mechanical damage. Specifically, a recursively staggered, temporally adaptive, Finite Difference Time Domain (FDTD) scheme is developed to resolve the internal microstructural thermal and mechanical fields, accounting for the simultaneous elasto-plasticity and damage. The time-step adaptation is constructed to allow the numerical scheme to iteratively resolve the changing physical fields by reducing the time-steps during phases of the process when the system is undergoing changes on relatively small timescales and also to enlarge the time-steps when the processes are relatively slow. The spatial discretization grids are uniform and dense, with the complex microstructure embedded into the mesh. The regular grid allows one to generate

T. I. Zohdi, *Modeling and Simulation of Functionalized Materials for Additive Manufacturing and 3D Printing: Continuous and Discrete Media*, Lecture Notes in Applied and Computational Mechanics 60, https://doi.org/10.1007/978-3-319-70079-3_5

a matrix-free iterative formulation which is amenable to rapid computation and minimal memory requirements, making it ideal for laptop computation. The presentation is broken into three main parts: (1) formulations for each field in the model problem, identifying the coupling terms, (2) iterative staggering schemes (including spatial and temporal discretizations), and (3) numerical examples for the model problem. The approach builds on work found in Zohdi [1–33] and then applies it to particle mixture deposition systems.

Remark In this section, we focus on the cooling and heating of a particulate mixture. The initial dynamic deposition process of multibody and inter-particle collisions is outside the scope of the present chapter. However, we mention in passing that to model the dynamics of particle systems, reduced-order particle-based or discrete element-based models, which treat such systems as multibody dynamical groups, are often used. They are advantageous in dealing with domains that break apart or coalesce, as compared to traditional continuum-based finite difference and finite element methods, which have limitations when dealing with dynamic discontinua. For reviews see, for example, Duran [34], Pöschel and Schwager [35], Onate et. al. [36, 37], Rojek et. al. [38], Carbonell et. al. [39], Labra and Onate [40], Leonardi et. al. [41], Cante et. al. [42], Rojek [43], Onate et. al. [79], Bolintineanu et. al. [44], Campello and Zohdi [45, 46], Avci and Wriggers [47], and Zohdi [1–33]. In many cases, the deposition of these materials is the first stage of a multistep process which may involve, among other processes, compaction. Compaction is also somewhat outside the scope of the present work, and we refer the reader to Akisanya et. al. [48], Anand and Gu [80], Brown and Abou-Chedid [49], Domas [50], Fleck [51], Gethin et. al., [52], Gu et. al. [53], Lewis et. al. [54], Ransing et. al. [55], Tatzel [56], and Zohdi [1–33].

5.1 Transient Thermo-Mechanical Coupled Fields

We consider a model problem consisting of a deposited set hot mixture of heterogeneous material which is in the cool-down phase of the process. The essential field equations and simplifying assumptions that will be used during the analysis are provided next.

Balance of linear momentum

We consider a balance of linear momentum governed by

$$\nabla_x \cdot \boldsymbol{\sigma} + \boldsymbol{f} = \rho \frac{d^2 \boldsymbol{u}}{dt^2}, \tag{5.1}$$

in regimes where *infinitesimal deformations are appropriate*, where $\boldsymbol{\sigma}$ is the Cauchy stress, $\boldsymbol{f}$ are body forces, ρ is the material density, and $\boldsymbol{u}$ is the displacement. Consistent with the infinitesimal deformation approximation, we write $\nabla_x \approx \nabla_X$ and

$\frac{d()}{dt} \approx \frac{\partial ()}{\partial t}|_X$, where $\boldsymbol{X}$ are the referential coordinates and $\boldsymbol{x}$ are the current coordinates. We consider a damaged, elasto-plastic, and isotropic constitutive law given by

$$\boldsymbol{\sigma} = \mathcal{D}I\!E_0 : (\epsilon - \epsilon_\theta - \epsilon_p), \tag{5.2}$$

where under the infinitesimal deformation framework the balance of linear momentum becomes ($\rho \approx \rho_o$)

$$\nabla_X \cdot (\mathcal{D}I\!E_0 : (\epsilon - \epsilon_\theta - \epsilon_p)) + \boldsymbol{f} = \rho \frac{\partial^2 \boldsymbol{u}}{\partial t^2} \tag{5.3}$$

with infinitesimal strains given by $\epsilon = \frac{1}{2}(\nabla_X \boldsymbol{u} + (\nabla_X \boldsymbol{u})^T)$, thermal strains given by $\epsilon_\theta \stackrel{\text{def}}{=} \gamma \cdot (\theta - \theta_0)\mathbf{1}$, where γ is the thermal expansion coefficient, and plastic strains given by ϵ_p, generated by the following unilateral conditions

$$\text{if} \quad ||\boldsymbol{\sigma}'|| > \sigma_y \Rightarrow \dot{\zeta} = a \left(\frac{||\boldsymbol{\sigma}'||}{\sigma_y} - 1 \right) \tag{5.4}$$

and

$$\text{if} \quad ||\boldsymbol{\sigma}'|| \leq \sigma_y \Rightarrow \dot{\zeta} = 0 \tag{5.5}$$

where $\dot{\epsilon}_p = \dot{\zeta} \frac{\boldsymbol{\sigma}'}{||\boldsymbol{\sigma}'||}$ and $\boldsymbol{\sigma}' = \boldsymbol{\sigma} - \frac{tr\boldsymbol{\sigma}}{3}\mathbf{1}$ is the deviatoric stress. Here, the (isotropic) damaged elasticity tensor is $I\!E = \mathcal{D}I\!E_0$, where $I\!E_0$ represents the "virgin" isotropic undamaged material, $0 \leq \mathcal{D} \leq 1$ is the scalar continuity (isotropic damage) parameter (Kachanov [57]), $\mathcal{D}(t = 0) = 1$ indicates the initial undamaged state, and $\mathcal{D} \to 0$ indicates a completely damaged state. The damage arising from mechanical and chemical sources is modeled as being governed by evolution over-stress functions of the form ($0 < \mathcal{D} \leq 1$)

$$||\boldsymbol{\sigma}'|| > \sigma_d \Rightarrow \dot{\mathcal{D}} = b \left(\frac{||\boldsymbol{\sigma}'||}{\sigma_d} - 1 \right) \tag{5.6}$$

and

$$||\boldsymbol{\sigma}'|| \leq \sigma_d \Rightarrow \dot{\mathcal{D}} = 0, \tag{5.7}$$

We note that the rate constants a and b and the critical stresses σ_y and σ_d are possibly spatially variable. Clearly, further evolution laws can be written for other material property changes, such as the thermal conductivity, although only changes in the mechanical property $I\!E$ are considered during the formulations to follow.[1] In the case of material isotropy,

[1] For further details on these types of phenomenological (damage) formulations, the interested reader is referred to the seminal work of Kachanov [57].

$$\boldsymbol{\sigma} = \mathcal{D}\left(\lambda_0 tr(\boldsymbol{\epsilon} - \boldsymbol{\epsilon}_\theta - \boldsymbol{\epsilon}_p)\mathbf{1} + 2\mu_0(\boldsymbol{\epsilon} - \boldsymbol{\epsilon}_\theta - \boldsymbol{\epsilon}_p)\right), \tag{5.8}$$

where λ_0 is the undamaged Lame parameter and μ_0 is the undamaged shear modulus.

Balance of energy

The interconversions of various forms of energy (mechanical, thermal, etc.) in a system are governed by the first law of thermodynamics,

$$\rho\dot{w} - \boldsymbol{\sigma} : \nabla_x \dot{\boldsymbol{u}} + \nabla_x \cdot \boldsymbol{q} - \rho z = 0, \tag{5.9}$$

where w is the stored energy per unit mass (which is a function of the temperature, θ and elastic strain, $\boldsymbol{\epsilon}^e = \boldsymbol{\epsilon} - \boldsymbol{\epsilon}_\theta - \boldsymbol{\epsilon}_p$), $\boldsymbol{q}$ is heat flux, and ρz is the rate of energy absorbed from sources. We employ the following for the stored energy (assuming infinitesimal deformations)

$$\rho w = W \approx \tfrac{1}{2}(\boldsymbol{\epsilon} - \boldsymbol{\epsilon}_\theta - \boldsymbol{\epsilon}_p) : \mathbb{E} : (\boldsymbol{\epsilon} - \boldsymbol{\epsilon}_\theta - \boldsymbol{\epsilon}_p) + \rho C\theta, \tag{5.10}$$

which implies (here assuming $C \neq C(t)$)

$$\rho\dot{w} = \dot{W} = (\dot{\boldsymbol{\epsilon}} - \dot{\boldsymbol{\epsilon}}_\theta - \dot{\boldsymbol{\epsilon}}_p) : \mathbb{E} : (\boldsymbol{\epsilon} - \boldsymbol{\epsilon}_\theta - \boldsymbol{\epsilon}_p) + \frac{1}{2}(\boldsymbol{\epsilon} - \boldsymbol{\epsilon}_\theta - \boldsymbol{\epsilon}_p) : \dot{\mathbb{E}} : (\boldsymbol{\epsilon} - \boldsymbol{\epsilon}_\theta - \boldsymbol{\epsilon}_p) + \rho C\dot{\theta}, \tag{5.11}$$

and thus the first law becomes

$$\rho C\dot{\theta} = \boldsymbol{\sigma} : (\dot{\boldsymbol{\epsilon}}_\theta + \dot{\boldsymbol{\epsilon}}_p) - \frac{1}{2}(\boldsymbol{\epsilon} - \boldsymbol{\epsilon}_\theta - \boldsymbol{\epsilon}_p) : \dot{\mathbb{E}} : (\boldsymbol{\epsilon} - \boldsymbol{\epsilon}_\theta - \boldsymbol{\epsilon}_p) + \nabla_X \cdot (\mathbb{K} \cdot \nabla_X \theta) + \rho z \tag{5.12}$$

where Fourier's law, $\boldsymbol{q} = -\mathbb{K} \cdot \nabla_X \theta$, has been used.

5.2 Iterative Staggering Scheme

We now develop a staggering solution framework to solve the coupled systems of interest. The general methodology is as follows (at a given time increment): (1) each field equation is solved individually, "freezing" the other (coupled) fields in the system, allowing only the primary field to be active and (2) after the solution of each field equation, the primary field variable is updated, and the next field equation is treated in a similar manner. For an "implicit" type of staggering, the process can be repeated in an iterative manner, while for an "explicit" type, one moves to the next time-step after one "passes" through the system. We will employ implicit staggering. Specifically, for the thermo-mechanical system under consideration, consider an abstract setting, whereby one solves for the mechanical field, assuming the thermal field is fixed (L is a time-step counter and K is a staggering-step counter),

$$\mathcal{A}_1(\underline{\boldsymbol{u}^{L+1,K}}, \theta^{L+1,K-1}) = \mathcal{B}_1(\boldsymbol{u}^{L+1,K-1}, \theta^{L+1,K-1}) \tag{5.13}$$

then one solves for the thermal fields, assuming the mechanical field fixed,

$$\mathcal{A}_2(\boldsymbol{u}^{L+1,K}, \underline{\theta^{L+1,K}}) = \mathcal{B}_2(\boldsymbol{u}^{L+1,K}, \theta^{L+1,K-1}) \tag{5.14}$$

where the only underlined variable is "active" at that stage of the process. Within the staggering (iterative) scheme, implicit time-stepping methods (with time-step size adaptivity) will be used throughout the upcoming analysis (described shortly). The process is driven by minimizing nondimensional relative-iteration-coupling error (of both fields) within a time-step (difference between successive iterations). A tolerance check determines whether the iterations should continue, or if the time-steps should be adaptively reduced to increase the rate of convergence. As alluded to earlier, the time-steps can be increased if convergence occurs too quickly, thus allowing larger time-steps and faster simulations for a given iterative error tolerance. The details of this process are discussed shortly. Generally speaking, if a recursive staggering process is not employed (an explicit coupling scheme), the staggering error can accumulate rapidly. However, simply employing extremely small time-steps, smaller than needed to control the discretization error, in order to suppress a (nonrecursive) staggering process error, can be computationally inefficient. Therefore, the objective of the next subsection is to develop a strategy to adaptively adjust, in fact maximize, the choice of the time-step size in order to control the staggering error, while simultaneously staying below a critical time-step size needed to control the discretization error. An important related issue is to simultaneously minimize the computational effort involved. We now develop a staggering scheme by extending an approach found in the work of Zohdi [1–33].

Remark 1 The symbol $||\cdot||$ will signify the $L_2(\Omega)$-norm throughout this work. The nondimensional error metric for the mechanical field is (where we assume that the denominator is nonzero)

$$\varpi_u^K \stackrel{\text{def}}{=} \frac{||\boldsymbol{u}^{L+1,K} - \boldsymbol{u}^{L+1,K-1}||}{||\boldsymbol{u}^{L+1,K} - \boldsymbol{u}^L||}, \tag{5.15}$$

and for the thermodynamic field

$$\varpi_\theta^K \stackrel{\text{def}}{=} \frac{||\theta^{L+1,K} - \theta^{L+1,K-1}||}{||\theta^{L+1,K} - \theta^L||}. \tag{5.16}$$

Thereafter, we select the maximum *nondimensionalized* error for adaptivity

$$\varpi^{*,K} \stackrel{\text{def}}{=} max(\varpi_u^K, \varpi_\theta^K). \tag{5.17}$$

Remark 2 Staggering schemes are widely used in the computational mechanics literature, dating back, at least, to Zienkiewicz [58] and Zienkiewicz et. al. [59]. For in-depth overviews, see the works of Lewis and Schrefler (Lewis et. al. [60] and Lewis

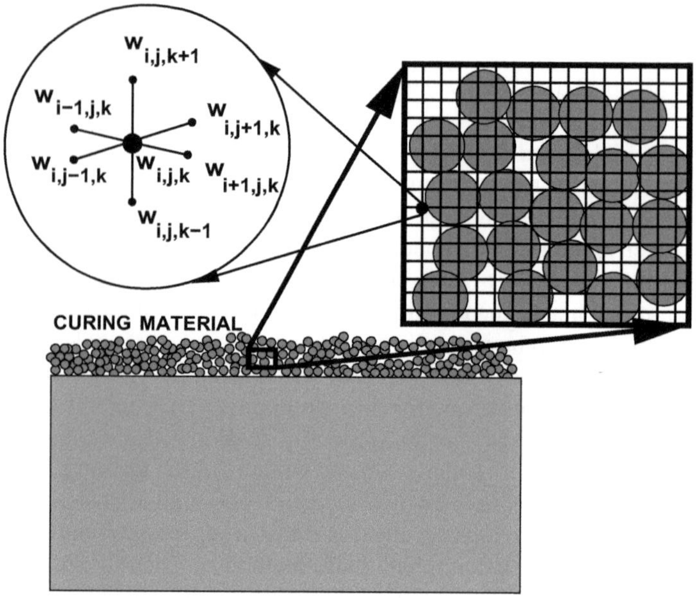

Fig. 5.1 A typical three-dimensional finite difference stencil for a field $w(x, y, z)$

and Schrefler [61]) and a series of works by Schrefler and collaborators: Schrefler [62], Turska and Schrefler [63], Biano et. al. [64], and Wang and Schrefler [65].

Spatial discretization of the fields

Numerically, the components of the gradient of functions such as $\boldsymbol{u}$ and θ are approximated by central finite difference stencils of the basic form (Fig. 5.1):

$$\frac{\partial u_i}{\partial x_j}|_x \approx \frac{u_i(x_j+\Delta x_j)-u_i(x_j - \Delta x_j)}{2\Delta x_j} \tag{5.18}$$

for each of the (x_1, x_2, x_3)-directions, in order to form the terms needed in $\nabla_x \boldsymbol{u}$ and $\nabla_x \cdot \boldsymbol{\sigma}$. This is a second-order accurate stencil. For a generic second-order scheme spatial derivative, such as

$$\frac{\partial \sigma}{\partial x}|_x \approx \frac{\sigma(x + \frac{\Delta x}{2}) - \sigma(x - \frac{\Delta x}{2})}{\Delta x}, \tag{5.19}$$

where generically, for example with an arbitrary material coefficient $a(x) = \lambda(x)$ or $a(x) = \mu(x)$:

$$\sigma(x+\frac{\Delta x}{2}) \approx a(x+\frac{\Delta x}{2})\underbrace{\frac{u(x+\Delta x)-u(x)}{\Delta x}}_{\frac{\partial u}{\partial x}|_{x+\frac{\Delta x}{2}}} \tag{5.20}$$

and

$$\sigma(x-\frac{\Delta x}{2}) \approx a(x-\frac{\Delta x}{2})\underbrace{\frac{u(x)-u(x-\Delta x)}{\Delta x}}_{\frac{\partial u}{\partial x}|_{x-\frac{\Delta x}{2}}}, \tag{5.21}$$

where

$$a(x+\frac{\Delta x}{2}) \approx \frac{1}{2}(a(x+\Delta x)+a(x)), \tag{5.22}$$

and

$$a(x-\frac{\Delta x}{2}) \approx \frac{1}{2}(a(x)+a(x-\Delta x)). \tag{5.23}$$

These approximations are made for all components and combinations in $\nabla_x \cdot \boldsymbol{\sigma}$. The mixed derivatives are derived in a similar manner, summarized in Chapter Appendix 1. Similarly, for a second-order scheme spatial derivatives in heat conduction

$$\frac{\partial q}{\partial x}|_x \approx \frac{q(x+\frac{\Delta x}{2})-q(x-\frac{\Delta x}{2})}{\Delta x}, \tag{5.24}$$

where (in conjunction with Fourier's Law)

$$q(x+\frac{\Delta x}{2}) \approx -\mathbb{K}(x+\frac{\Delta x}{2})\underbrace{\frac{\theta(x+\Delta x)-\theta(x)}{\Delta x}}_{\frac{\partial \theta}{\partial x}|_{x+\frac{\Delta x}{2}}} \tag{5.25}$$

and

$$q(x-\frac{\Delta x}{2}) \approx -\mathbb{K}(x-\frac{\Delta x}{2})\underbrace{\frac{\theta(x)-\theta(x-\Delta x)}{\Delta x}}_{\frac{\partial \theta}{\partial x}|_{x-\frac{\Delta x}{2}}}, \tag{5.26}$$

where

$$\mathbb{K}(x+\frac{\Delta x}{2}) \approx \frac{1}{2}(\mathbb{K}(x+\Delta x)+\mathbb{K}(x)), \tag{5.27}$$

and

$$\mathbb{K}(x - \frac{\Delta x}{2}) \approx \frac{1}{2}(\mathbb{K}(x) + \mathbb{K}(x - \Delta x)). \tag{5.28}$$

These approximations are made for $\frac{\partial q_1}{\partial x_1}$, $\frac{\partial q_2}{\partial x_2}$, and $\frac{\partial q_3}{\partial x_3}$, in order to form the terms needed in $\nabla_x \cdot \boldsymbol{q}$. This is done at each node in the grid. See Chapter Appendix 1 for more details.

5.3 Temporal Discretization of Fields

Mechanical fields

For the mechanical field (infinitesimal deformation formulation), we write

$$\frac{d\boldsymbol{v}}{dt} = \frac{\partial \boldsymbol{v}}{\partial t} = \frac{1}{\rho}(\nabla_X \cdot \boldsymbol{\sigma} + \boldsymbol{f}) \stackrel{\text{def}}{=} \boldsymbol{\Psi}. \tag{5.29}$$

We discretize for time $= t + \phi\Delta t$, and using a trapezoidal "$\phi - scheme$" ($0 \leq \phi \leq 1$, see Chapter Appendix 2)

$$\frac{\boldsymbol{v}(t + \Delta t) - \boldsymbol{v}(t)}{\Delta t} \approx \boldsymbol{\Psi}(t + \phi\Delta t) \approx \phi\boldsymbol{\Psi}(t + \Delta t) + (1 - \phi)\boldsymbol{\Psi}(t). \tag{5.30}$$

Rearranging yields

$$\boldsymbol{v}(t + \Delta t) \approx \boldsymbol{v}(t) + \Delta t\,(\phi\boldsymbol{\Psi}(t + \Delta t) + (1 - \phi)\boldsymbol{\Psi}(t)) \tag{5.31}$$

where the previously introduced spatial discretization is applied to the terms in $\boldsymbol{\Psi}$ ($\nabla_x \cdot \boldsymbol{\sigma}$). Since this is a second-order system, the procedure is then repeated to determine the displacement field $\boldsymbol{u}$ (see Chapter Appendix 2)

$$\begin{aligned} \boldsymbol{u}(t + \Delta t) &= \boldsymbol{u}(t) + \boldsymbol{v}(t + \phi\Delta t)\Delta t \\ &= \boldsymbol{u}(t) + (\phi\boldsymbol{v}(t + \Delta t) + (1 - \phi)\boldsymbol{v}(t))\,\Delta t, \end{aligned} \tag{5.32}$$

or more explicitly

$$\boldsymbol{u}(t + \Delta t) = \boldsymbol{u}(t) + \boldsymbol{v}(t)\Delta t + \phi(\Delta t)^2\boldsymbol{\Psi}(t + \phi\Delta t). \tag{5.33}$$

The term $\boldsymbol{\Psi}(t + \phi\Delta t)$ can be handled in two main ways:

- $\boldsymbol{\Psi}(t + \phi\Delta t) \approx \boldsymbol{\Psi}(\phi\boldsymbol{u}(t + \Delta t) + (1 - \phi)\boldsymbol{u}(t))$ or
- $\boldsymbol{\Psi}(t + \phi\Delta t) \approx \phi\boldsymbol{\Psi}(\boldsymbol{u}(t + \Delta t)) + (1 - \phi)\boldsymbol{\Psi}(\boldsymbol{u}(t))$.

The differences are quite small between either of the above; thus, for brevity, we choose the latter. Therefore,

$$\boldsymbol{u}(t+\Delta t)=\boldsymbol{u}(t)+\boldsymbol{v}(t)\Delta t+\phi(\Delta t)^2\left(\phi\boldsymbol{\Psi}(t+\Delta t)+(1-\phi)\boldsymbol{\Psi}(t)\right). \quad (5.34)$$

When $\phi=1$, then this approach can be considered to be a (implicit) Backward Euler scheme, which is very stable (very dissipative) and $\mathcal{O}((\Delta t)^2)$ locally in time, while if $\phi=0$, the scheme can be considered as a (explicit) Forward Euler scheme, which is conditionally stable and $\mathcal{O}((\Delta t)^2)$ locally in time and if $\phi=0.5$, then the scheme can be considered as a (implicit) Midpoint scheme, which is marginally stable and $\hat{\mathcal{O}}((\Delta t)^2)=\mathcal{O}((\Delta t)^3)$ locally in time. The dependent plastic and damage variables are also integrated in a similar manner

$$\epsilon(t+\Delta t)=\epsilon(t)+\Delta t\left(\phi\dot{\epsilon}(t+\Delta t)+(1-\phi)\dot{\epsilon}(t)\right) \quad (5.35)$$

and

$$\mathcal{D}(t+\Delta t)=\mathcal{D}(t)+\Delta t\left(\phi\dot{\mathcal{D}}(t+\Delta t)+(1-\phi)\dot{\mathcal{D}}(t)\right). \quad (5.36)$$

Thermal fields
For the thermal field, we write

$$\frac{\partial\theta}{\partial t}=\frac{1}{\rho C}\left(\sigma:(\dot{\epsilon}_\theta+\dot{\epsilon}_p)-\frac{1}{2}(\epsilon-\epsilon_\theta-\epsilon_p):\dot{\mathbb{E}}:(\epsilon-\epsilon_\theta-\epsilon_p)+\nabla_X\cdot(\mathbb{K}\cdot\nabla_X\theta)+\rho z\right)\overset{\text{def}}{=}\boldsymbol{Y}. \quad (5.37)$$

We discretize for around the time $=t+\phi\Delta t$, yielding

$$\theta(t+\Delta t)\approx\theta(t)+\Delta t\left(\phi\boldsymbol{Y}(t+\Delta t)+(1-\phi)\boldsymbol{Y}(t)\right), \quad (5.38)$$

where the previously introduced spatial discretization is applied to the terms in $\boldsymbol{Y}$.

5.4 The Overall Solution Scheme

In order to construct a solution, the algorithm is as follows:

- (1) *Spatiotemporal discretization*: Construct derivative terms such as

$$\frac{\partial u(x)}{\partial x}\approx\frac{u(x+\Delta x)-u(x-\Delta x)}{2\Delta x},\quad etc., \quad (5.39)$$

  and insert into them the governing equations. This leads to a system of coupled equations, for each node $((i,j,k)$ in Fig. 5.2), which are cast in the following (implicit/recursive) form (which are a recasting of the abstract system (Eqs. 5.13

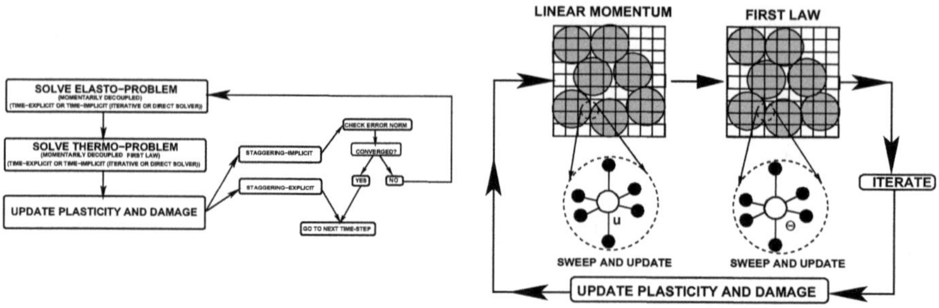

Fig. 5.2 Overall coupled staggering (left) solution and the matrix-free approach (right)

and 5.14))

$$\boldsymbol{u}(t+\Delta t) = \mathcal{F}(\boldsymbol{u}(t+\Delta t), \theta(t+\Delta t), ...), \tag{5.40}$$

and

$$\theta(t+\Delta t) = \mathcal{Y}(\boldsymbol{u}(t+\Delta t), \theta(t+\Delta t), ...). \tag{5.41}$$

- (2) *System staggering*: Compute $\boldsymbol{u}$-field with θ-fields fixed, then compute θ-field with $\boldsymbol{u}$-fields fixed, etc., and iterate at time interval $L+1$, $K = 1, 2, ...$ for

$$\boldsymbol{u}^{L+1,K} = \mathcal{F}(\underline{\boldsymbol{u}^{L+1,K-1}}, \theta^{L+1,K-1}), \tag{5.42}$$

and

$$\theta^{L+1,K} = \mathcal{Y}(\boldsymbol{u}^{L+1,K}, \underline{\theta^{L+1,K-1}}), \tag{5.43}$$

Solving each of the above equations (5.42 and 5.43), with the respective other fields fixed, can be achieved in a variety of ways, for example iteratively or by direct (Gaussian-type) solution methods (Fig. 5.2). For example, an interior iterative loop, within the staggering loop (within a time-step), can be used to update the solution to solve the individual field, for example the mechanical field, before moving to the next field (e.g., the thermal field). Those internal iterations can be performed until that individual field converges. This can then be repeated for the next field. This would then complete one staggering iteration. There are of course many possible variants of this process. In theory, one could even simply perform an explicit update (no recursion). This is discussed further in the remarks that follow.

- (3) Compute error measures: $\varpi^{*,K} \stackrel{\text{def}}{=} max(\varpi_u^K, \varpi_\theta^K)$, $i = 1, \ldots, nodes$ in the system.
- (4a) If tolerance is met, $\varpi^{*,K} \leq C_{tol}$ and $K \leq K_d$, then:

(i) Increment time forward: $t = t + \Delta t$,

(ii) Construct new time-step: $\Delta t^{new} = \Phi_K \Delta t^{old}$, where $\Phi_K \stackrel{\text{def}}{=} \left(\frac{(\frac{C_{tol}}{\varpi^{*,0}})^{\frac{1}{pK_d}}}{(\frac{\varpi^{*,K}}{\varpi^{*,0}})^{\frac{1}{pK}}} \right)$

(iii) Select $\Delta t = min(\Delta t^{lim}, \Delta t)$ and go to (1)

- (4b) If tolerance is not met, $\varpi^{*,K} > C_{tol}$ and $K = K_d$, then construct (refine) new time-step: $\Delta t^{\text{new}} \stackrel{\text{def}}{=} \Phi_K \Delta t^{\text{old}}$

$$\Phi_K \stackrel{\text{def}}{=} \left(\frac{\left(\frac{C_{tol}}{\varpi^{*,0}}\right)^{\frac{1}{pK_d}}}{\left(\frac{\varpi^{*,K}}{\varpi^{*,0}}\right)^{\frac{1}{pK}}} \right) \tag{5.44}$$

and go to (1). This time-scaling relation is derived in Chapter Appendix 3.

At a given time, once the process is complete, then the time is incremented forward and the process is repeated. The overall goal is to deliver solutions where the iterative error is controlled and the temporal discretization accuracy dictates the upper limit on the time-step size (Δt^{lim}). Clearly, there are various combinations of solution methods that one can choose. For example, for the overall field coupling, one may choose implicit or explicit staggering and, within the staggering process, either implicit ($0 < \phi \leq 1$) or explicit time-stepping ($\phi = 0$). As mentioned previously in the case of implicit time-stepping, one can use iterative or direct solvers for the balance of linear momentum and the first law of thermodynamics (Fig. 5.2).

Algorithmic observation 1

It is important to emphasize that one should use the previous (converged) time-step's solution as the starting guess for the next time-step to obtain a "head-start" ($\boldsymbol{u}^{K=0}(t + \Delta t) = \boldsymbol{u}(t)$). When selecting a time-step, one must balance accuracy concerns and, simultaneously, stability issues.[2] Clearly, the smaller the time-step, the more stable the solution process; however, more time-steps implies more system evaluations. Since the multifield staggering scheme iterates anyway, implicit methods are preferred for the applications of interest. As the physics changes, the field that is most sensitive (exhibits the largest amount of relative nondimensional change) dictates the time-step size. Because the internal system solvers within the staggering scheme are also iterative and use the previously converged solution as their starting value to solve the system of equations, a field that is relatively insensitive at a given stage of the simulation will converge in a very few internal iterations (perhaps even one).

Algorithmic observation 2

Generally speaking, the solution to the individual field equations progresses in a node-by-node fashion whereby, at a node (i, j, k), for example for the mechanical field calculations, one has an abstract form

$$\boldsymbol{u}(t + \Delta t) \approx \mathcal{F}(\boldsymbol{u}(t), \boldsymbol{u}(t + \Delta t), \theta(t), \theta(t + \Delta t)), \tag{5.45}$$

where the term on the left-hand side is updated and the terms on the right-hand side are previously iterated (old) values. This entails using the old values for all finite dif-

[2]Typically, the number of iterations needed to solve the coupled system, if an iterative scheme is used, increases with the time-step size and the value of ϕ.

ference stencils that eventually become updated only after the algorithm completely traverses through the system, updating values, node by node (*no matrices need to be formed*, Fig. 5.2). There exist many methods to accelerate such computations, such as successive over-relaxation, based on the pioneering work of Young [66]. For reviews, see Ames [67] or Axelsson [68]. Note that for the mechanical field calculations the thermal field is instantaneously fixed and is updated only when it is to be solved, in the staggered manner (fixing the mechanical variables). At the algebraic equation solution level, after the individual field has been solved, the entire solution is passed to the next field equation, as described in the previous algorithm (Fig. 5.2). This is a Jacobi-type scheme, whereby the updates are made only after one complete system iteration, which is easier to address theoretically, as opposed to a Gauss–Seidel-type method, which involves immediately using the most current field values, when they become available. The Jacobi method is easily parallelizable, if desired. In other words, the calculation for each node is (momentarily) uncoupled, with the updates only coming at the end of an iteration. Gauss–Seidel, since it requires the most current updates, couples the nodal calculations immediately.

5.5 Numerical Examples

As a model problem, we consider a group of particles with a smaller-scale interstitial (compacted very fine particle) material that is assumed to be a continuous phase. We generated a group of N_p randomly dispersed spherical particles, of equal size, embedded in a cubical domain of dimensions, $D \times D \times D$. The particle size was determined by a particle/sample size ratio, which was defined via a subvolume size $V \stackrel{\text{def}}{=} \frac{D\times D\times D}{N_p}$. The nondimensional ratio between the particle radii (R) and the subvolume was denoted by $\mathcal{L} \stackrel{\text{def}}{=} \frac{R}{V^{\frac{1}{3}}}$. The volume fraction occupied by the particles consequently can be written as $v_p \stackrel{\text{def}}{=} \frac{4\pi\mathcal{L}^3}{3}$. Thus, the total volume occupied by the particles denoted ζ can be written as $\zeta = v_p N_p V$. Large values of $\zeta > 0.5$ allow for overlap. We used $N_p = 100$ particles (Fig. 5.9). This sample size was arrived at by successively enlarging sample until there were no significant changes in the overall system response for further enlargements. The classical random sequential addition algorithm was used to place nonoverlapping particles randomly into the domain of interest (RSA; Widom [69]). The particles were then enlarged from those locations and allowed to overlap (Fig. 5.3).

Remark For higher volume fractions, during the first phase of this algorithm (particle placement), more sophisticated algorithms, such as the equilibrium-driven Metropolis algorithm, can be used or methods based on simultaneous particle flow and growth found in Torquato [70], Kansaal et. al. [71], and Donev et. al. [72–74].

Sample size selection

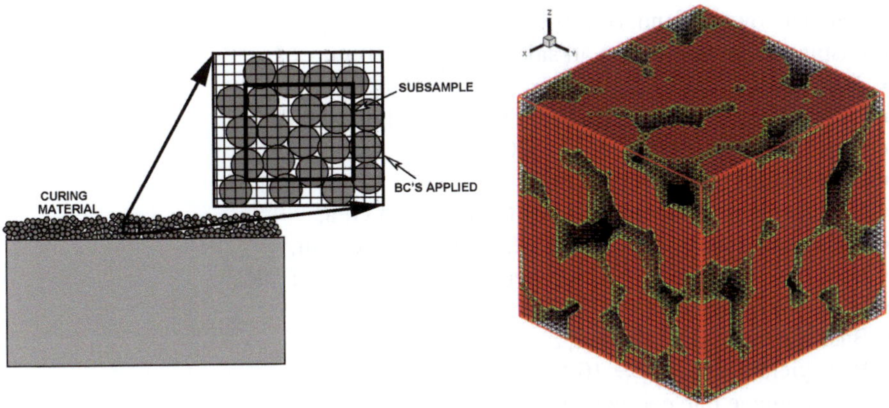

Fig. 5.3 Left: Hot deposited particles on a surface. With the framing method, a sample is probed with interior subsamples, within the larger sample, in order to avoid boundary layer effects that occur from imposing boundary conditions on the large-sample exterior. Right: A mesh of the curing subsample (showing only one of the particle phases for illustration purposes)

In order to select a suitably sized sample that is statistically representative (a RVE), we employ a "framing" method, whereby the boundary conditions are applied ($\boldsymbol{u}$ and θ) to the boundary of a sample (Fig. 5.9), and an interior subsample is used to probe what the material would experience without the direct influence of the applied boundary conditions. This approach avoids introducing boundary layer effects into the interior response. For more details, see Zohdi [1–33]. An implementation of a "framing" approach is as follows:

- *STEP (1)*: Generate a sample with a certain number of particles in its interior,
- *STEP (2)*: For the effective property calculation (averaging), select a subsample ("a sub-box," Fig. 5.9) in the interior (to avoid boundary layer effects that arise from the imposition of boundary conditions),
- *STEP (3)*: Repeat STEPS (1) and (2) for different random realizations for a given sample size, and average the resulting response to determine a mean value,
- *STEP (4)*: Repeat STEPS (1)–(3) for a larger sample,
- *STEP (5)*: Continue the process (STEPS (1)–(4)) until the response ceases to change to within an acceptable tolerance.

For a more in-depth discussion on size-effect issues, see the works of Zohdi [1–33].

Numerical examples

As an example, the following parameters were used:

- Reference temperature, $\theta_r = 600\,^\circ\mathrm{K}$,
- Initial temperature, $\theta_0 = 600\,^\circ\mathrm{K}$,
- Total time, $T = 10^{-5}$ s,
- Initial time-step size, $\Delta t = 10^{-10}$ s,

- Damage lower bound, $\alpha = 0.1$,
- Displacement loading on all sides $\boldsymbol{u} = (0, 0, 0)$ m,
- Temperature of all sides, $\theta(t) = -300^{\circ}\frac{t}{T} + 600\,^{\circ}$K,
- Dimensions of the sample, 0.001 m $\times$ 0.001 m $\times$ 0.001 m,
- Particles in the sample, $N_p = 100$,
- Base density, $\rho = 1000\,\mathrm{kg/m^3}$,
- Base Lame parameters, $\lambda_o = 10$ GPa, $\mu_o = 3$ GPa,
- Base conductivity, $\boldsymbol{I\!K}_o = K_o\mathbf{1}$, $K_o = 100$ W/°K − m,
- Base thermal expansion coefficient, $\beta_o = 0.000001$, 1/°K,
- Base heat capacity, $C_o = 10$ J/K° − kg,
- Base plastic rate coefficient, $a_o = 0.001$,
- Base yield stress, $\sigma_{yo} = 10$ MPa,
- Base damage rate coefficient, $b_o = -10000000$,
- Base damage flow stress, $\sigma_{do} = 10$ MPa,
- Relative densities, $\rho_{1r} = \frac{\rho_1}{\rho} = 1$, $\rho_{2r} = \frac{\rho_2}{\rho} = 2$,
- Relative Lame parameters, $\lambda_{1r} = \frac{\lambda_1}{\lambda_o}$, $\lambda_{2r} = \frac{\lambda_2}{\lambda_o} = 5$,
- Relative Lame parameters, $\mu_{1r} = \frac{\mu_1}{\mu_o} = 1$, $\mu_{2r} = \frac{\mu_2}{\mu_o} = 5$,
- Relative conductivity, $K_{1r} = \frac{\boldsymbol{I\!K}_1}{\boldsymbol{I\!K}_o} = 1$, $K_{2r} = \frac{\boldsymbol{I\!K}_2}{\boldsymbol{I\!K}_o} = 5$,
- Relative thermal expansion, $\beta_{1r} = \frac{\beta_1}{\beta_o} = 1$, $\beta_{2r} = \frac{\beta_2}{\beta_o} = 10$,
- Relative heat capacity, $C_{1r} = \frac{C_1}{C_o} = 1$, $C_{2r} = \frac{C_2}{C_o} = 2$,
- Relative plastic rate coefficient, $a_{1r} = \frac{a_1}{a_o} = 1$, $a_{2r} = \frac{a_2}{a_o} = 1$,
- Relative damage rate coefficient, $b_{1r} = \frac{b_1}{b_o} = 1$, $b_{2r} = \frac{b_2}{b_o} = 1$,
- Relative plastic yield, $\sigma_{y1r} = \frac{\sigma_{y1}}{\sigma_{yo}} = 1$, $\sigma_{y2r} = \frac{\sigma_{y2}}{\sigma_{yo}} = 5$,
- Relative damage threshold, $\sigma_{d1r} = \frac{\sigma_{d1}}{\sigma_{do}} = 1$, $\sigma_{d1r} = \frac{\sigma_{d2}}{\sigma_{do}} = 5$,
- A time-stepping factor $\phi = 0.5$ (midpoint rule),
- An overlapping length scale of the particles of $\zeta = 0.625$,
- The number of desired iterations per time-step set to $K_d = 5$, along with a coupling/staggering tolerance of $C_{tol} = 10^{-2}$,
- Weights for the iterative error norm, $w_1 = 0.5$ and $w_2 = 0.5$.

Throughout the computations, the spatial discretization meshes were repeatedly refined until the solutions did not exhibit any more sensitivity to further refinement of the grid-spacing. We started with meshes such as a 21 × 21 × 21 mesh, arising from having a cubical mesh with 10 nodes from the centerline plane of symmetry and one node in the middle, and then repeatedly refined in the following sequential manner:

1. **Mesh # 1**: a 21 × 21 × 21 mesh, which has 9,261 degrees of freedom (DOF) for the thermal field and 27,783 DOF for the mechanical field, for a total of 37,044 DOF.
2. **Mesh # 2**: a 41 × 41 × 41 mesh, which has 68921 DOF for the thermal field and 206783 DOF for the mechanical field, for a total of 275,684 DOF.
3. **Mesh # 3**: a 61 × 61 × 61 mesh, which has 226,981 DOF for the thermal field and 680,943 DOF for the mechanical field, for a total of 907,924 DOF.

4. **Mesh # 4**: a 81 × 81 × 81 mesh, which has 531,441 DOF for the thermal field and 1,594,322 DOF for the mechanical field, for a total of 2,125,764 DOF.

Approximately between a 61-level and a 81-level mesh, the results stabilized, indicating that the results are essentially free of any appreciable numerical error. *All numerical results are shown in* Figs. 5.4, 5.5, 5.6, and 5.7. At the length scales of interest, it is questionable whether the ideas of a sharp material interface are justified. Accordingly, we simulated the system with and without Laplacian smoothing, whereby one smooths the material data by post-processing the material data, node by node, to produce a smoother material representation, for example, for the thermal conductivity, $\hat{\mathbb{K}}$ (using the stencil in Fig. 5.8 in Chapter Appendix 1)

$$\nabla_X^2 \mathbb{K} = \mathbf{0} \Rightarrow \hat{\mathbb{K}}_{i,j,k} = \frac{1}{6}\left(\mathbb{K}_{i+1,j,k} + \mathbb{K}_{i-1,j,k} + \mathbb{K}_{i,j+1,k} + \mathbb{K}_{i,j-1,k} + \mathbb{K}_{i,j,k+1} + \mathbb{K}_{i,j,k-1}\right). \quad (5.46)$$

The same was done for the mechanical properties by enforcing $\nabla_X^2 \lambda_0 = \mathbf{0}$ and $\nabla_X^2 \mu_0 = \mathbf{0}$ and as well as other material data. The simulations were run with and without data smoothing, with the results being negligibly different for sufficiently fine meshes (Fig. 5.9). In particular, Fig. 5.9 depicts a typical microstructure showing the contact area (only illustrating one of the phases), while Fig. 5.4 shows successive frames of the deviatoric stress, 25% into the interior of the sample. Figures 5.5, 5.6, and 5.7 illustrate the various metrics that a materials designer would be interested in quantifying. In Fig. 5.7, the variation of the time-step size (normalized by the starting time-step size) is depicted. The size of the time-steps was purposely started quite small ($\Delta t = 10^{-10}$ s) and given an enlargement cap of 50 times in magnitude. This allows the system to slowly evolve to capture the quite transient behavior. During the bulk of the computation, the large steps were warranted (the time-step size evolved), as dictated by the physics and the adaptive algorithm. For other material selections and loading regimes, other adaptivity modes can occur. All simulations were run on a standard laptop requiring minimal memory requirements. It is important to stress that it is virtually impossible to determine a priori whether the initial time step is adequate to meet a tolerance and whether adaptivity is needed. *Obviously, we can use this scheme for any (trapezoidal) value of* $0 \le \phi \le 1$. Time-step size adaptivity is important, since the solution can dramatically change over the course of time, possibly requiring quite different time-step sizes to control the iterative (staggering) error. However, to maintain the accuracy of the time-stepping scheme, one must respect an upper bound dictated by the discretization error, i.e., $\Delta t \le \Delta t^{lim}$. The example shown was simply to illustrate the overall process.

Remark While our stated focus is on the evolution of stresses in deposited hot powders/particles, a by-product of the analysis is the post-processing of the overall effective properties such as, for the mechanical fields

$$\langle \boldsymbol{\sigma} \rangle_{\Omega} = \mathcal{F}^*(\langle \boldsymbol{\epsilon} \rangle_{\Omega}), \quad (5.47)$$

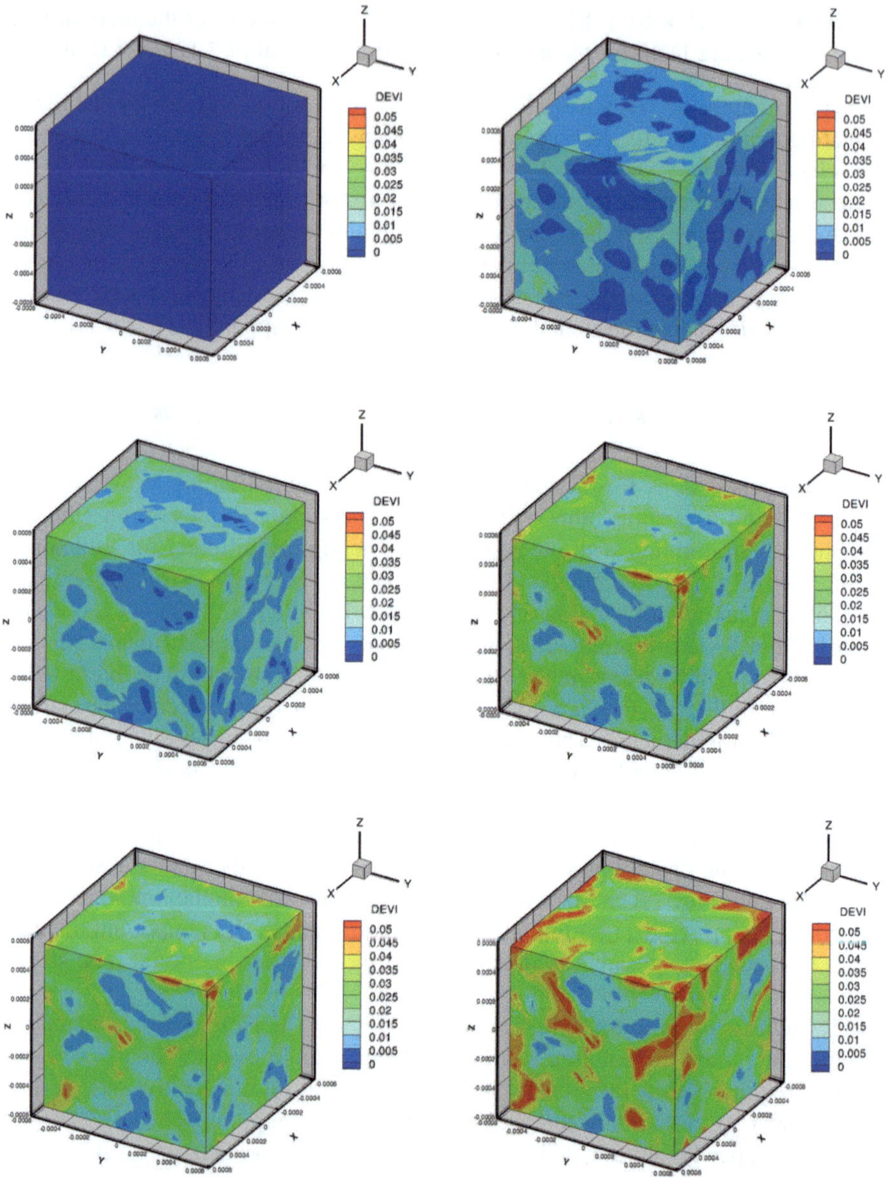

Fig. 5.4 From left to right and top to bottom: the deviatoric stress (in gigapascals). The morphology is shown in Fig. 5.9

where $\langle \cdot \rangle_\Omega \stackrel{\text{def}}{=} \frac{1}{|\Omega|}\int_\Omega \cdot d\Omega$. Similarly, one can generate effective thermal responses through

$$\langle \boldsymbol{q} \rangle_\Omega = \mathcal{G}^*(\langle \nabla\theta \rangle_\Omega). \tag{5.48}$$

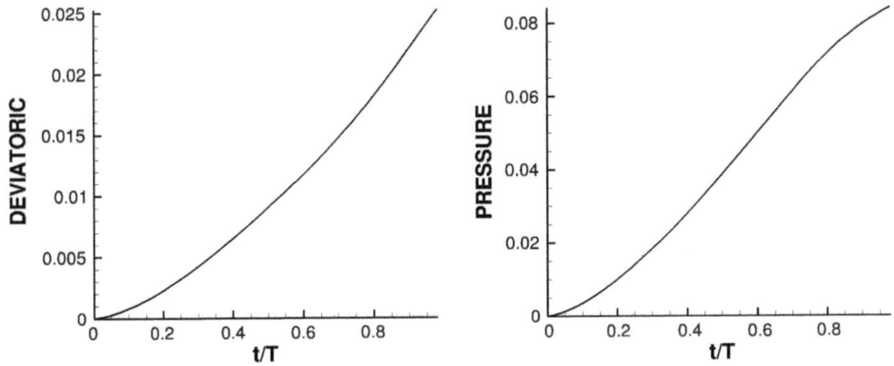

Fig. 5.5 Left: The volume averaged normed deviator $||\sigma'||$ (in gigapascals). Right: The volume averaged pressure $p \stackrel{\text{def}}{=} \frac{tr\sigma}{3}$ (in gigapascals)

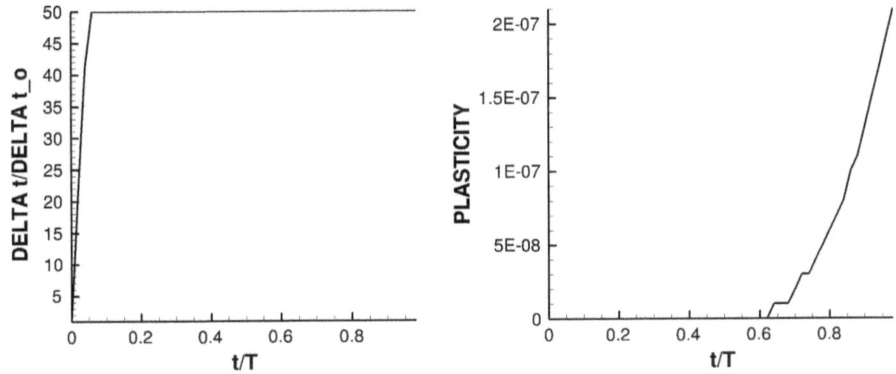

Fig. 5.6 Left: The volume averaged temperature $\langle\theta\rangle_{\Omega}$ (in Kelvin) Right: The volume averaged norm of the plastic strain $||\epsilon_p||$

See Zohdi [1–33] for more details. *As we have introduced earlier, there are a variety of estimates for effective responses. The numerical procedures in this chapter augment those classical approaches and are a more "brute-force" and robust route, since they can provide time-transient and nonlinear behavior directly.*

5.6 Summary and Extensions

The spatial discretization grids used were uniform and dense, and the deposited microstructure was embedded in spatial discretization. The regular grid allows one to generate a matrix-free iterative formulation which is amenable to rapid calculation and minimal memory requirements, making it ideal for laptop computation. Variants

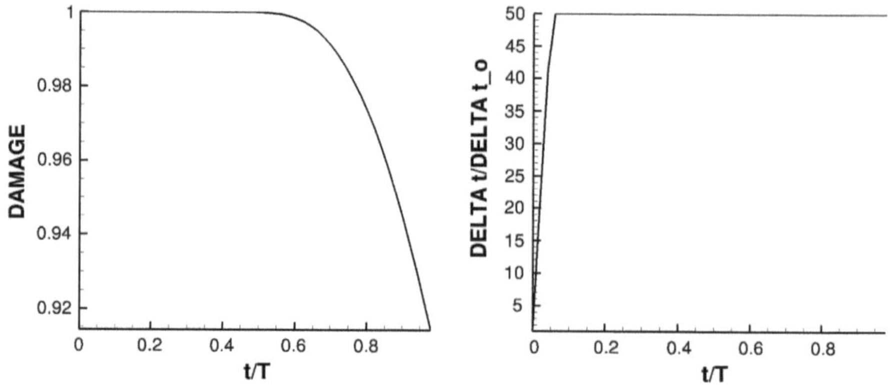

Fig. 5.7 Left: The volume averaged damage $||\mathcal{D}||$. Right: The time-step size variation

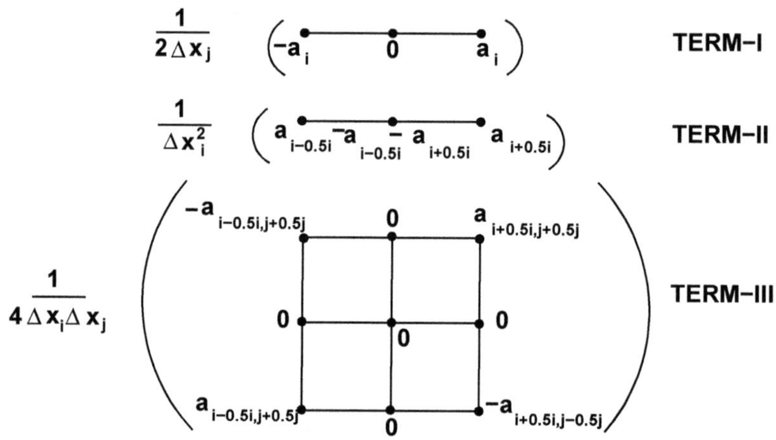

Fig. 5.8 Various finite difference stencils in "computational molecule" form (centered at (x_i, x_j, x_k)), where (1) TERM-I: $a\frac{\partial u}{\partial x_i}$, (2) TERM-II: $\frac{\partial}{\partial x_i}\left(a\frac{\partial u}{\partial x_i}\right)$, and (3) TERM-III: $\frac{\partial}{\partial x_j}\left(a\frac{\partial u}{\partial x_i}\right)$

of the technique have been applied to related problems involving more coupled multiphysics, such as electro-magneto-thermo-mechano-chemo effects, in Zohdi [1–33] whereby one computes the electrical field $\boldsymbol{E}$ with the magnetic field $\boldsymbol{H}$, thermal field θ, displacement field $\boldsymbol{u}$, and chemical field c fixed, then computes $\boldsymbol{H}$-field with $\boldsymbol{E}, \theta, \boldsymbol{u}$, and c fields fixed, etc., and iterates at time interval $L+1$, $K = 1, 2, \ldots$ for (written directly in iterative implicit form)

$$\boldsymbol{E}^{L+1,K} = \mathcal{F}(\underline{\boldsymbol{E}^{L+1,K-1}}, \boldsymbol{H}^{L+1,K-1}, \theta^{L+1,K-1}, \boldsymbol{u}^{L+1,K-1}, c^{L+1,K-1}), \quad (5.49)$$

and

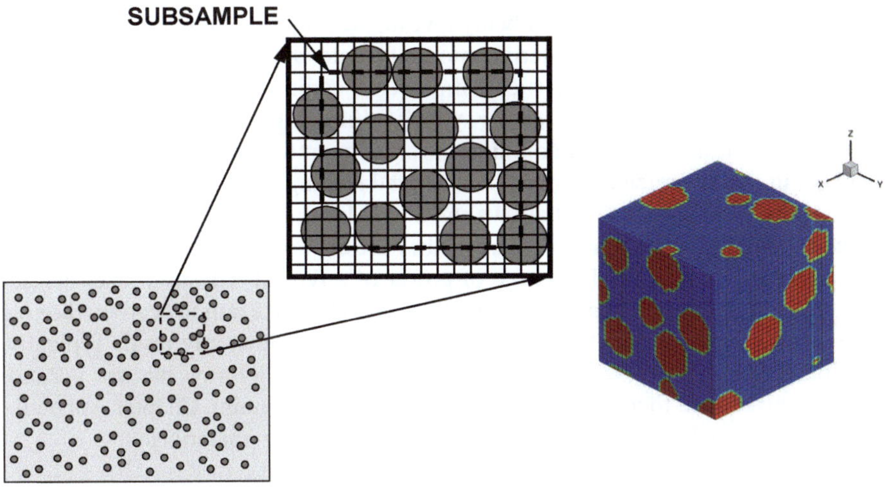

Fig. 5.9 Left: With the framing method, a sample is probed with interior subsamples, within the larger sample, in order to avoid boundary layer effects that occur from imposing boundary conditions on the large-sample exterior. Right: A mesh of the subsample

$$\boldsymbol{H}^{L+1,K} = \mathcal{G}(\boldsymbol{E}^{L+1,K}, \underline{\boldsymbol{H}^{L+1,K-1}}, \theta^{L+1,K-1}, \boldsymbol{u}^{L+1,K-1}, c^{L+1,K-1}), \tag{5.50}$$

and

$$\theta^{L+1,K} = \mathcal{Y}(\boldsymbol{E}^{L+1,K}, \boldsymbol{H}^{L+1,K}, \underline{\theta^{L+1,K-1}}, \boldsymbol{u}^{L+1,K-1}, c^{L+1,K-1}), \tag{5.51}$$

and

$$\boldsymbol{u}^{L+1,K} = \mathcal{L}(\boldsymbol{E}^{L+1,K}, \boldsymbol{H}^{L+1,K}, \theta^{L+1,K}, \underline{\boldsymbol{u}^{L+1,K-1}}, c^{L+1,K-1}), \tag{5.52}$$

and

$$c^{L+1,K} = \mathcal{C}(\boldsymbol{E}^{L+1,K}, \boldsymbol{H}^{L+1,K}, \theta^{L+1,K}, \boldsymbol{u}^{L+1,K}, \underline{c^{L+1,K-1}}), \tag{5.53}$$

where the only underlined variable is active at that stage of the process. One then computes the maximum of the error measures $\varpi^{*,K} \stackrel{\text{def}}{=} max(\varpi_E^K, \varpi_H^K, \varpi_\theta^K, \varpi_u^K, \varpi_c^K)$ in order to determine if time-step adaptivity is necessary, as introduced earlier for the thermo-mechanical scheme. Generally, the methods discussed in this work can be combined to create hybrid block-partitioned approaches, whereby the entire domain is partitioned into subdomains and within each subdomain an iterative method is applied. In other words, for a subdomain, the values at all nodes from outside are initially frozen, as far as calculations involving members of the group are concerned. After each isolated subdomain's solution (nodal values) has converged (computed in parallel), then all nodal values are updated; i.e., the most current values become available to all members of the grid, and the isolated subdomain calculations are repeated. Although parallel computation of the introduced algorithms was not pursued in this work, it is currently being investigated by the author. Finally, in addition to the deposition of such material onto a substrate, there are a number of related post-processes which make up a successful overall additive manufacturing process. *One key component is laser processing*, which utilizes high-intensity beams to heat the material to desired temperatures either to subsequently bond, soften, sinter, melt, or ablate, in a very targeted manner. Laser-based heating is quite attractive because of the degree of precision that it allows. As alluded to earlier, because of the monochromatic and collimated nature of lasers, they are an attractive, highly controllable way to process materials.[3] The range of power of a typical industrial laser is relatively wide, ranging from approximately 100–10000 W. For example, carbon dioxide (CO_2) and yttrium aluminum garnet (YAG) lasers are commonly used. Typically, the initial beam produced is in the form of collimated (parallel) rays that are 1–2 mm apart, which are then focused with a lens onto a small focal point. Chapter Appendix 4 provides an example of laser simulation for particle-laden materials.

5.7 Chapter Appendix 1: Summary of Spatial Finite Difference Stencils

The following standard approximations are used:

1. For the first derivative of a primal variable u at (x_1, x_2, x_3):

$$\frac{\partial u}{\partial x_1} \approx \frac{u(x_1 + \Delta x_1, x_2, x_3) - u(x_1 - \Delta x_1, x_2, x_3)}{2\Delta x_1} \tag{5.54}$$

2. For the derivative of a flux at (x_1, x_2, x_3), with an arbitrary material coefficient a:

$$\begin{aligned}\frac{\partial}{\partial x_1}\left(a\frac{\partial u}{\partial x_1}\right) &\approx \frac{\left(a\frac{\partial u}{\partial x_1}\right)|_{x_1+\frac{\Delta x_1}{2},x_2,x_3} - \left(a\frac{\partial u}{\partial x_1}\right)|_{x_1-\frac{\Delta x_1}{2},x_2,x_3}}{\Delta x_1} \\ &= \frac{1}{\Delta x_1}\left[a(x_1 + \frac{\Delta x_1}{2}, x_2, x_3)\left(\frac{u(x_1 + \Delta x_1, x_2, x_3) - u(x_1, x_2, x_3)}{\Delta x_1}\right)\right] \\ &- \frac{1}{\Delta x_1}\left[a(x_1 - \frac{\Delta x_1}{2}, x_2, x_3)\left(\frac{u(x_1, x_2, x_3) - u(x_1 - \Delta x_1, x_2, x_3)}{\Delta x_1}\right)\right],\end{aligned} \tag{5.55}$$

where we have used

$$a(x_1 + \frac{\Delta x_1}{2}, x_2, x_3) \approx \frac{1}{2}\left(a(x_1 + \Delta x_1, x_2, x_3) + a(x_1, x_2, x_3)\right) \tag{5.56}$$

and

[3] In particular with pulsing, via continuous beam chopping or modulation of the voltage.

$$a(x_1 - \frac{\Delta x_1}{2}, x_2, x_3) \approx \frac{1}{2}(a(x_1, x_2, x_3) + a(x_1 - \Delta x_1, x_2, x_3)) \quad (5.57)$$

3. For the cross-derivative of a flux at (x_1, x_2):

$$\begin{aligned}\frac{\partial}{\partial x_2}\left(a\frac{\partial u}{\partial x_1}\right) &\approx \frac{\partial}{\partial x_2}\left(a(x_1, x_2, x_3)\left(\frac{u(x_1+\Delta x_1, x_2, x_3) - u(x_1 - \Delta x_1, x_2, x_3)}{2\Delta x_1}\right)\right)\\ &\approx \frac{1}{4\Delta x_1 \Delta x_2}(a(x_1, x_2+\Delta x_2, x_3)\left[u(x_1+\Delta x_1, x_2+\Delta x_2, x_3) - u(x_1-\Delta x_1, x_2+\Delta x_2, x_3)\right]\\ &- a(x_1, x_2-\Delta x_2, x_3)\left[u(x_1+\Delta x_1, x_2-\Delta x_2, x_3) - u(x_1-\Delta x_1, x_2-\Delta x_2, x_3)\right]), \end{aligned} \quad (5.58)$$

Remark To illustrate second-order accuracy, consider a Taylor series expansion for an arbitrary function u

$$u(x+\Delta x) = u(x) + \frac{\partial u}{\partial x}|_x \Delta x + \frac{1}{2}\frac{\partial^2 u}{\partial x^2}|_x (\Delta x)^2 + \frac{1}{6}\frac{\partial^3 u}{\partial x^3}|_x (\Delta x)^3 + \mathcal{O}((\Delta x)^4) \quad (5.59)$$

and

$$u(x-\Delta x) = u(x) - \frac{\partial u}{\partial x}|_x \Delta x + \frac{1}{2}\frac{\partial^2 u}{\partial x^2}|_x (\Delta x)^2 - \frac{1}{6}\frac{\partial^3 u}{\partial x^3}|_x (\Delta x)^3 + \mathcal{O}((\Delta x)^4) \quad (5.60)$$

Subtracting the two expressions yields

$$\frac{\partial u}{\partial x}|_x = \frac{u(x+\Delta x) - u(x-\Delta x)}{2\Delta x} + \mathcal{O}((\Delta x)^2). \quad (5.61)$$

5.8 Chapter Appendix 2: Second-Order Temporal Discretization

Discretization of temporally second-order equations can be illustrated by considering (with general variables $\boldsymbol{U}$ and $\boldsymbol{V}$)

$$\ddot{\boldsymbol{U}} = \dot{\boldsymbol{V}} = \boldsymbol{\Psi}(\boldsymbol{U}). \quad (5.62)$$

Expanding the field $\boldsymbol{V}$ in a Taylor series about $t + \phi\Delta t$, we obtain

$$\boldsymbol{V}(t+\Delta t) = \boldsymbol{V}(t+\phi\Delta t) + \frac{d\boldsymbol{V}}{dt}|_{t+\phi\Delta t}(1-\phi)\Delta t + \frac{1}{2}\frac{d^2\boldsymbol{V}}{dt^2}|_{t+\phi\Delta t}(1-\phi)^2(\Delta t)^2 + \mathcal{O}((\Delta t)^3) \quad (5.63)$$

and

$$\boldsymbol{V}(t) = \boldsymbol{V}(t+\phi\Delta t) - \frac{d\boldsymbol{V}}{dt}|_{t+\phi\Delta t}\phi\Delta t + \frac{1}{2}\frac{d^2\boldsymbol{V}}{dt^2}|_{t+\phi\Delta t}\phi^2(\Delta t)^2 + \mathcal{O}((\Delta t)^3) \quad (5.64)$$

Subtracting the two expressions yields

$$\frac{d\boldsymbol{V}}{dt}|_{t+\phi\Delta t} = \frac{\boldsymbol{V}(t+\Delta t) - \boldsymbol{V}(t)}{\Delta t} + \hat{\mathcal{O}}(\Delta t), \tag{5.65}$$

where $\hat{\mathcal{O}}(\Delta t) = \mathcal{O}((\Delta t)^2)$, when $\phi = \frac{1}{2}$. Thus, inserting this into the governing equation yields

$$\boldsymbol{V}(t+\Delta t) = \boldsymbol{V}(t) + \Delta t \boldsymbol{\Psi}(t+\phi\Delta t) + \hat{\mathcal{O}}((\Delta t)^2). \tag{5.66}$$

Note that adding a weighted sum of Eqs. 5.63 and 5.64 yields

$$\boldsymbol{V}(t+\phi\Delta t) = \phi\boldsymbol{V}(t+\Delta t) + (1-\phi)\boldsymbol{V}(t) + \mathcal{O}((\Delta t)^2), \tag{5.67}$$

which will be useful shortly. Now expanding the field $\boldsymbol{U}$ in a Taylor series about $t + \phi\Delta t$, we obtain

$$\boldsymbol{U}(t+\Delta t) = \boldsymbol{U}(t+\phi\Delta t) + \frac{d\boldsymbol{U}}{dt}|_{t+\phi\Delta t}(1-\phi)\Delta t + \frac{1}{2}\frac{d^2\boldsymbol{U}}{dt^2}|_{t+\phi\Delta t}(1-\phi)^2(\Delta t)^2 + \mathcal{O}((\Delta t)^3) \tag{5.68}$$

and

$$\boldsymbol{U}(t) = \boldsymbol{U}(t+\phi\Delta t) - \frac{d\boldsymbol{U}}{dt}|_{t+\phi\Delta t}\phi\Delta t + \frac{1}{2}\frac{d^2\boldsymbol{U}}{dt^2}|_{t+\phi\Delta t}\phi^2(\Delta t)^2 + \mathcal{O}((\Delta t)^3). \tag{5.69}$$

Subtracting the two expressions yields

$$\frac{\boldsymbol{U}(t+\Delta t) - \boldsymbol{U}(t)}{\Delta t} = \boldsymbol{V}(t+\phi\Delta t) + \hat{\mathcal{O}}(\Delta t). \tag{5.70}$$

Inserting Eq. 5.67 yields

$$\boldsymbol{U}(t+\Delta t) = \boldsymbol{U}(t) + (\phi\boldsymbol{V}(t+\Delta t) + (1-\phi)\boldsymbol{V}(t))\,\Delta t + \hat{\mathcal{O}}((\Delta t)^2). \tag{5.71}$$

Thus, using Eq. 5.66 yields

$$\boldsymbol{U}(t+\Delta t) = \boldsymbol{U}(t) + \boldsymbol{V}(t)\Delta t + \phi(\Delta t)^2\boldsymbol{\Psi}(\boldsymbol{U}(t+\phi\Delta t)) + \hat{\mathcal{O}}((\Delta t)^2). \tag{5.72}$$

The term $\boldsymbol{\Psi}(\boldsymbol{U}(t+\phi\Delta t))$ can be handled in two main ways:

- $\boldsymbol{\Psi}(t+\phi\Delta t) \approx \boldsymbol{\Psi}(\phi\boldsymbol{U}(t+\Delta t) + (1-\phi)\boldsymbol{U}(t))$ or
- $\boldsymbol{\Psi}(t+\phi\Delta t) \approx \phi\boldsymbol{\Psi}(\boldsymbol{U}(t+\Delta t)) + (1-\phi)\boldsymbol{\Psi}(\boldsymbol{U}(t))$.

The differences are quite minute between either of the above; thus, for brevity, we choose the latter. In summary, we have the following:

$$\boldsymbol{U}(t+\Delta t) = \boldsymbol{U}(t) + \boldsymbol{V}(t)\Delta t + \phi(\Delta t)^2\left(\phi\boldsymbol{\Psi}(\boldsymbol{U}(t+\Delta t)) + (1-\phi)\boldsymbol{\Psi}(\boldsymbol{U}(t))\right) + \hat{\mathcal{O}}((\Delta t)^2). \tag{5.73}$$

We note that

- When $\phi = 1$, then this is the (implicit) Backward Euler scheme, which is very stable (very dissipative) and $\mathcal{O}((\Delta t)^2)$ locally in time,
- When $\phi = 0$, then this is the (explicit) Forward Euler scheme, which is conditionally stable and $\mathcal{O}((\Delta t)^2)$ locally in time,
- When $\phi = 0.5$, then this is the (implicit) "Midpoint" scheme, which is stable and $\hat{\mathcal{O}}((\Delta t)^2) = \mathcal{O}((\Delta t)^3)$ locally in time.

In summary, we have for the velocity[4]

$$\boldsymbol{V}(t + \Delta t) = \boldsymbol{V}(t) + \Delta t \left(\phi \boldsymbol{\Psi}(\boldsymbol{U}(t + \Delta t)) + (1 - \phi)\boldsymbol{\Psi}(\boldsymbol{U}(t))\right) \tag{5.74}$$

and for the position

$$\begin{aligned} \boldsymbol{U}(t + \Delta t) &= \boldsymbol{U}(t) + \boldsymbol{V}(t + \phi\Delta t)\Delta t \\ &= \boldsymbol{U}(t) + \left(\phi \boldsymbol{V}(t + \Delta t) + (1 - \phi)\boldsymbol{V}(t)\right)\Delta t, \end{aligned} \tag{5.75}$$

or more explicitly,

$$\boldsymbol{U}(t + \Delta t) = \boldsymbol{U}(t) + \boldsymbol{V}(t)\Delta t + \phi(\Delta t)^2 \left(\phi \boldsymbol{\Psi}(\boldsymbol{U}(t + \Delta t)) + (1 - \phi)\boldsymbol{\Psi}(\boldsymbol{U}(t))\right). \tag{5.76}$$

In iterative (recursion) form,

$$\boldsymbol{U}^{L+1,K} = \underbrace{(\phi\Delta t)^2 \boldsymbol{\Psi}(\boldsymbol{U}^{L+1,K-1})}_{\mathcal{G}(\boldsymbol{U}^{L+1,K-1})} + \underbrace{\boldsymbol{U}^L + \boldsymbol{V}^L \Delta t + (\Delta t)^2 \phi(1 - \phi)\boldsymbol{\Psi}(\boldsymbol{U}^L)}_{\mathcal{R}} \tag{5.77}$$

Remark Applying this scheme to the balance of linear momentum continuum formulation, under infinitesimal deformations, $\nabla_X \cdot \boldsymbol{\sigma} + \boldsymbol{f} = \rho \frac{\partial^2 \boldsymbol{u}}{\partial t^2}$, we use $\boldsymbol{\Psi}(\boldsymbol{u}(t)) = \frac{\nabla_X \cdot \boldsymbol{\sigma} + \boldsymbol{f}}{\rho}$ and must apply the (iterative) process introduced earlier to all nodes in the system.

5.9 Chapter Appendix 3: Temporally Adaptive Iterative Methods

Implicit time-stepping methods, with time-step size adaptivity, built on approaches found in Zohdi [1–33] were used throughout the analysis in the body of this work. In order to introduce basic concepts, we consider a first-order differential equation for a field $\boldsymbol{W}$:

---

[4]In order to streamline the notation, we drop the cumbersome $\mathcal{O}(\Delta t)$-type terms.

$$\dot{\boldsymbol{W}} = \boldsymbol{\Lambda}(\boldsymbol{W}), \tag{5.78}$$

which after being discretized using a trapezoidal "ϕ-method" ($0 \leq \phi \leq 1$)

$$\boldsymbol{W}^{L+1} = \boldsymbol{W}^{L} + \Delta t \left(\phi \boldsymbol{\Lambda}(\boldsymbol{W}^{L+1}) + (1-\phi)\boldsymbol{\Lambda}(\boldsymbol{W}^{L})\right). \tag{5.79}$$

Generally, for systems of equations of this form, a straightforward iterative scheme can be written as

$$\boldsymbol{W}^{L+1,K} = \mathcal{G}(\boldsymbol{W}^{L+1,K-1}) + \mathcal{R}, \tag{5.80}$$

where $\mathcal{R}$ is a remainder term that does not depend on the solution, i.e., $\mathcal{R} \neq \mathcal{R}(\boldsymbol{W}^{L+1})$, and $K = 1, 2, 3, \ldots$ is the index of iteration within time-step $L+1$. The convergence of such a scheme is dependent on the behavior of $\mathcal{G}$. Namely, a sufficient condition for convergence is that $\mathcal{G}$ is a contraction mapping for all $\boldsymbol{W}^{L+1,K}$, $K = 1, 2, 3, \ldots$. In order to investigate this further, we define the iteration error as

$$\varpi^{L+1,K} \stackrel{\text{def}}{=} ||\boldsymbol{W}^{L+1,K} - \boldsymbol{W}^{L+1}||. \tag{5.81}$$

A necessary restriction for convergence is iterative self-consistency, i.e., the "exact" (discretized) solution must be represented by the scheme

$$\mathcal{G}(\boldsymbol{W}^{L+1}) + \mathcal{R} = \boldsymbol{W}^{L+1}. \tag{5.82}$$

Enforcing this restriction, a sufficient condition for convergence is the existence of a contraction mapping

$$\varpi^{L+1,K} = ||\boldsymbol{W}^{L+1,K} - \boldsymbol{W}^{L+1}|| = ||\mathcal{G}(\boldsymbol{W}^{L+1,K-1}) - \mathcal{G}(\boldsymbol{W}^{L+1})|| \tag{5.83}$$

$$\leq \eta^{L+1,K} ||\boldsymbol{W}^{L+1,K-1} - \boldsymbol{W}^{L+1}||, \tag{5.84}$$

where if $0 \leq \eta^{L+1,K} < 1$ for each iteration K, then $\varpi^{L+1,K} \to 0$ for any arbitrary starting value $\boldsymbol{W}^{L+1,K=0}$, as $K \to \infty$. This type of contraction condition is sufficient, but not necessary, for convergence. Inserting these approximations into $\dot{\boldsymbol{W}} = \boldsymbol{\Lambda}(\boldsymbol{W})$ leads to

$$\boldsymbol{W}^{L+1,K} \approx \underbrace{\Delta t \left(\phi \boldsymbol{\Lambda}(\boldsymbol{W}^{L+1,K-1})\right)}_{\mathcal{G}(\boldsymbol{W}^{L+1,K-1})} + \underbrace{\Delta t (1-\phi)\boldsymbol{\Lambda}(\boldsymbol{W}^{L}) + \boldsymbol{W}^{L}}_{\mathcal{R}}, \tag{5.85}$$

whose contraction constant is scaled by $\eta \propto \phi \Delta t$. Therefore, if convergence is slow within a time-step, the time step size, which is adjustable, can be reduced by an appropriate amount to increase the rate of convergence. Decreasing the time-step size improves the convergence; however, we want to simultaneously maximize the time-step sizes to decrease overall computing time, while still meeting an error tolerance on the numerical solution's accuracy. In order to achieve this goal, we follow an approach found in Zohdi [1–33] originally developed for continuum thermo-chemical multifield problems in which one firstly approximates

$$\eta^{L+1,K} \approx S(\Delta t)^{p} \tag{5.86}$$

(S is a constant) and secondly one assumes the error within an iteration to behave according to

$$(S(\Delta t)^{p})^{K}\varpi^{L+1,0} = \varpi^{L+1,K}, \tag{5.87}$$

$K = 1, 2, \ldots$, where $\varpi^{L+1,0}$ is the initial norm of the iterative error and S is intrinsic to the system.[5] Our goal is to meet an error tolerance in exactly a preset number of iterations. To this end, one writes

$$(S(\Delta t_{\text{tol}})^{p})^{K_d}\varpi^{L+1,0} = C_{tol}, \tag{5.88}$$

where C_{tol} is a (coupling) tolerance and K_d is the number of desired iterations.[6] If the error tolerance is not met in the desired number of iterations, the contraction constant $\eta^{L+1,K}$ is too large. Accordingly, one can solve for a new smaller step size, under the assumption that S is constant,

$$\Delta t_{\text{tol}} = \Delta t \left(\frac{\left(\frac{C_{tol}}{\varpi^{L+1,0}}\right)^{\frac{1}{pK_d}}}{\left(\frac{\varpi^{L+1,K}}{\varpi^{L+1,0}}\right)^{\frac{1}{pK}}} \right). \tag{5.89}$$

The assumption that S is constant is not critical, since the time-steps are to be recursively refined and unrefined throughout the simulation. Clearly, the expression in Eq. 5.89 can also be used for time-step enlargement, if convergence is met in less than K_d iterations.[7]

5.10 Chapter Appendix 4: Laser Processing

One concern of manufacturers is the microstructural defects generated in additively manufactured products, created by imprecisely controlled heat-affected zones, and brought on by miscalibration of the laser power needed for a specific goal. For example, due to the rise of one particular branch of additive manufacturing, printed flexible electronics, involving sensitive substrates, it has become important to pre-

---

[5] For the class of problems under consideration, due to the linear dependency on Δt, $p \approx 1$.

[6] Typically, K_d is chosen to be between five to ten iterations.

[7] At the implementation level, since the exact solution is unknown, the following relative error term is used, $\varpi^{L+1,K} \stackrel{\text{def}}{=} ||\boldsymbol{w}^{L+1,K} - \boldsymbol{w}^{L+1,K-1}||$.

cisely understand how much laser input is needed. With minimal additional modeling and simulation effort, one can add laser input to the previously introduced model.

5.10.1 Formulations for Particulate-Laden Continua

The specific manufacturing scenario that we are interested in modeling is a particulate-functionalized composite material which experiences laser-pulsing in a targeted region. This covers a wide range of applications in additive manufacturing. Of particular interest is the resolution of thermal and residual stresses. Accordingly, this section is concerned with the computational characterization of the evolution of thermal and stress fields, in materials with particulate-functionalized microstructure using direct methods.

Laser input

Selective laser processing/sintering was pioneered by Householder [75] in 1979 and Deckard [76] in the mid-1980s. Generally, an overall technological goal is to develop computational tools to accelerate the manufacturing of printed electronics. *Lasers can play a central role in precisely processing these systems.* To describe the *laser–target* interaction, the following must be accounted for: (a) absorption of laser energy input, (b) beam interference (attenuation) from the heterogeneous media, and (c) heat transfer by conduction. There are varying degrees of sophistication for modeling the heat input from a laser, for example (Zohdi [1–33]). We utilize the Beer–Lambert law, whereby one performs an overall power balance (Fig. 6.5):

$$(I + \Delta I) - I + \mathcal{L}\Delta\zeta = 0 \Rightarrow \frac{dI}{d\zeta} = \mathcal{L} \approx -\alpha I, \tag{5.90}$$

where $\mathcal{L} \approx \alpha I$ is the absorbed irradiance (per unit area) from the laser and $\zeta = \zeta^*$ is the penetration location depth, which is solved to yield

$$I(\zeta^*) = I_o e^{-\int_0^{\zeta^*} \alpha d\zeta}. \tag{5.91}$$

This is then incorporated into the first law of thermodynamics.

Remark We could perform a discretization of the irradiant beam into rays and perform a full-blown ray-tracking scheme (see Zohdi [1–33]), or perform a discretization of the beam into its electromagnetic field components via Maxwell's equations (see Zohdi [1–33]). While such approaches provide extremely detailed field information at the smallest scales, they are extraordinarily computationally expensive and, for the applications in this work, unwarranted. The Beer–Lambert framework provides an approach that is useful for practical computation. In such an approach, one can easily introduce nonuniform beam profiles, for example

$$I(r) = I(r = 0)e^{-c||\boldsymbol{r}-\boldsymbol{r}_o||}, \tag{5.92}$$

where $||\boldsymbol{r} - \boldsymbol{r}_o||$ is the distance from the center of the incident beam line. In the case of $c = 0$, we recapture a flat beam, $I(r) = I(r = 0)$. The previous thermo-mechanical formulation in this chapter is then used. Following Zohdi [1–33], we consider a model problem of a particulate composite. The essential field equations and simplifying assumptions that will be used during the analysis are provided next. We employ the same coupled balance of linear momentum–first law framework as before, but add a laser source term: (where here we assume $C = C(t)$)

$$\rho C\dot{\theta} = \boldsymbol{\sigma} : (\dot{\boldsymbol{\epsilon}}_\theta + \dot{\boldsymbol{\epsilon}}_p) - \frac{1}{2}(\boldsymbol{\epsilon} - \boldsymbol{\epsilon}_\theta - \boldsymbol{\epsilon}_p) : \dot{\mathbb{E}} : (\boldsymbol{\epsilon} - \boldsymbol{\epsilon}_\theta - \boldsymbol{\epsilon}_p) + \nabla_X \cdot (\mathbb{K} \cdot \nabla_X \theta) + \rho z - \rho\dot{C}\theta + LASER - SOURCES \quad (5.93)$$

where Fourier's law, $\boldsymbol{q} = -\mathbb{K} \cdot \nabla_X \theta$, has been used.

Remark As before, following Zohdi [1–33], we now develop a staggering solution framework to solve the coupled systems of interest, where, at a given time increment: (1) each field equation is solved individually, "freezing" the other (coupled) fields in the system, allowing only the primary field to be active and (2) after the solution of each field equation, the primary field variable is updated, and the next field equation is treated in a similar manner. For an "implicit" type of staggering, the process can be repeated in an iterative manner, while for an "explicit" type, one moves to the next time-step after one "passes" through the system. As before, we will employ implicit staggering.

5.10.2 A Specific Numerical Example—Controlled Heating

As mentioned at the outset of this chapter, the specific manufacturing scenario that we are interested in modeling is a mixture which experiences laser-pulsing in targeted regions to induce a desired temperature field locally. This covers a wide range of applications in additive manufacturing. For illustration purposes, in this example, we control the lateral and side boundary conditions, set to be constant (thus laterally drawing heat out of the overall system). The top boundary is flux-type according to the following (Fig. 5.10):

- (a) if in the laser zone

$$(\mathbb{K} \cdot \nabla\theta) \cdot \boldsymbol{n} = I \quad (5.94)$$

and

- (b) otherwise

$$(\mathbb{K} \cdot \nabla\theta) \cdot \boldsymbol{n} = 0. \quad (5.95)$$

In addition to the previously defined parameters, we have the following:

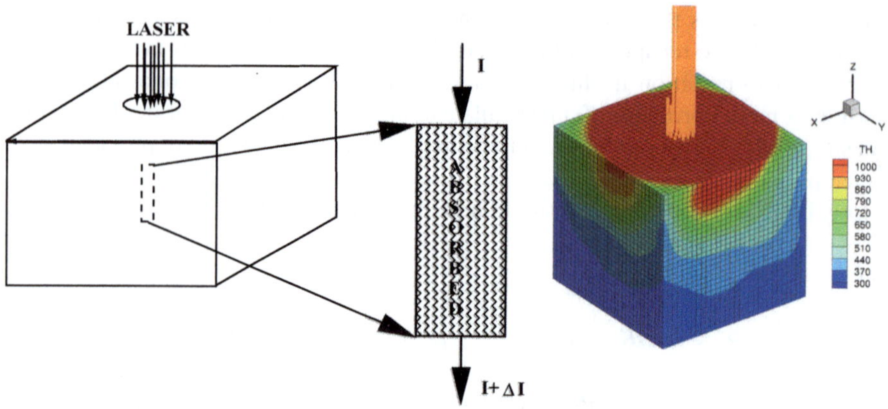

Fig. 5.10 Left: representations of laser input and absorption. Right: upcoming results showing mesh, absorption of energy, and temperature

- Laser beam radius: $R_L = 0.1L = 0.00001$ m and
- Laser strength ($I_o = 10^9\ W/m^2$), where

$$I(\boldsymbol{r}, t) = I_o e^{-c_o||\boldsymbol{r}-\boldsymbol{r}_o||}\left((c_1 + c_2\frac{t}{T})(1 + sin(2\pi\omega t/T))\right), \tag{5.96}$$

  and $\omega = 10\ s^{-1}$, $c_o = 0$, $c_1 = 0$, $c_2 = 100$, and $T = 10^{-4}$s.
- Absorption for the Beer–Lambert law:

$$I(\boldsymbol{r}, \zeta, t) = I(\boldsymbol{r}, t)e^{-\int_0^{\zeta^*} \alpha\, d\zeta}, \tag{5.97}$$

  where $\alpha = \alpha_1 e^{-\alpha_2 \frac{(\theta(x)-\theta_o)}{\theta_o}}$, for the matrix material: $\alpha_1 = 1$ and $\alpha_2 = 0$ and for this particulate material: $\alpha_1 = 100$ and $\alpha_2 = 0$.

As in the previous model problem, we consider a group of particles with a smaller-scale interstitial material that is assumed to be a continuous phase.

5.10.3 Numerical Examples

As in the previous examples, the same system parameters were used. Throughout the computations, the spatial discretization meshes were repeatedly refined until the solutions did not exhibit any more sensitivity to further refinement of the grid-spacing. In Figs. 5.11, 5.12, 5.13, and 5.14, frames of the sample and cross-sectional temperature profiles are shown for targeted laser heating. As before, we started with meshes such as a 21 × 21 × 21 mesh, arising from having a cubical mesh with 10 nodes from the centerline plane of symmetry and one node in the middle, and

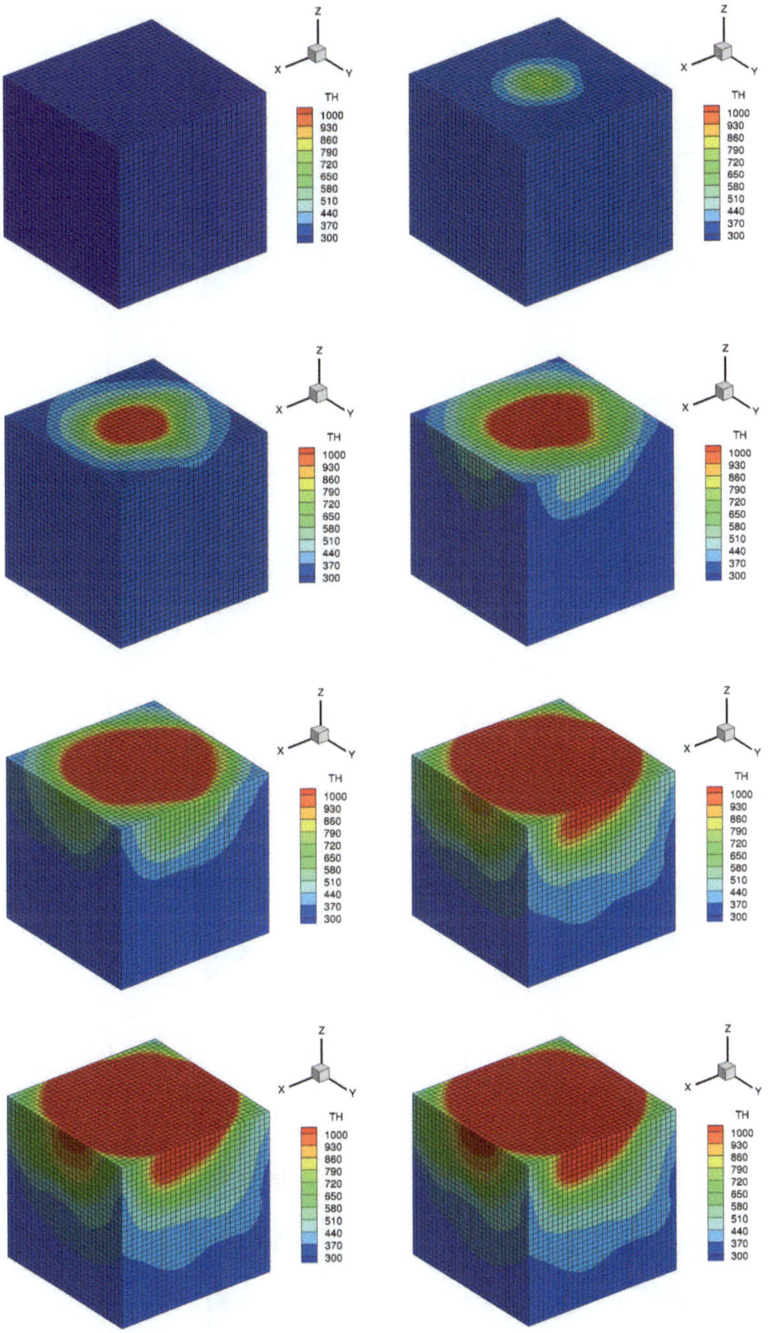

Fig. 5.11 From left to right and top to bottom: the temperature (in Kelvin), **with pulsed laser input**. The morphology is shown in Fig. 5.9

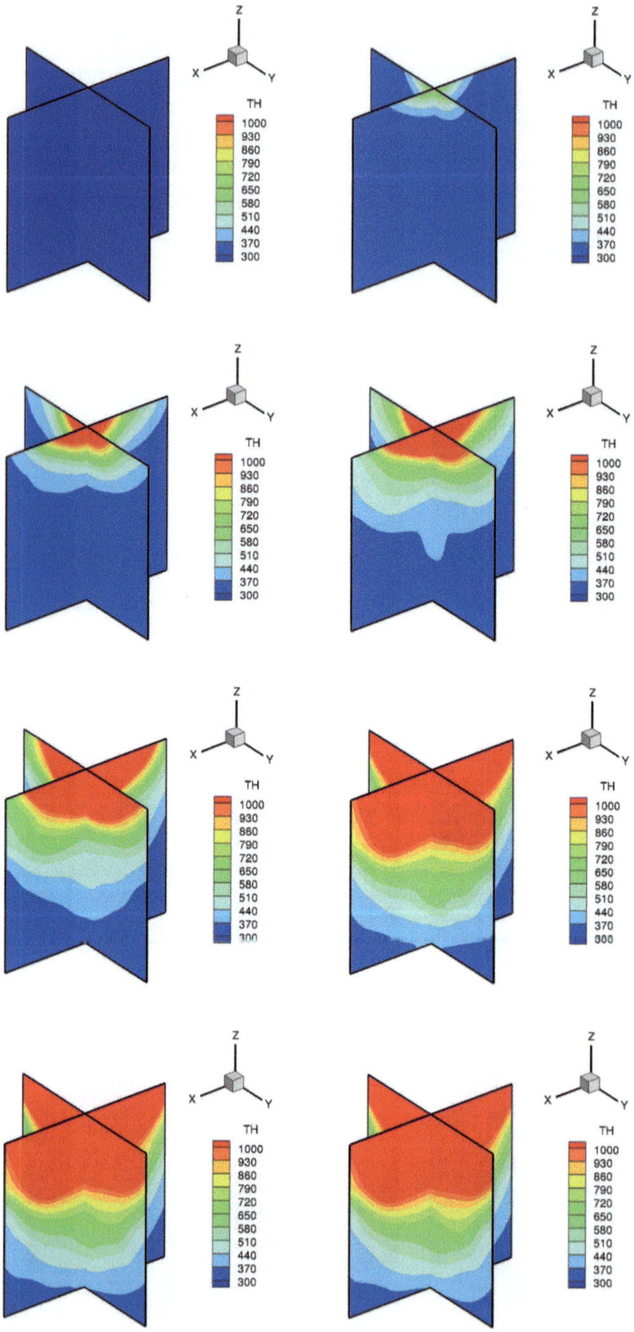

Fig. 5.12 From left to right and top to bottom: orthogonal slices through the microstructure for the temperature (in Kelvin), **with pulsed laser input**

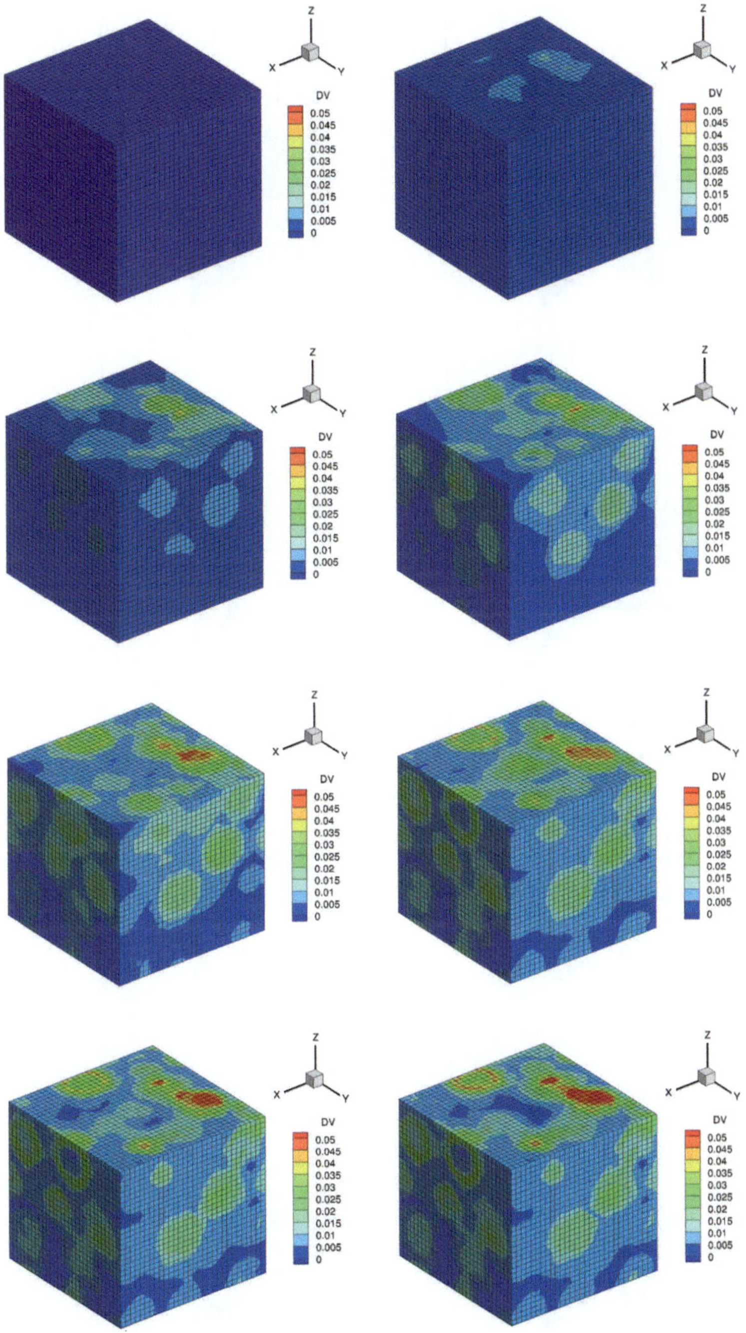

Fig. 5.13 From left to right and top to bottom: the norm of the deviatoric stress (in GPa), **with pulsed laser input**

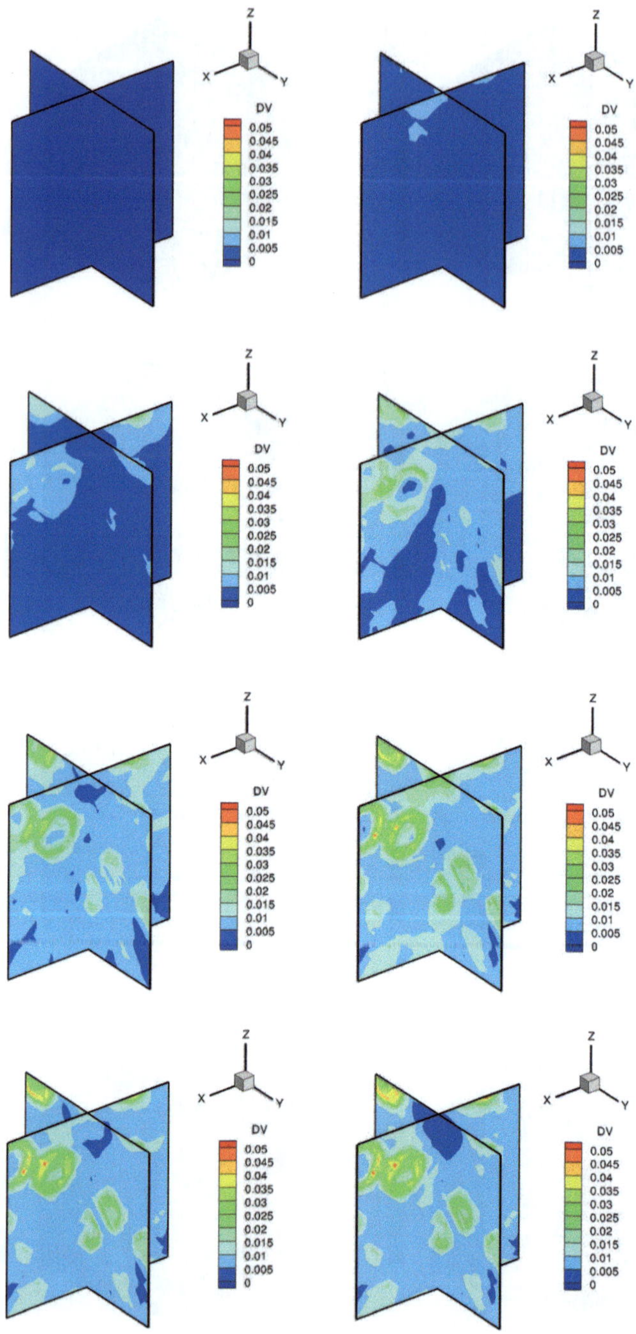

Fig. 5.14 From left to right and top to bottom: orthogonal slices through the microstructure for the norm of the deviatoric stress (in GPa), **with pulsed laser input**

then repeatedly refined. Approximately between a 61-level and a 81-level mesh, the results stabilized, indicating that the results are essentially free of any appreciable numerical error. The example shown was simply to illustrate the overall process. The simulation of targeted heating is a subject of current research by the author, extending this model to include phase transformations involving melting and vaporization and the debris ejecta, which involves multiple stages of nonmonotone evaporative heating and cooling, and mass transfer.

5.10.4 *Extensions: Advanced Models for Conduction Utilizing Thermal Relaxation*

Within the last decade, technological advances have enabled the reliable control of ultrafast pulsed lasers to activate small timescale heat wave effects. These effects are often referred to as thermally relaxed "second-sound" effects, because of their mathematical similarity to wave propagation in acoustics, although normal sound waves are fluctuations in the density of molecules in a substance while thermally relaxed second-sound waves are fluctuations in the density of phonons. Such phenomena are predicted by models which introduce thermal relaxation terms into heat-conduction relations. The thermally relaxed second-sound is a quantum mechanical phenomenon in which heat transfer occurs by wave-like motion, rather than by the more usual mechanism of diffusion. This leads to a very high confinement of thermal energy in very targeted zones. Thermally relaxed phenomena can be observed in any system, in which most phonon–phonon collisions conserve momentum, and can play a role when the timescale of heat input is quite small. More advanced models, based on thermal relaxation, which are a key feature of fast-pulsing laser technologies, are discussed in Zohdi [30, 33] and are outlined briefly here.

Thermally relaxed continuum model

The thermally relaxed second-sound-type model can be motivated by a Jeffreys-type relation between the conductive flux and temperature gradient (Joseph and Preziosi [77] and Ignaczak and Ostoja-Starzewski [78]):

$$\tau \frac{\partial \boldsymbol{q}_k}{\partial t} + \boldsymbol{q}_k = -\boldsymbol{I\!K} \cdot \nabla \theta, \tag{5.98}$$

where τ is the relaxation time, θ is the temperature, t is time, $\boldsymbol{q}_k$ is the conductive heat flux, and $\boldsymbol{I\!K}$ is the thermal conductivity. To clearly illustrate the character of this model, relative to standard head conduction, we ignore stress-power effects, yielding

$$\rho C \frac{\partial \theta}{\partial t} = -\nabla \cdot \boldsymbol{q}_k + \mathcal{S}, \tag{5.99}$$

where ρ is the mass density, C is the heat capacity, and $\mathcal{S}$ represents other sources, **such as laser energy input.** By taking the partial derivative with respect to time of the above yields, assuming no material changes,

$$\rho C \frac{\partial^2 \theta}{\partial t^2} = -\frac{\partial \nabla \cdot \boldsymbol{q}_k}{\partial t} + \frac{\partial \mathcal{S}}{\partial t} = -\nabla \cdot \frac{\partial \boldsymbol{q}_k}{\partial t} + \frac{\partial \mathcal{S}}{\partial t}. \tag{5.100}$$

Inserting Eq. 5.98 into 5.99 and 5.100 yields

$$\frac{\partial^2 \theta}{\partial t^2} + \frac{1}{\tau}\frac{\partial \theta}{\partial t} = \frac{1}{\rho C \tau} \nabla \cdot (\mathbb{K} \cdot \nabla \theta) + \frac{1}{\rho C \tau}\mathcal{S} + \frac{1}{\rho C}\frac{\partial \mathcal{S}}{\partial t}. \tag{5.101}$$

This produces attenuating heat waves. In the case of a homogeneous medium, the wave speed is $\sqrt{\frac{\mathbb{K}}{\rho C \tau}}$.

Extreme cases

We have the following extreme parameter cases:

- In the special case of $\tau \rightarrow 0$, one obtains the classical heat-conduction equation

$$\rho C \frac{\partial \theta}{\partial t} = \nabla \cdot \mathbb{K} \cdot \nabla \theta + \mathcal{S}. \tag{5.102}$$

- In the special case of $\tau \rightarrow \infty$, one obtains

$$\rho C \frac{\partial^2 \theta}{\partial t^2} = \frac{\partial \mathcal{S}}{\partial t} \Rightarrow \rho C \frac{\partial \theta}{\partial t} = \mathcal{S}, \tag{5.103}$$

which eliminates heat losses due to conduction, thus confining heat input.
- In the special case when $\tau \rightarrow \infty$, $\mathbb{K} \rightarrow \infty$, and $\frac{\mathbb{K}}{\tau}$ remaining finite, then a purely hyperbolic equation arises

$$\frac{\partial^2 \theta}{\partial t^2} = \frac{1}{\rho C \tau} \nabla \cdot \mathbb{K} \cdot \nabla \theta. \tag{5.104}$$

An excellent review of a wide range of heat transfer models can be found in the seminal review paper of Joseph and Preziosi [77] or the text of Ignaczak and Ostoja-Starzewski [78]. In the general case, we couple Eq. 5.98 to the more general first law:

$$\rho \dot{w} = \boldsymbol{\sigma} : \nabla \dot{\boldsymbol{u}} - \nabla \cdot \boldsymbol{q}_k + \rho z. \tag{5.105}$$

Algorithm for thermal relaxation

The thermally relaxed heat flux, since it is governed by its own PDE, requires simultaneous solution with the other governing equations. We approach this from a staggering point of view, at a time-step, at every node in the system:

- Solve for $\boldsymbol{q}$ fixing θ,

- Solve for θ (using the just updated $\boldsymbol{q}$) and
- Repeat until convergence.

At each iteration, $\boldsymbol{q}$ can be solved analytically (for fixed θ). We proceed by solving the following ODE (spatially fixed), for each component $\boldsymbol{q} = (q_1, q_2, q_3)$, over the interval $t_o \leq t \leq t_o + \phi \Delta t$

$$\tau \frac{\partial q_i}{\partial t} + q_i = -(\mathbb{K} \cdot \nabla \theta)_i, \tag{5.106}$$

with initial condition $q_i = q_i(t = t_o)$. Since we are "freezing" θ, this yields an ODE in time with solution (defining $\tilde{t} = t - t_o$)

$$q_i(\tilde{t}) = \underbrace{\left(q_i(\tilde{t} = 0) + (\mathbb{K} \cdot \nabla \theta(\tilde{t} + t_o))_i\right) e^{-\frac{\tilde{t}}{\tau}}}_{slows\ conduction} - \underbrace{(\mathbb{K} \cdot \nabla \theta(\tilde{t} + t_o))_i}_{regular\ conduction}, \tag{5.107}$$

which yields at time t

$$\boldsymbol{q}(t) = \underbrace{(\boldsymbol{q}(t_o) + (\mathbb{K} \cdot \nabla \theta(t)))\, e^{-\frac{t-t_o}{\tau}}}_{slows\ conduction} - \underbrace{(\mathbb{K} \cdot \nabla \theta(t))}_{regular\ conduction}. \tag{5.108}$$

For the spatial discretization (needed in the first law of thermodynamics), this yields:

$$\begin{aligned}\nabla \cdot \boldsymbol{q} &= \nabla \cdot (\boldsymbol{q}(t_o) + (\mathbb{K} \cdot \nabla \theta(t_o + \phi \Delta t)))\, e^{-\frac{t-t_o}{\tau}} \\ &+ (\boldsymbol{q}(t_o) + (\mathbb{K} \cdot \nabla \theta(t_o + \phi \Delta t))) \cdot \nabla e^{-\frac{t-t_o}{\tau}} - \nabla \cdot (\mathbb{K} \cdot \nabla \theta(t_o + \phi \Delta t)),\end{aligned} \tag{5.109}$$

where

$$\nabla e^{-\frac{t-t_o}{\tau}} = (t - t_o)\tau^{-2} e^{-\frac{t-t_o}{\tau}} \nabla \tau. \tag{5.110}$$

Remark 1 This yields at time $t = t_o + \phi \Delta t$

$$\boldsymbol{q}(t_o + \phi \Delta t) = \underbrace{(\boldsymbol{q}(t_o) + (\mathbb{K} \cdot \nabla \theta(t_o + \phi \Delta t)))\, e^{-\frac{\phi \Delta t}{\tau}}}_{slows\ conduction} - \underbrace{(\mathbb{K} \cdot \nabla \theta(t_o + \phi \Delta t))}_{regular\ conduction}, \tag{5.111}$$

which is needed in the trapezoidal time-stepping scheme.

Remark 2 We note that for a for a Discrete Element Method formulation,

$$m_i C_i \dot{\theta}_i = -\boldsymbol{q}_i \cdot \boldsymbol{n}_i A_i = q_i^* A_i, \tag{5.112}$$

this becomes[8]

$$q^*(t)_i = -\left(q(t_o) + \left(I\!K_i \frac{\theta_j - \theta_i}{||\boldsymbol{r}_j - \boldsymbol{r}_i||}\right)|_t\right) e^{-\frac{t-t_o}{\tau}} + \left(I\!K_i \frac{\theta_j - \theta_i}{||\boldsymbol{r}_j - \boldsymbol{r}_i||}\right)|_t. \quad (5.114)$$

References

1. Zohdi, T.I.: Genetic design of solids possessing a random-particulate microstructure. Philos. Trans. R. Soc. Math. Phys. Eng. Sci. **361**(1806), 1021–1043 (2003)
2. Zohdi, T.I.: On the compaction of cohesive hyperelastic granules at finite strains. Proc. R. Soc. **454**(2034), 1395–1401 (2003)
3. Zohdi, T.I.: Constrained inverse formulations in random material design. Comput. Methods Appl. Mech. Eng. 1–20. **192**(18), 28–30, 3179–3194 (2003)
4. Zohdi, T.I.: Staggering error control for a class of inelastic processes in random microheterogeneous solids. Int. J. Nonlinear Mech. **39**, 281–297 (2004)
5. Zohdi, T.I.: Modeling and simulation of a class of coupled thermo-chemo-mechanical processes in multiphase solids. Comput. Methods Appl. Mech. Eng. **193**(6–8), 679–699 (2004)
6. Zohdi, T.I.: Statistical ensemble error bounds for homogenized microheterogeneous solids. J. Appl. Math. Phys. (Zeitschrift für Angewandte Mathematik und Physik) **56**(3), 497–515 (2005)
7. Zohdi, T.I., Kachanov, M.: A note on the micromechanics of plastic yield of porous solids. Int. J. Fract. Lett. Micromech. **133**, L31–L35 (2005)
8. Zohdi, T.I.: On the optical thickness of disordered particulate media. Mech. Mater. **38**, 969–981 (2006)
9. Zohdi, T.I., Kuypers, F.A.: Modeling and rapid simulation of multiple red blood cell light scattering. Proc. R. Soc. Interface **3**(11), 823–831 (2006)
10. Zohdi, T.I.: Computation of the coupled thermo-optical scattering properties of random particulate systems. Comput. Methods Appl. Mech. Eng. **195**, 5813–5830 (2006)
11. Zohdi, T.I.: Computation of strongly coupled multifield interaction in particle-fluid systems. Comput. Methods Appl. Mech. Eng. **196**, 3927–3950 (2007)
12. Zohdi, T.I.: On the computation of the coupled thermo-electromagnetic response of continua with particulate microstructure. Int. J. Numer. Methods Eng. **76**, 1250–1279 (2008)
13. Zohdi, T.I., Kuypers, F.A., Lee, W.C.: Estimation of Red Blood Cell volume fraction from overall permittivity measurement. Int. J. Eng. Sci. **48**, 1681–1691 (2010)
14. Zohdi, T.I.: Simulation of coupled microscale multiphysical-fields in particulate-doped dielectrics with staggered adaptive FDTD. Comput. Methods Appl. Mech. Eng. **199**, 79–101 (2010)
15. Zohdi, T.I.: Joule-heating field phase-amplification in particulate-doped dielectrics. Int. J. Eng. Sci. **49**, 30–40 (2011)
16. Zohdi, T.I.: Estimation of electrical-heating load-shares for sintering of powder mixtures. Proc. R. Soc. **468**, 2174–2190 (2012)
17. Zohdi, T.I.: Modeling and simulation of the optical response rod-functionalized reflective surfaces. Comput. Mech. **50**(2), 257–268 (2012)

---

[8] For example, for a trapezoidal time-stepping scheme

$$q^*(t_o + \phi\Delta t)_i = -\left(q(t_o) + \left(I\!K_i \frac{\theta_j - \theta_i}{||\boldsymbol{r}_j - \boldsymbol{r}_i||}\right)|_{t_o+\phi\Delta t}\right) e^{-\frac{\phi\Delta t}{\tau}} + \left(I\!K_i \frac{\theta_j - \theta_i}{||\boldsymbol{r}_j - \boldsymbol{r}_i||}\right)|_{t_o+\phi\Delta t}. \quad (5.113)$$

18. Zohdi, T.I.: On the reduction of heat generation in lubricants using microscale additives. Int. J. Eng. Sci. **62**, 84–89 (2013)
19. Zohdi, T.I.: Numerical simulation of charged particulate cluster-droplet impact on electrified surfaces. J. Comput. Phys. **233**, 509–526 (2013)
20. Zohdi, T.I.: On inducing compressive residual stress in microscale print-lines for flexible electronics. Int. J. Eng. Sci. **62**, 157–164 (2013)
21. Zohdi, T.I.: Rapid simulation of laser processing of discrete particulate materials. Arch. Comput. Methods Eng. **20**, 309–325 (2013)
22. Zohdi, T.I.: A direct particle-based computational framework for electrically-enhanced thermo-mechanical sintering of powdered materials. Math. Mech. Solids **19**(1), 93–113 (2014)
23. Zohdi, T.I.: On cross-correlation between thermal gradients and electric fields. Int. J. Eng. Sci. **74**, 143–150 (2014)
24. Zohdi, T.I.: Mechanically-driven accumulation of microscale material at coupled solid-fluid interfaces in biological channels. Proc. R. Soc. Interface **11**, 20130922 (2014)
25. Zohdi, T.I.: A computational modeling framework for heat transfer processes in laser-induced dermal tissue removal. Comput. Mech. Eng. Sci. **98**(3), 261–277 (2014)
26. Zohdi, T.I.: Additive particle deposition and selective laser processing-a computational manufacturing framework. Comput. Mech. **54**, 171–191 (2014)
27. Zohdi, T.I.: Embedded electromagnetically sensitive particle motion in functionalized fluids. Comput. Part. Mech. **1**, 27–45 (2014)
28. Zohdi, T.I.: Rapid computation of statistically-stable particle/feature ratios for consistent substrate stresses in printed flexible electronics. J. Manuf. Sci. Eng. ASME, MANU-14-1476 (2015). https://doi.org/10.1115/1.4029327
29. Zohdi, T.I.: A computational modelling framework for high-frequency particulate obscurant cloud performance. Int. J. Eng. Sci. **89**, 75–85 (2015)
30. Zohdi, T. I.: On the thermal response of a laser-irradiated powder particle in additive manufacturing. CIRP J. Manuf. Sci. Technol. **10**, 77–83 (2015)
31. Zohdi, T.I.: Modeling and simulation of cooling-induced residual stresses in heated particulate mixture depositions. Comput. Mech. **56**, 613–630 (2015)
32. Zohdi, T.I.: Modeling and efficient simulation of the deposition of particulate flows onto compliant substrates. Int. J. Eng. Sci. **99**, 74–91 (2015). https://doi.org/10.1016/j.ijengsci.2015.10.012
33. Zohdi, T.I.: Modeling and simulation of laser processing of particulate-functionalized materials. Arch. Comput. Methods Eng. 1–25 (2015). https://doi.org/10.1007/s11831-015-9160-1
34. Duran, J.: Sands, Powders and Grains. An introduction to the Physics of Granular Matter. Springer (1997)
35. Pöschel, T., Schwager, T.: Computational Granular Dynamics. Springer (2004)
36. Onate, E., Idelsohn, S.R., Celigueta, M.A., Rossi, R.: Advances in the particle finite element method for the analysis of fluid-multibody interaction and bed erosion in free surface flows. Comput. Methods Appl. Mech. Eng. **197**(19–20), 1777–1800 (2008)
37. Onate, E., Celigueta, M.A., Idelsohn, S.R., Salazar, F., Surez, B.: Possibilities of the particle finite element method for fluid-soil-structure interaction problems. Comput. Mech. **48**, 307–318 (2011)
38. Rojek, J., Labra, C., Su, O., Onate, E.: Comparative study of different discrete element models and evaluation of equivalent micromechanical parameters. Int. J. Solids Struct. **49**, 1497–1517 (2012). https://doi.org/10.1016/j.ijsolstr.2012.02.032
39. Carbonell, J.M., Onate, E., Suarez, B.: Modeling of ground excavation with the particle finite element method. J. Eng. Mech. ASCE **136**, 455–463 (2010)
40. Labra, C., Onate, E.: High-density sphere packing for discrete element method simulations. Commun. Numer. Methods Eng. **25**(7), 837–849 (2009)
41. Leonardi, A., Wittel, F.K., Mendoza, M., Herrmann, H.J.: Coupled DEM-LBM method for the free-surface simulation of heterogeneous suspensions. Comput. Part. Mech. **1**(1), 3–13 (2014)
42. Cante, J., Davalos, C., Hernandez, J.A., Oliver, J., Jonsen, P., Gustafsson, G., Haggblad, H.A.: PFEM-based modeling of industrial granular flows. Comput. Part. Mech. **1**(1), 47–70 (2014)

43. Rojek, J.: Discrete element thermomechanical modelling of rock cutting with valuation of tool wear. Comput. Part. Mech. **1**(1), 71–84 (2014)
44. Bolintineanu, D.S., Grest, G.S., Lechman, J.B., Pierce, F., Plimpton, S.J., Schunk, P.R.: Particle dynamics modeling methods for colloid suspensions. Comput. Part. Mech. **1**(3), 321–356 (2014)
45. Campello, E.M.B., Zohdi, T.I.: A computational framework for simulation of the delivery of substances into cells. Int. J. Numer. Methods Biomed. Eng. **30**(11), 1132–1152 (2014)
46. Campello, E.M.B., Zohdi, T.I.: Design evaluation of a particle bombardment system to deliver substances into cells. Comput. Mech. Eng. Sci. **98**(2), 221–245 (2014)
47. Avci, B., Wriggers, P.: A DEM-FEM coupling approach for the direct numerical simulation of 3D particulate flows. J. Appl. Mech. **79**, 010901–1–7 (2012)
48. Akisanya, A.R., Cocks, A.C.F., Fleck, N.A.: The yield behavior of metal powders. Int. J. Mech. Sci. **39**, 1315–1324 (1997)
49. Brown, S., Abou-Chedid, G.: Yield behavior of metal powder assemblages. J. Mech. Phys. Solids **42**, 383–398 (1994)
50. Domas, F.: Eigenschaft profile und Anwendungsübersicht von EPE und EPP. Technical Report of the BASF Company (1997)
51. Fleck, N.A.: On the cold compaction of powders. J. Mech. Phys. Solids **43**, 1409–1431 (1995)
52. Gethin, D.T., Lewis, R.W., Ransing, R.S.: A discrete deformable element approach for the compaction of powder systems. Model. Simul. Mater. Sci. Eng. **11**(1), 101–114 (2003)
53. Gu, C., Kim, M., Anand, L.: Constitutive equations for metal powders: application to powder forming processes. Int. J. Plast. **17**, 147–209 (2001)
54. Lewis, R.W., Gethin, D.T., Yang, X.S.S., Rowe, R.C.: A combined finite-discrete element method for simulating pharmaceutical powder tableting. Int. J. Numer. Methods Eng. **62**, 853869 (2005)
55. Ransing, R.S., Lewis, R.W., Gethin, D.T.: Using a deformable discrete-element technique to model the compaction behaviour of mixed ductile and brittle particulate systems. Philosoph. Trans. R. Soc. Ser. A Math. Phys. Eng. Sci. **362**(1822), 1867–1884 (2004)
56. Tatzel, H.: Grundlagen der Verarbeitungstechnik von EPP-Bewährte und neue Verfahren. Technical Report of the BASF Company (1996)
57. Kachanov, L.M.: Introduction to Continuum Damage Mechanics. Martinus Nijoff, Dordricht (1986)
58. Zienkiewicz, O.C.: Coupled problems & their numerical solution. In: Lewis, R.W., Bettes, P., Hinton, E. (eds.) Numerical Methods in Coupled Systems, pp. 35–58. Wiley, Chichester (1984)
59. Zienkiewicz, O.C., Paul, D.K., Chan, A.H.C.: Unconditionally stable staggered solution procedure for soil-pore fluid interaction problems. Int. J. Numer. Methods Eng. **26**, 1039–1055 (1988)
60. Lewis, R.W., Schrefler, B.A., Simoni, L.: Coupling versus uncoupling in soil consolidation. Int. J. Num. Anal. Metho. Geomech. **15**, 533–548 (1992)
61. Lewis, R.W., Schrefler, B.A.: The Finite Element Method in the Static and Dynamic Deformation and Consolidation of Porous Media, 2nd edn. Wiley press (1998)
62. Schrefler, B.A.: A partitioned solution procedure for geothermal reservoir analysis. Comm. Appl. Num. Meth. **1**, 53–56 (1985)
63. Turska, E., Schrefler, B.A.: On consistency, stability and convergence of staggered solution procedures. Rend. Mat. acc. Lincei, Rome, S. 9, **5**, 265–271 (1994)
64. Bianco, M., Bilardi, G., Pesavento, F., Pucci, G., Schrefler, B.A.: A frontal solver tuned for fully coupled non-linear hygro-thermo-mechanical problems. Int. J. Numer. Meth. Eng. **57**, 18011818 (2003)
65. Wang, X., Schrefler, B.A.: A multifrontal parallel algorithm for coupled thermo-hydro-mechanical analysis of deforming porous media. Int. J. Numer. Meth. Eng. **43**, 10691083 (1998)
66. Young. D.M.: Iterative methods for solving partial difference equations of elliptic type. Doctoral Thesis, Harvard University (1950)

67. Ames, W.F.: Numerical Methods for Partial Differential Equations, 2nd edn. Academic Press (1977)
68. Axelsson, O.: Iterative Solution Methods. Cambridge University Press (1994)
69. Widom, B.: Random sequential addition of hard spheres to a volume. J. Chem. Phys. **44**, 3888–3894 (1966)
70. Torquato, S.: Random Heterogeneous Materials: Microstructure and Macroscopic Properties. Springer, New York (2002)
71. Kansaal, A., Torquato, S., Stillinger, F.: Diversity of order & densities in jammed hard-particle packings. Phys. Rev. E **66**, 041109 (2002)
72. Donev, A., Cisse, I., Sachs, D., Variano, E.A., Stillinger, F., Connelly, R., Torquato, S., Chaikin, P.: Improving the density of jammed disordered packings using ellipsoids. Science **303**, 990-993 (2004)
73. Donev, A., Stillinger, F.H., Chaikin, P.M., Torquato, S.: Unusually dense crystal ellipsoid packings. Phys. Rev. Lett. **92**, 255506 (2004)
74. Donev, A., Torquato, S., Stillinger, F.: Neighbor list collision-driven molecular dynamics simulation for nonspherical hard particles-I Algorithmic details. J. Comput. Phys. **202**, 737 (2005)
75. Householder, R.: Molding Process. U.S. Patent 4,247,508 (1979)
76. Deckard, C.: Method and apparatus for producing parts by selective sintering U.S. Patent 4,863,538 (1986)
77. Joseph, D.D., Preziosi, L.: Heat waves. Rev. Mod. Phys. **61**, 41–74 (1989)
78. Ignaczak, J., Ostoja-Starzewski, M.: Thermoelasticity with Finite Wave Speeds. Oxford Mathematical Monographs (2010)
79. Onate, E., Celigueta, M.A., Latorre, S., Casas, G., Rossi, R., Rojek, J.: Lagrangian analysis of multiscale particulate flows with the particle finite element method. Comput. Part. Mech. **1**(1), 85–102 (2014)
80. Anand, L., Gu, C.: Granular materials: constitutive equations and shear localization. J. Mech. Phys. Solids **48**, 1701–1733 (2000)

Chapter 6
PART II—Discrete Element Method (DEM) Approaches: Dynamic Powder Deposition

Dry powders require different modeling and simulation tools to characterize their behavior. One family of methods that is ideally suited to this task are discrete element methods. This chapter introduces the reader to this type of modeling. The key for next-generation manufacturers to succeed in developing additive manufacturing into a robust reliable approach is to draw upon rigorous theory and high-fidelity computation to guide and simultaneously develop design rules for scaling up to industrial-level manufacturing. Because of the extremely tight profit margins and short turnaround times in manufacturing of new materials, there is an industrial need for numerical simulation of these types of processes. However, continuum-based simulation methods, such as the finite element methods, are ill-suited to simulate systems comprised of discrete units (particles). A relatively new modeling and simulation paradigm for such advanced manufacturing systems are discrete element/particle-based mechanics and methods. Particle-based mechanics and numerical methods have become widespread in the natural sciences, industrial applications, engineering, biology, applied mathematics, and many other areas. The term "particle mechanics/methods" has now come to imply several different areas of research in the twenty-first century, for example: (1) particles as a physical unit in granular media, particulate flows, plasmas, swarms, etc., (2) particles representing material phases in continua at the meso-, micro- and nanoscale, and (3) particles as a discretization unit in continua and discontinua in numerical methods. The application areas of particle-based methods are quite wide ranging, for example: (1) particulate and granular flow problems, motivated by high-tech industrial processes such as those stemming from spray, deposition, and printing processes, (2) fluid–structure interaction problems accounting for free surface flow effects on civil and marine structures (water jets, wave loads, ship hydrodynamics and sea keeping situations, debris flows, etc.), (3) coupled multiphysical phenomena involving solid, fluid, thermal, electromagnetic, and optical systems, (4) material design/functionalization using particles to modify base materials, (5) manufacturing processes involving forming, cutting, compaction, and material processing, (6) biomedical engineering, involving cell mechanics, molecular dynamics, and scale-bridging, (7) multifracture and fragmentation of materials

T. I. Zohdi, *Modeling and Simulation of Functionalized Materials for Additive Manufacturing and 3D Printing: Continuous and Discrete Media*, Lecture Notes in Applied and Computational Mechanics 60, https://doi.org/10.1007/978-3-319-70079-3_6

and structures under impact and blast loads, and (8) excavation and drilling problems in the oil/gas industry and tunneling processes. Particle or discrete element-based computation has emerged in multiple fields and is ideal for simulation of additive manufacturing processes, since the physical systems are inherently discontinuous. They are advantageous in dealing with domains that break apart or come together, as compared to traditional continuum-based finite difference and finite element methods which have severe limitations when dealing with discontinua. For reviews see, for example, Duran [44], Pöschel and Schwager [45], Onate et al. [46, 47], Rojek et al. [48], Carbonell et al. [49], Labra and Onate [50], Mukherjee and Zohdi [51–54], and Zohdi [1–43].[1]

Following Zohdi [1–43], the objective of this chapter is to introduce a multiparticle-based Discrete Element Method (DEM) computational framework which captures the following main physical events:

- *Particle dynamics*, which primarily entails: (a) the movement of the particles induced by contact with the surface, (b) particle-to-particle contact forces, and (c) near-field interaction and external electromagnetic fields,
- *Laser input*, which primarily entails: (a) absorption of laser energy input and (b) beam interference (attenuation) from particles, and
- *Particle thermodynamics*, which primarily entails: (a) heat transfer between particles in contact by conduction and (b) subsequent thermal softening of the particles.

We remark that the inclusion of electromagnetic effects stems from the fact that in many emerging processes, the deposited particles are endowed with charges and guided to the surface with an electromagnetic field, in order to obtain superior deposition control, relative to a charge-free system. The charges are achieved through a variety of possible methods, such as: (1) post-atomization charging—whereby the particles come into contact with an electrostatic field (produced by electrostatic induction or by electrodes) downstream of the outlet nozzle, (2) direct charging—whereby an electrode is immersed in the coating supply, and (3) tribological charging—whereby the friction in the nozzle induces an electrostatic charge on the particles as they rub the surface. There are a variety of industrial deposition techniques, and we refer the reader to the surveys of the state of the art found in Martin [65, 66], as well as the extensive works of Choi et al. [67–69] and Demko et al. [70].

The overall multiphysical system is strongly coupled, since the dynamics controls which particles are in mechanical contact and, consequently, the induced thermal fields, spurred on by selective laser input, which softens and binds the material. The approach taken in the present work is to construct a submodel for each primary physical process. These submodels are coupled to one another. In order to resolve the coupling, a recursive multiphysical staggering scheme is constructed. As before,

[1]There has been considerable research activity in processing of powders, in particular by compaction, for example, see Akisanya et al. [55], Anand and Gu [56], Brown and Abou-Chedid [57], Domas [58], Fleck [59], Gethin et al. [60], Gu et al. [61], Lewis et al. [62], Ransing et al. [63], Tatzel [64], and Zohdi [1–43]. The study of "granular" or "particulate" media is wide ranging. Classical examples include the study of natural materials, such as sand and gravel, associated with coastal erosion, landslides, and avalanches.

the general methodology is as follows (at a given time increment): (1) Each field equation is solved individually, "freezing" the other (coupled) fields in the system, allowing only the primary field to be active and (2) after the solution of each field equation, the primary field variable is updated, and the next field equation is treated in a similar manner. As the physics evolves, the field that is most sensitive (exhibiting the largest amount of relative nondimensional change) dictates the time-step size. This approach can be classified as an implicit, staggered, time-stepping scheme, in conjunction with an iterative solution method that automatically adapts the time-step sizes to control the rates of convergence within a time-step. If the process does not converge (below an error tolerance) within a preset number of iterations, the time-step is adapted (reduced) by utilizing an estimate of the spectral radius of the coupled system. The modular approach allows for easy replacement of submodels if needed. For more details, the reader is referred to Zohdi [1–43]. Toward the end of the work, extensions are also addressed for two advanced processing scenarios involving phase transformations and subsequent multiphase dynamics whereby: (1) material on the system surface is melted and penetrates the subsurface and (2) material on the surface is melted, vaporized, and vacuumed away (Fig. 6.1).

Remark The study of "granular" or "particulate" media is wide ranging. Classical examples include the study of natural materials, such as sand and gravel, associated with coastal erosion, landslides, and avalanches. For reviews see, for example, Duran [44], Pöschel and Schwager [45], the works of Torquato and collaborators: Torquato [71], Kansaal et al. [72], and Donev et al. [73–77], the works of Onate and collaborators: Onate et al. [46, 47], Rojek et al. [48], Carbonell et al. [49] and Labra and Onate [50], Jaeger and Nagel [78, 79], Nagel [80], Liu et al. [81], Liu and Nagel [82], Jaeger and Nagel [83], Jaeger et al. [84–86], Jaeger and Nagel [87], the extensive works of Hutter and collaborators: Tai et al. [88–90], Gray et al. [91], Wieland et al. [92], Berezin et al. [93], Gray and Hutter [94], Gray [95], Hutter [96], Hutter et al. [97], Hutter and Rajagopal [98], Koch et al. [99], Greve and Hutter [100], and Hutter et al. [101]; the works of Behringer and collaborators: Behringer [102], Behringer and Baxter [103], Behringer and Miller [104], and Behringer et al. [105] and the works

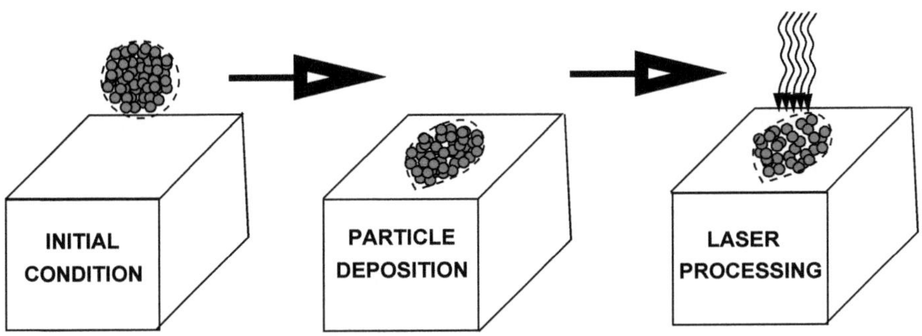

Fig. 6.1 An example of powder-based processes: (*1*) particle deposition and (*2*) selective laser processing Zohdi [1–43]

of Jenkins and collaborators: Jenkins and Strack [106], Jenkins and La Ragione [90], Jenkins and Koenders [107], and Jenkins et al. [108].

6.1 Direct Particle Representation/Calculations

We consider a group of nonintersecting particles ($i = 1, 2, ..., N_p$). The objects in the system are assumed to be small enough to be considered (idealized) as particles, spherical in shape, and that the effects of their rotation with respect to their mass center are unimportant to their overall motion, although we will make further remarks on these effects shortly. The equation of motion for the ith particle in system is

$$m_i \ddot{\boldsymbol{r}}_i = \boldsymbol{\Psi}_i^{tot}(\boldsymbol{r}_1, \boldsymbol{r}_2, ..., \boldsymbol{r}_{N_p}) = \boldsymbol{\Psi}_i^{con} + \boldsymbol{\Psi}_i^{wall} + \boldsymbol{\Psi}_i^{bond} + \boldsymbol{\Psi}_i^{damp} + \boldsymbol{\Psi}_i^{e+m}, \quad (6.1)$$

where $\mathbf{r_i}$ is the position vector of the ith particle and $\boldsymbol{\Psi}_i^{tot}$ represents all forces acting on particle i, which is decomposed into the sum of forces due to:

- Inter-particle forces ($\boldsymbol{\Psi}_i^{con}$) generated by contact with other particles,
- Wall forces ($\boldsymbol{\Psi}_i^{wall}$) generated by contact with constraining surfaces,
- Adhesive bonding forces ($\boldsymbol{\Psi}_i^{bond}$) with other particles and walls,
- Damping forces arising from the surrounding interstitial environment ($\boldsymbol{\Psi}_i^{damp}$) occurring from potentially viscous, surrounding, interstitial fluids, surfactants, and
- External electromagnetic forces ($\boldsymbol{\Psi}_i^{e+m}$) which can play a key role in small charged or magnetized particles.

In the next sections, we examine each of the types of forces in the system in detail.

6.1.1 *Comments on Rolling*

The introduction of rolling and spin is questionable for a small object, idealized by a particle, in particular because of rolling resistance. In addition to the balance of linear momentum, $m_i \dot{\boldsymbol{v}}_i = \boldsymbol{\Psi}_i^{tot}$, where the $\boldsymbol{v}_i$ is the velocity of the center of mass, the equations of angular momentum read $\dot{\boldsymbol{H}}_{i,cm} = \frac{d(\overline{\boldsymbol{I}}_i \cdot \boldsymbol{\omega}_i)}{dt} = \boldsymbol{M}_{i,cm}^{tot}$. For spheres, we have $\boldsymbol{H}_{i,cm} = \overline{I}_{i,s} \omega_i = \frac{2}{5} m_i R_i^2 \omega_i$ and for the time discretization

$$\omega_i(t + \Delta t) = \omega_i(t) + \frac{\Delta t}{\overline{I}_{i,s}} \left(\phi \boldsymbol{M}_{i,cm}^{tot}(t + \Delta t) + (1 - \phi) \boldsymbol{M}_{i,cm}^{tot}(t) \right), \quad (6.2)$$

where $\boldsymbol{M}_{i,cm}^{tot}$ are the total moments generated by interaction forces, such as contact forces, rolling resistance. For the applications at hand, the effects of rolling are generally negligible, in particular because the particles are small. However, nonetheless, we formulate the system with rotations where $\boldsymbol{r}_i$ is the position of the center of mass, $\boldsymbol{v}_i$ is the velocity of the center of mass, and $\boldsymbol{\omega}_i$ is the angular velocity. An important quantity of interest is the velocity on the surface of the "particles," which is a potential contact point with other particles, denoted $\boldsymbol{v}_i^c$

$$\boldsymbol{v}_i^c = \boldsymbol{v}_i + \boldsymbol{\omega}_i \times \boldsymbol{r}_{i\to c}, \tag{6.3}$$

where $\boldsymbol{r}_{i\to c}$ is the relative position vector from the center to the possible point of contact. This is discussed further later.

6.1.2 Particle-to-particle Contact Forces

Following Zohdi [1–43], we employ a simple particle overlap model to determine the normal contact force contributions from the surrounding particles (N_{ci}) in contact, $\boldsymbol{\Psi}_i^{con,n} = \sum_{j=1}^{N_{ci}} \psi_{ij}^{con,n}$, based on separation distance between particles in contact (Fig. 6.2). Generally,

$$\boldsymbol{\Psi}_{ij}^{con,n} = \mathcal{F}(||\boldsymbol{r}_i - \boldsymbol{r}_j||, R_i, R_j, material\ parameters). \tag{6.4}$$

There is no shortage of contact models, of varying complexity, to generate a contact interaction force. Throughout this work, we will utilize a particularly simple relation whereby contact force is proportional to the relative *normalized* proximity of particles i and j in contact, detected by the distance between centers being less that the sum of the radii

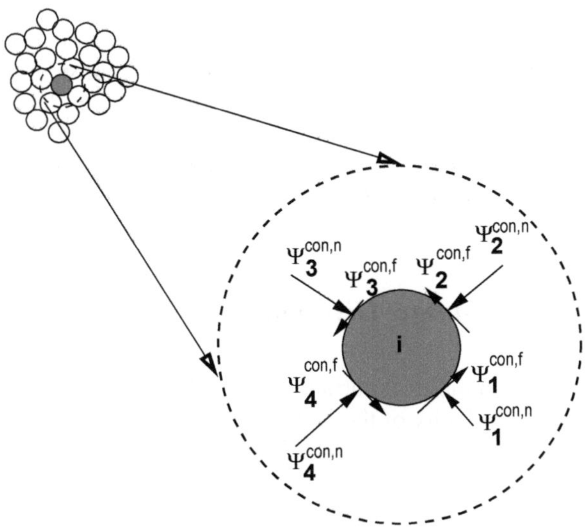

Fig. 6.2 Normal contact and friction forces induced by neighboring particles in contact (after Zohdi [1–43])

$$\text{If } ||\boldsymbol{r}_i - \boldsymbol{r}_j|| \leq R_i + R_j \Rightarrow \text{activate contact}, \tag{6.5}$$

where we define the overlap as

$$\delta_{ij} \stackrel{\text{def}}{=} (R_i + R_j) - ||\boldsymbol{r}_i - \boldsymbol{r}_j||. \tag{6.6}$$

Accordingly, we consider the following

$$\boldsymbol{\Psi}_{ij}^{con,c} \propto -K_{pij}|\mathcal{E}_{ij}|^{p_p}\boldsymbol{n}_{ij}A_{ij}^c, \tag{6.7}$$

where $0 < K_{pij} < \infty$ is a particle-to-particle contact compliance constant, p_p is a material parameter, $\mathcal{E}_{ij}$ is normalized/nondimensional (strain-like) deformation metric

$$\mathcal{E}_{ij} = \frac{(R_i + R_j) - ||\boldsymbol{r}_i - \boldsymbol{r}_j||}{(R_i + R_j)} = \frac{\delta_{ij}}{(R_i + R_j)} \tag{6.8}$$

and

$$\boldsymbol{n}_{ij} = -\frac{\boldsymbol{r}_i - \boldsymbol{r}_j}{||\boldsymbol{r}_i - \boldsymbol{r}_j||} = \frac{\boldsymbol{r}_j - \boldsymbol{r}_i}{||\boldsymbol{r}_i - \boldsymbol{r}_j||}, \tag{6.9}$$

where the R_i and R_j are the radii of particles i and j, respectively. The term A_{ij}^c is a contact area parameter, which is discussed in Chapter Appendix 1. Chapter Appendix 1 also provides a brief review of alternative models, such as the classical Hertzian contact model.

6.1.3 Particle-Wall Contact

Contact of a particle-to-wall contact is handled in the identical manner to particle-to-particle, except that the wall displacement is considered given (externally controlled), and independent of the action with the particles. The contact between the wall and the particles is handled exactly in the same manner as the particle-to-particle contact, with the amount of overlap of the particle with the wall position, dictates the force (see Fig. 11.2) (Fig. 6.3).

6.1.4 Contact Dissipation

Phenomenological particle contact dissipation can be incorporated by tracking the relative velocity of the particles in contact. A simple model to account for this is

$$\boldsymbol{\Psi}_{ij}^{con,d} = c^{cd}(\boldsymbol{v}_{nj} - \boldsymbol{v}_{ni})A_{ij}^c \tag{6.10}$$

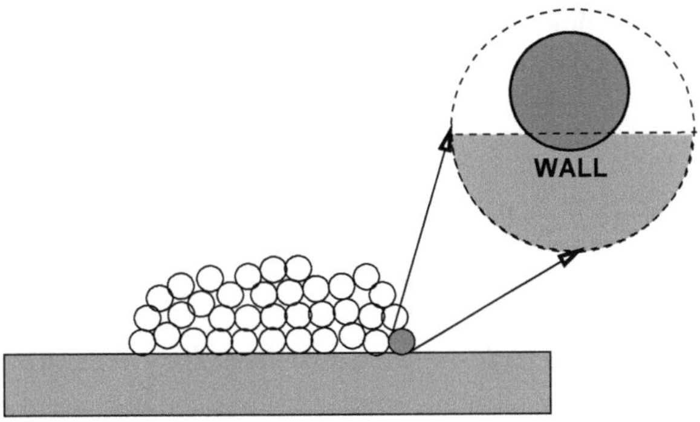

Fig. 6.3 An example of overlap contact between a wall and a particle. The amount of overlap of the particle with the wall position dictating the force

6.1.5 Regularized Contact Friction Models

Frictional stick is modeled via the following regularized friction algorithm: (at the point of contact)

- Check static friction threshold (K^f is a tangential contact friction compliance constant):

$$K^f ||\boldsymbol{v}^c_{j,T} - \boldsymbol{v}^c_{i,T}|| A^c_{ij} \Delta t \text{ against } \mu_s ||\boldsymbol{\Psi}^{con,n}||, \tag{6.11}$$

  where $||\boldsymbol{v}^c_{j,T} - \boldsymbol{v}^c_{i,T}||\Delta t$ (of dimensions of length) is the relative tangential velocity at a point of contact, Δt is the time-step used later in the numerical discretization, μ_s is the static friction coefficient. *This step replaces ("regularizes") a more rigorous, and difficult, step of first assuming no slip, generating the no-slip contact forces, by solving an entire multibody/multisurface contact problem, $\boldsymbol{\Psi}^{ns}$, and checking $\boldsymbol{\Psi}^{ns}$ against the threshold $\mu_s||\boldsymbol{\Psi}^{con,n}||$ on each surface.*
- If the threshold not met ($K^f ||\boldsymbol{v}^c_{j,T} - \boldsymbol{v}^c_{i,T}|| A^c_{ij} \Delta t < \mu_s ||\boldsymbol{\Psi}^{con,n}||$), then

$$\boldsymbol{\Psi}^{con,f} = K^f ||\boldsymbol{v}^c_{j,T} - \boldsymbol{v}^c_{i,T}|| A^c_{ij} \Delta t \boldsymbol{\tau}^c_{ij} \tag{6.12}$$

  where

$$\boldsymbol{\tau}^c_{ij} = -\frac{\boldsymbol{v}^c_{i,T} - \boldsymbol{v}^c_{j,T}}{||\boldsymbol{v}^c_{j,T} - \boldsymbol{v}^c_{i,T}||} = \frac{\boldsymbol{v}^c_{j,T} - \boldsymbol{v}^c_{i,T}}{||\boldsymbol{v}^c_{j,T} - \boldsymbol{v}^c_{i,T}||}, \tag{6.13}$$

  where the subscripts indicate the tangential components of velocity. The tangential velocity at the contact point is obtained by subtracting away the normal component of the velocity

$$\boldsymbol{v}^c_t = \boldsymbol{v}^c - (\boldsymbol{v}^c \cdot \boldsymbol{n})\boldsymbol{n}. \tag{6.14}$$

- If the threshold is met or exceeded ($K^f||\boldsymbol{v}^c_{j,\tau} - \boldsymbol{v}^c_{i,\tau}||A^c_{ij}\Delta t \geq \mu_s||\boldsymbol{\Psi}^{con,n}||$), then one adopts a slip model of the form

$$\boldsymbol{\Psi}^{con,f}_{ij} = \mu_d||\boldsymbol{\Psi}^{con,n}_{ij}||\boldsymbol{\tau}^c_{ij}, \tag{6.15}$$

where μ_d is the dynamic friction coefficient.

6.1.6 Particle-to-particle Bonding Relation

For the particles to bond, we adopt a criterion based on exceeding a critical interpenetration distance. Explicitly:

- Recall, if $||\boldsymbol{r}_i - \boldsymbol{r}_j|| \leq (R_i + R_j)$, then the particles are in contact and $\mathcal{E}_{ij} = \frac{\delta_{ij}}{(R_i+R_j)}$.
- If the particles are in contact and $\mathcal{E}_{ij} \geq \mathcal{E}^*$, then an (adhesive/attractive) normal bond is activated between the particles of the form

$$\boldsymbol{\Psi}^{bond,n}_{ij} = K^{nb}_{ij}|\mathcal{E}_{ij}|^{p_b}\boldsymbol{n}_{ij}A^c_{ij}, \tag{6.16}$$

  where $0 \leq K^{nb}_{ij}$ is a bonding constant and p_b is a material parameter.
- If the particles have an activated normal bond, then the particles automatically have a rotational bond equivalent in form to stick friction

$$\boldsymbol{\Psi}^{bond,r}_{ij} = K^{rb}_{ij}||\boldsymbol{v}^c_{j,\tau} - \boldsymbol{v}^c_{i,\tau}||A^c_{ij}\Delta t\boldsymbol{\tau}^c_{ij}, \tag{6.17}$$

  as well as a torsional bond of the form

$$\boldsymbol{M}^{bond,t}_{ij} = K^{rt}_{ij}\left((\boldsymbol{\omega}_i - \boldsymbol{\omega}_j)a^2 \cdot \boldsymbol{n}\right)A^c_{ij}\Delta t\boldsymbol{n}, \tag{6.18}$$

  where a is the contact area radius (see Chapter Appendix 1). The same model is also used for torsional frictional moments.

Note: We assume that torsional friction is due to relative spinning along axis connecting the particle centers. The effect is generally small, unless the particles are bonded to one another.

6.1.7 Electromagnetic Forces

The electromagnetic forces are decomposed into three contributions, (1) Lorentz forces (for charged particles), (2) magnetic forces (for magnetic particles), and (3) inter-particle near-field forces. We will utilize the decomposition of the electromagnetic forces generated into a (inter-particle) near-field interaction and the external electromagnetic field

$$\boldsymbol{\Psi}_i^{e+m} = \boldsymbol{\Psi}_i^{lor,e+m} + \boldsymbol{\Psi}_i^{mag} + \underbrace{\sum_{j \neq i}^{N} \psi_{ij}^{nf}}_{\boldsymbol{\Psi}_i^{nf}} = \underbrace{q_i(\boldsymbol{E}^{ext} + \boldsymbol{v}_i \times \boldsymbol{B}^{ext})}_{\boldsymbol{\Psi}_i^{lor,e+m}} + \boldsymbol{\Psi}_i^{mag} + \boldsymbol{\Psi}_i^{nf}, \quad (6.19)$$

where $\sum_{j \neq i}^{N} \psi_{ij}^{nf}$ represents the interaction between particle i and all other particles $j = 1, 2, ..., N$ $(j \neq i)$ and $\boldsymbol{\Psi}_i^{lor,e+m}$ represents external Lorentz-induced forces from the surrounding environment, for example, comprised of $\boldsymbol{E}^{ext}$ and $\boldsymbol{B}^{ext}$, which are externally controlled fields that are independent of the response of the system. The terms $\boldsymbol{E}^{ext}$ and $\boldsymbol{B}^{ext}$ can be considered as static (or extremely slowly varying), and thus mutually uncoupled and independently controllable. The self-induced magnetic fields developed between particles are insignificant for the velocity ranges of interest here (well below the speed of light). For the Lorentz force, we recall the following important observations in conjunction with electromagnetic phenomena (Jackson [109]):

- If a point charge q experiences a force $\boldsymbol{\Psi}^{lor,e}$, the electric field, $\boldsymbol{E}^{ext}$, at the location of the charge is defined by $\boldsymbol{\Psi}^{lor,e} = q\boldsymbol{E}^{ext}$.
- If the charge is moving, another force may arise, $\boldsymbol{\Psi}^{lor,m}$, which is proportional to its velocity $\boldsymbol{v}$. This other (induced) field is denoted as the "magnetic induction" or just the "magnetic field", $\boldsymbol{B}^{ext}$, such that $\boldsymbol{\Psi}^{lor,m} = q\boldsymbol{v} \times \boldsymbol{B}^{ext}$.
- If the forces occur concurrently (the charge is moving through the region possessing both electric and magnetic fields), then the electromagnetic force is $\boldsymbol{\Psi}^{lor,e+m} = q\boldsymbol{E}^{ext} + q\boldsymbol{v} \times \boldsymbol{B}^{ext}$.

6.1.8 Inter-particle Near-Field Interaction

Following Zohdi [1–43], a simple form that captures the essential near-field effects is

$$\boldsymbol{\Psi}_i^{nf} = \sum_{j \neq i}^{N_p} \left(\underbrace{\alpha_{1ij}||\boldsymbol{r}_i - \boldsymbol{r}_j||^{-\beta_1}}_{\text{attraction}} - \underbrace{\alpha_{2ij}||\boldsymbol{r}_i - \boldsymbol{r}_j||^{-\beta_2}}_{\text{repulsion}} \right) \boldsymbol{n}_{ij}, \quad (6.20)$$

where the α's (attraction and repulsion parameters) and β's (decay exponents) are empirical material parameters. The various representations (decompositions) of the coefficients that appear in Eq. 6.20 are with $c_i = \pm 1$ (a positive/negative identifier)

- Mass-based (m=mass): $\alpha_{ij} = \bar{\alpha}_{ij} m_i m_j c_i c_j$,
- Surface area-based (a=surface area): $\alpha_{ij} = \bar{\alpha}_{ij} a_i a_j c_i c_j$,

- Volume-based (V=volume): $\alpha_{ij} = \bar{\alpha}_{ij} V_i V_j c_i c_j$, and
- Charge-based: $\alpha_{ij} = \bar{\alpha}_{ij} q_i q_j c_i c_j$,

where the $\bar{\alpha}_{ij}$ are empirical material parameters. There are vast numbers of empirical representations, for example, found in the field of "molecular dynamics" (MD), which typically refers to mathematical models of systems of atoms or molecules where each atom (or molecule) is represented by a material point and is treated as a point mass. The overall motion of such mass-point systems is dictated by Newtonian mechanics. For an extensive survey of MD-type interaction forces, which includes comparisons of the theoretical and computational properties of a variety of interaction laws, we refer the reader to Frenklach and Carmer [110]. In the usual MD approach (see Haile [111], for example), the motion of individual atoms is described by Newton's second law with the forces computed from differentiating a prescribed potential energy function, with applications to solids, liquids, and gases, as well as biological systems (Hase [112], Schlick [113], and Rapaport [114]). The interaction functions usually take the form of the familiar Mie, Lennard-Jones, and Morse potentials (Moelwyn-Hughes [115]); however, three-body terms can be introduced directly into the interaction functions (Stillinger [116]), or alternatively, "local" modifications can be made to two-body representations (Tersoff [117]).

6.1.9 Magnetic Forces

An additional force can be exerted on magnetic particles, independent of the electrodynamically induced Lorentz forces. A relatively simple model for the characterization of this force is given by

$$\boldsymbol{\Psi}^{mag} = \nabla(\gamma \boldsymbol{B}^{ext} \cdot \boldsymbol{B}^{ext}), \tag{6.21}$$

where γ is a material parameter that is related to the magnetization of the particle, and which is dependent on the magnetic dipole properties, the magnetic susceptibility, the magnetic permeability, and the internal magnetic moment density of the material (see Feynman et al. [118], Cullity and Graham [119], Boyer [120] or Jackson [109]). For the specific applications in this monograph, $\boldsymbol{\Psi}^{mag}$ is considered small.

6.1.10 Interstitial Damping

Finally, we note that damping from interstitial fluid (or even smaller-scale particles, solvents) between particles, such as binding enhancers, surfactants, and lubricants, is possible. A simple model to account for this is (a very low Reynolds number "Stokesian" model)

$$\boldsymbol{\Psi}_i^{damp} = c^e 6\pi R_i(\boldsymbol{v}^e - \boldsymbol{v}_i), \tag{6.22}$$

where $\boldsymbol{v}^e$ is the local average velocity of the external interstitial medium, which one may assume $\boldsymbol{v}^e \approx \mathbf{0}$, for most applications of interest here, and c^e is the viscosity. The mechanics of the interstitial fluid is unimportant in problems of interest here. However, for other applications, such as high-speed flow, the motion of the fluid can be important, necessitating more sophisticated drag laws and/or fully coupled (two-way) particle–fluid interaction models. This is outside the scope of the present work. Generally, this requires the use of solid–fluid staggering-type schemes (e.g., see Zohdi [1–43] and Avci and Wriggers [121]).

6.2 Time-Stepping

Integrating Eq. 6.1 leads to (using a trapezoidal rule with variable integration metric, $0 \leq \phi \leq 1$)

$$\begin{aligned} \boldsymbol{v}_i(t+\Delta t) &= \boldsymbol{v}_i(t) + \frac{1}{m_i}\int_t^{t+\Delta t} \boldsymbol{\Psi}_i^{tot}\, dt \\ &\approx \boldsymbol{v}_i(t) + \frac{\Delta t}{m_i}\left(\phi \boldsymbol{\Psi}_i^{tot}(t+\Delta t) + (1-\phi)\boldsymbol{\Psi}_i^{tot}(t)\right), \end{aligned} \tag{6.23}$$

where $\boldsymbol{\Psi}_i^{tot} = \boldsymbol{\Psi}_i^{con} + \boldsymbol{\Psi}_i^{wall} + \boldsymbol{\Psi}_i^{bond} + \boldsymbol{\Psi}_i^{damp} + \boldsymbol{\Psi}_i^{e+m}$. The position can be computed via application of the trapezoidal rule again:

$$\boldsymbol{r}_i(t+\Delta t) \approx \boldsymbol{r}_i(t) + \Delta t(\phi \boldsymbol{v}_i(t+\Delta t) + (1-\phi)\boldsymbol{v}_i(t)), \tag{6.24}$$

which can be consolidated into

$$\boldsymbol{r}_i(t+\Delta t) = \boldsymbol{r}_i(t) + \boldsymbol{v}_i(t)\Delta t + \frac{\phi(\Delta t)^2}{m_i}\left(\phi \boldsymbol{\Psi}_i^{tot}(t+\Delta t) + (1-\phi)\boldsymbol{\Psi}_i^{tot}(t)\right). \tag{6.25}$$

This leads to a coupled system of equations, which are solved using an adaptive iterative scheme, building on approaches found in various forms in Zohdi [1–43]. We note that the material contact compliance constants in the various force terms are functions of temperature, $K = K(\theta)$, and the temperature is, in turn, a function of the laser input. We will discuss the thermal effects shortly, but first indicate how the preceding dynamical particle system can be solved.

6.2.1 Iterative (Implicit) Solution Method

As before in CM approaches, following the basic framework in Zohdi [1–43], we write Eq. 6.25 in a slightly more streamlined form for particle i

$$\boldsymbol{r}_i^{L+1} = \boldsymbol{r}_i^L + \boldsymbol{v}_i^L \Delta t + \frac{\phi(\Delta t)^2}{m_i} \left(\phi(\boldsymbol{\Psi}_i^{tot,L+1}) + (1-\phi)(\boldsymbol{\Psi}^{tot,L}) \right), \quad (6.26)$$

which leads to a coupled set equations for $i = 1, 2, ..., N_p$ particles, where the superscript L is a time interval counter. The set of equations represented by Eq. 6.26 can be solved recursively by recasting the relation as

$$\boldsymbol{r}_i^{L+1,K} = \boldsymbol{r}_i^L + \boldsymbol{v}_i^L \Delta t + \frac{(\phi \Delta t)^2}{m_i} \boldsymbol{\Psi}_i^{tot,L+1,K-1} + \frac{\phi(\Delta t)^2}{m_i}(1-\phi)\boldsymbol{\Psi}_i^{tot,L}, \quad (6.27)$$

which is of the form

$$\boldsymbol{r}_i^{L+1,K} = \mathcal{G}(\boldsymbol{r}_i^{L+1,K-1}) + \mathcal{R}_i, \quad (6.28)$$

where $K = 1, 2, 3, ...$ is the index of iteration within time-step $L + 1$ and

- $\boldsymbol{\Psi}_i^{tot,L+1,K-1} \stackrel{\text{def}}{=} \boldsymbol{\Psi}_i^{tot,L+1,K-1}(\boldsymbol{r}_1^{L+1,K-1}, \boldsymbol{r}_2^{L+1,K-1} ... \boldsymbol{r}_N^{L+1,K-1})$,
- $\boldsymbol{\Psi}_i^{tot,L} \stackrel{\text{def}}{=} \boldsymbol{\Psi}_i^{tot,L}(\boldsymbol{r}_1^L, \boldsymbol{r}_2^L ... \boldsymbol{r}_N^L)$,
- $\mathcal{G}(\boldsymbol{r}_i^{L+1,K-1}) = \frac{(\phi\Delta t)^2}{m_i} \boldsymbol{\Psi}_i^{tot,L+1,K-1}$, and
- $\mathcal{R}_i = \boldsymbol{r}_i^L + \boldsymbol{v}_i^L \Delta t + \frac{\phi(\Delta t)^2}{m_i}(1-\phi)\boldsymbol{\Psi}_i^{tot,L}$.

The term $\mathcal{R}_i$ is a remainder term that does not depend on the solution. The convergence of such a scheme is dependent on the behavior of $\mathcal{G}$. Namely, a sufficient condition for convergence is that $\mathcal{G}$ is a contraction mapping for all $\boldsymbol{r}_i^{L+1,K}$, $K = 1, 2, 3, ...$ In order to investigate this further, we define the iteration error as

$$\varpi_i^{L+1,K} \stackrel{\text{def}}{=} \boldsymbol{r}_i^{L+1,K} - \boldsymbol{r}_i^{L+1}. \quad (6.29)$$

A necessary restriction for convergence is iterative self-consistency, i.e., the "exact" (discretized) solution must be represented by the scheme, $\boldsymbol{r}_i^{L+1} = \mathcal{G}(\boldsymbol{r}_i^{L+1}) + \mathcal{R}_i$. Enforcing this restriction, a sufficient condition for convergence is the existence of a contraction mapping

$$\underbrace{||\boldsymbol{r}_i^{L+1,K} - \boldsymbol{r}_i^{L+1}||}_{\varpi_i^{L+1,K}} = ||\mathcal{G}(\boldsymbol{r}_i^{L+1,K-1}) - \mathcal{G}(\boldsymbol{r}_i^{L+1})|| \leq \eta^{L+1,K} ||\boldsymbol{r}_i^{L+1,K-1} - \boldsymbol{r}_i^{L+1}||, \quad (6.30)$$

where, if $0 \leq \eta^{L+1,K} < 1$ for each iteration K, then $\varpi_i^{L+1,K} \rightarrow \mathbf{0}$ for any arbitrary starting value $\boldsymbol{r}_i^{L+1,K=0}$, as $K \rightarrow \infty$, which is a contraction condition that is sufficient, but not necessary, for convergence. The convergence of Eq. 6.27 is scaled by $\eta \propto \frac{(\phi\Delta t)^2}{m_i}$. Therefore, we see that the contraction constant of $\mathcal{G}$ is:

- Directly dependent on the magnitude of the interaction forces ($||\boldsymbol{\Psi}||$),
- Inversely proportional to the masses m_i, and
- Directly proportional to $(\Delta t)^2$.

Thus, decreasing the time-step size improves the convergence. *In order to maximize the time-step sizes (to decrease overall computing time) and still meet an error tolerance on the numerical solution's accuracy*, we build on an approach originally developed for continuum thermochemical multifield problems (Zohdi [1–43]), where one assumes: (1) $\eta^{L+1,K} \approx S(\Delta t)^p$, ($S$ is a constant) and (2) the error within an iteration behaves according to $(S(\Delta t)^p)^K \varpi^{L+1,0} = \varpi^{L+1,K}$, $K = 1, 2, ...$, where $\varpi^{L+1,0} = \boldsymbol{r}^{L+1,K=1} - \boldsymbol{r}^L$ is the initial norm of the iterative (relative) error and S is intrinsic to the system. For example, for second-order problems, due to the quadratic dependency on Δt, $p \approx 2$. The objective is to meet an error tolerance in exactly a preset (the analyst sets this) number of iterations. To this end, one writes $(S(\Delta t_{\text{tol}})^p)^{K_d} \varpi^{L+1,0} = TOL$, where TOL is a tolerance and K_d is the number of desired iterations. If the error tolerance is not met in the desired number of iterations, the contraction constant $\eta^{L+1,K}$ is too large. Accordingly, one can solve for a new smaller step size, under the assumption that S is constant,

$$\Delta t_{\text{tol}} = \Delta t \left(\frac{\left(\frac{TOL}{\varpi^{L+1,0}}\right)^{\frac{1}{pK_d}}}{\left(\frac{\varpi^{L+1,K}}{\varpi^{L+1,0}}\right)^{\frac{1}{pK}}} \right) \stackrel{\text{def}}{=} \Delta t \Lambda_K. \tag{6.31}$$

The assumption that S is constant is not critical, since the time-steps are to be recursively refined and unrefined throughout the simulation. Clearly, the expression in Eq. 6.31 can also be used for time-step enlargement if convergence is met in less than K_d iterations (typically chosen to be between five to ten iterations).

6.2.2 Algorithm

The solution steps are, within a time-step:

- (1): Start a global fixed iteration (set $i = 1$ (particle counter) and $K = 0$ (iteration counter))
- (2): If $i > N_p$ then go to (4)
- (3): If $i \leq N_p$ then:

(a) Compute the position $\boldsymbol{r}_i^{L+1,K}$
(b) Go to (2) for the next particle ($i = i + 1$)

- (4): Measure error (normalized) quantities

(a) $\varpi_K \stackrel{\text{def}}{=} \dfrac{\sum_{i=1}^{N_p} ||\boldsymbol{r}_i^{L+1,K} - \boldsymbol{r}_i^{L+1,K-1}||}{\sum_{i=1}^{N_p} ||\boldsymbol{r}_i^{L+1,K} - \boldsymbol{r}_i^{L}||}$

(b) $Z_K \stackrel{\text{def}}{=} \dfrac{\varpi_K}{TOL_r}$

(c) $\Lambda_K \stackrel{\text{def}}{=} \left(\dfrac{\left(\frac{TOL}{\varpi_0}\right)^{\frac{1}{pK_d}}}{\left(\frac{\varpi_K}{\varpi_0}\right)^{\frac{1}{pK}}} \right)$

- (5): If the tolerance is met: $Z_K \leq 1$ and $K < K_d$ then
  (a) Increment time: $t = t + \Delta t$
  (b) Construct the next time-step: $\Delta t^{new} = \Lambda_K \Delta t^{old}$,
  (c) Select the minimum size: $\Delta t = \mathbf{MIN}(\Delta t^{lim}, \Delta t^{new})$ and go to (1)
- (6): If the tolerance is not met: $Z_K > 1$ and $K < K_d$ then
  (a) Update the iteration counter: $K = K + 1$
  (b) Reset the particle counter: $i = 1$
  (c) Go to (2)
- (7): If the tolerance is not met ($Z_K > 1$) and $K = K_d$ then
  (a) Construct a new time-step: $\Delta t^{new} = \Lambda_K \Delta t^{old}$
  (b) Restart at time t and go to (1)

Time-step size adaptivity is critical, since the system's dynamics and configuration can dramatically change over the course of time, possibly requiring quite different time-step sizes to control the iterative error. However, to maintain the accuracy of the time-stepping scheme, one must respect an upper bound dictated by the discretization error, i.e., $\Delta t \leq \Delta t^{lim}$. Note that in Step (5), Λ_K may enlarge the time-step if the error is lower than the preset tolerance. The algorithm will be modified shortly to account for coupled thermal effects. As mentioned, the parameters such as K_{pij} can be thermally dependent since the particles can thermally soften. For example, the compliance constant for the particles in the contact law can be written as:

$$K_{pi} = \mathbf{MAX}(K_{pio}\left(e^{-a_i \frac{\theta_i}{\theta_i^*} - 1}\right), K_{pi}^{lim}), \tag{6.32}$$

where K_{pio} is the reference value and for particle j

$$K_{pj} = \mathbf{MAX}(K_{pjo}\left(e^{-a_j \frac{\theta_j}{\theta_j^*} - 1}\right), K_{pj}^{lim}), \tag{6.33}$$

and take the average at the interface, and the value in the contact law becomes:

$$K_{pij} = \frac{1}{2}\left(K_{pi} + K_{pj}\right). \tag{6.34}$$

There is a multitude of possible representations, and it is relatively easy to select one or the other, and to embed in the staggering framework developed.

Remark We next include thermal fields using the Discrete Element Method, following Zohdi [1–43] which developed a computational model and corresponding solution algorithm for the rapid simulation of the laser processing and targeted localized heating of materials composed of multiple discrete particle systems (Fig. 6.4).

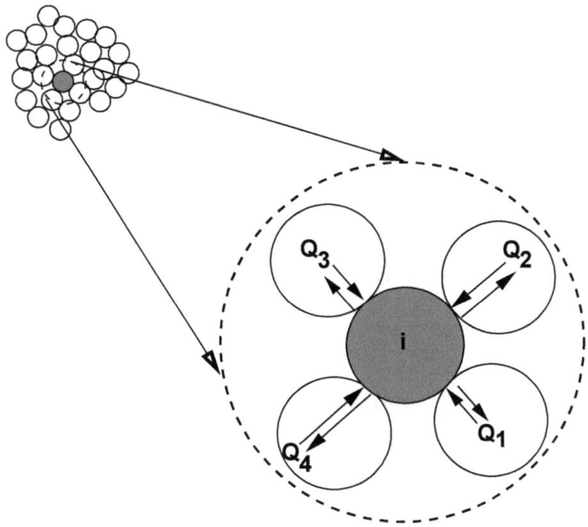

Fig. 6.4 Heat flux exchange for a particle induced by neighboring particles in contact

6.3 Thermal Fields

6.3.1 Heat Transfer Model

We assume that (infrared) radiative, convective, and strain-rate effects are negligible for the particle's thermodynamics. Only heating by laser input and conduction are considered important. Thus, for each particle $i = 1, 2, ..., N_p$,

$$m_i C_i \dot{\theta}_i = \mathcal{Q}_i + \mathcal{H}_i, \tag{6.35}$$

where $\mathcal{Q}_i$ represents the conductive contribution from surrounding particles in contact (including walls) and $\mathcal{H}_i$ represents the external heating term. It is assumed that the temperature fields are uniform within the (small) particles. This assumption is justified, i.e., a lumped thermal model, ignoring temperature gradients and assuming a uniform temperature within a particle, when the Biot number is small. The Biot number for spheres scales with the ratio of the particle volume (V) to the particle surface area (A_s), $\frac{V}{A_s} = \frac{\frac{4}{3}\pi R^3}{4\pi R^2} = \frac{R}{3}$, which indicates that a uniform temperature distribution is appropriate, since the particles, by definition, are small. Assuming that the fields are uniform in each particle allows for the following (for particle i)

$$\mathcal{Q}_i = -\int_{\partial\omega} \boldsymbol{Q} \cdot \boldsymbol{n} \, dA \approx \sum_{j=1}^{N_{pc}} I\!K_{ij} \frac{\theta_j - \theta_i}{||\boldsymbol{r}_i - \boldsymbol{r}_j||} A^c_{ij}, \tag{6.36}$$

where the summation extends over all particles $j = 1, 2, 3, ..., N_{pc}$ that are in contact with particle i (Fig. 6.4).[2] This yields

$$m_i C_i \frac{d\theta_i}{dt} = \underbrace{\sum_{j=1}^{N_{pc}} I\!K_{ij} \frac{\theta_j - \theta_i}{||\boldsymbol{r}_i - \boldsymbol{r}_j||} A^c_{ij}}_{\mathcal{Q}_i} + \mathcal{H}_i, \tag{6.37}$$

where the specific form of laser-induced heating is

$$\mathcal{H}_i \approx a_i I_i V_i, \tag{6.38}$$

where aI_i is the absorbed laser input for particle i and $0 \leq a_i \leq 1$ is an absorption constant. More remarks on laser input will be given next.

6.3.2 Lasers—Various Levels of Description

There are varying degrees of sophistication for modeling the heat input from a laser, for example:

- **Method 1**-A *very rough* overall energy balance whereby the total energy input $\mathcal{W}^{tot}$ (given by integrating the power output from the laser) and performing an energy balance

$$MC\Delta\theta = \mathcal{W}^{tot} \Rightarrow \Delta\theta = \frac{\mathcal{W}^{tot}}{MC} \tag{6.39}$$

  where the total mass is $M = \sum_{i=1}^{N} m_i$ and C is the heat capacity. This provides no information about conductive losses, the distribution of the heat, etc. It simply gives a rough estimate of the overall effects of laser power input.
- **Method 2**-Utilization of the Beer–Lambert law, whereby one performs an overall power balance (Fig. 6.5):

$$(I + \Delta I) - I + \mathcal{L}\Delta\zeta = 0 \Rightarrow \frac{dI}{d\zeta} =\approx -aI, \tag{6.40}$$

  where aI is the absorbed irradiance (per unit area) from the laser and ζ is the penetration depth, which is solved to yield

$$I(\zeta) = I_o e^{-a\zeta}. \tag{6.41}$$

[2] $I\!K_{ij}$ can be approximated by an average interfacial value of the $i - j$ pair, $I\!K_{ij} \approx \frac{I\!K_i + I\!K_j}{2}$. If the materials are the same, this collapses to simply $I\!K$. As for the mechanical contact, A^c_{ij} is the contact area associated with the particle pair (ij).

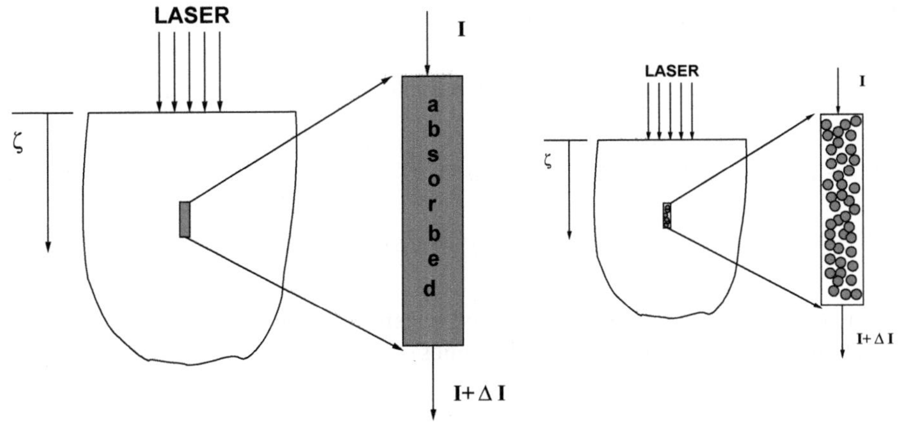

Fig. 6.5 Representations of laser penetration based on modification (right) of the classical Beer–Lambert relation (left)

One can modify this relation to account for a particle at depth ζ, which is blocked by the particle interference from above it by determining the intersections with particles above the particle, computing $\zeta^* = \sum_{i=1}^{N} \Delta\zeta_i$, and modifying the Beer–Lambert relation to read

$$I(z) = I_o e^{-a\zeta^*}. \tag{6.42}$$

- **Method 3**-Discretization of the irradiant beam into rays and performing a full-blown ray-tracking scheme (see Zohdi [1–43]).
- **Method 4**-Discretization of the beam into its electromagnetic field components via Maxwell's equations (see Zohdi [1–43]).

Method 1 is rather simplistic and can be used more as a rough semianalytical guide to laser input selection. Methods 3 and 4 provide extremely detailed field information at the smallest scales. They are extraordinarily computationally expensive, and for the applications in this work, unwarranted. This leaves Method 2, which we will employ for the remainder of the work. In such an approach, one can easily introduce nonuniform beam profiles, for example

$$I(d) = I(d = 0)e^{-bd}, \tag{6.43}$$

where d is the distance from the center of the incident beam line. In the case of $b = 0$, we recapture a flat beam, $I(d) = I(d = 0)$.

6.3.3 Numerical Integration

Integrating the energy equation for each particle ($i = 1, 2, ..., N_p$) yields

$$\theta_i(t+\Delta t) = \theta_i(t) + \frac{1}{m_i C_i}\left(\int_t^{t+\Delta t} \mathcal{Q}_i \, dt + \int_t^{t+\Delta t} \mathcal{H}_i \, dt\right) \tag{6.44}$$

$$\approx \theta_i(t) + \frac{\Delta t}{m_i C_i}\left(\phi(\mathcal{Q}_i(t+\Delta t) + \mathcal{H}_i(t+\Delta t)) + (1-\phi)(\mathcal{Q}_i(t) + \mathcal{H}_i(t))\right).$$

We note that Eq. 6.44 represents a coupled system of the general form (similar to the equation that arises for the particle dynamics)

$$\theta_i(t+\Delta t) = \mathcal{G}_i(\theta_i(t+\Delta t)) + \mathcal{R}_i, \tag{6.45}$$

where for the "remainder" term, $\mathcal{R}_i \neq \mathcal{R}_i(\theta_i(t+\Delta t))$, and where $\mathcal{G}_i$'s behavior is controlled by the magnitude of Δt. Clearly, the temperature is coupled to the mechanical behavior of the system. Shortly, we develop a multiphysical staggering scheme to solve the overall system.

6.4 Total System Coupling: Multiphysical Staggering Scheme

We now extend the iterative solution process introduced earlier for the particle dynamics to a multifield setting by (at a given time increment): (1) solving each field equation individually, "freezing" the other (coupled) fields in the system, allowing only the primary field to be active and (2) updating the primary field variable after the solution of each field equation. The next field equation is treated in a similar manner where, as the physics changes, the field that is most sensitive (exhibits the largest amount of relative nondimensional change) dictates the time-step size. This is an implicit, staggered, adaptive time-stepping scheme. Such approaches have a long history in the computational mechanics community, for example, Zienkiewicz [122], Zienkiewicz et al. [123], Lewis et al. [124], Lewis and Schrefler [125], Park and Felippa [126], Farhat et al. [127], Farhat and Lesoinne [128], Farhat et al. [129], Piperno [130], Piperno et al. [131], Piperno and Farhat [132] and Michopoulos et al. [133], Steuben et al. [134–136] Lesoinne and Farhat [137], and Le Tallec and Muoro [138].

6.4.1 A General Iterative Framework

The staggering scheme introduced earlier utilizes approaches found in Zohdi [1–43] and proceeds by considering an abstract setting, whereby one solves for the particle positions, assuming the thermal fields fixed,

$$\mathcal{A}_1(\underline{\boldsymbol{r}^{L+1,K}}, \theta^{L+1,K-1}) = \mathcal{F}_1(\boldsymbol{r}^{L+1,K-1}, \theta^{L+1,K-1}), \tag{6.46}$$

then one solves for the thermal fields, assuming the particle positions fixed,

$$\mathcal{A}_2(\boldsymbol{r}^{L+1,K}, \underline{\theta^{L+1,K}}) = \mathcal{F}_2(\boldsymbol{r}^{L+1,K}, \theta^{L+1,K-1}), \tag{6.47}$$

where only the underlined variable is "active," L indicates the time-step, and K indicates the iteration counter. Within the staggering scheme, implicit time-stepping methods, with time-step size adaptivity, will be used throughout the upcoming analysis. We define the normalized errors within each time-step, for the two fields,

$$\varpi_{rK} \stackrel{\text{def}}{=} \frac{||\boldsymbol{r}^{L+1,K} - \boldsymbol{r}^{L+1,K-1}||}{||\boldsymbol{r}^{L+1,K} - \boldsymbol{r}^{L}||} \quad \text{and} \quad \varpi_{\theta K} \stackrel{\text{def}}{=} \frac{||\theta^{L+1,K} - \theta^{L+1,K-1}||}{||\theta^{L+1,K} - \theta^{L}||}. \tag{6.48}$$

We define maximum "violation ratio," i.e., as the larger of the ratios of each field variable's error to its corresponding tolerance, by $Z_K \stackrel{\text{def}}{=} \mathbf{MAX}(z_{rK}, z_{\theta K})$, where

$$z_{rK} \stackrel{\text{def}}{=} \frac{\varpi_{rK}}{TOL_r} \quad \text{and} \quad z_{\theta K} \stackrel{\text{def}}{=} \frac{\varpi_{\theta K}}{TOL_\theta}, \tag{6.49}$$

with the minimum scaling factor defined as $\Lambda_K \stackrel{\text{def}}{=} \mathbf{MIN}(\Lambda_{rK}, \Lambda_{\theta K})$, where

$$\Lambda_{rK} \stackrel{\text{def}}{=} \left(\frac{(\frac{TOL_r}{\varpi_{r0}})^{\frac{1}{pK_d}}}{(\frac{\varpi_{rK}}{\varpi_{r0}})^{\frac{1}{pK}}} \right), \quad \Lambda_{\theta K} \stackrel{\text{def}}{=} \left(\frac{(\frac{TOL_\theta}{\varpi_{\theta 0}})^{\frac{1}{pK_d}}}{(\frac{\varpi_{\theta K}}{\varpi_{\theta 0}})^{\frac{1}{pK}}} \right). \tag{6.50}$$

6.4.2 Overall Solution Algorithm

The algorithm is as follows:

$$
\begin{array}{l}
(1)\ \text{GLOBAL FIXED} - \text{POINT ITERATION}:\ (\text{SET i} = 1 \text{ AND K} = 0):\\
(2)\ \text{IF i} > \text{N}_\text{p} \text{ THEN GO TO (4)}\\
(3)\ \text{IF i} \leq \text{N}_\text{p} \text{ THEN}:\ (\text{FOR PARTICLE i})\\
\quad (a)\ \text{COMPUTE POSITION}: \boldsymbol{r}_i^{L+1,K}\\
\quad (b)\ \text{COMPUTE TEMPERATURE}: \theta_i^{L+1,K}\\
\quad (c)\ \text{GO TO (2) AND NEXT PARTICLE (i} = \text{i} + 1)\\
(4)\ \text{ERROR MEASURES(normalized)}:\\
\quad (a) \varpi_{rK} \stackrel{\text{def}}{=} \dfrac{\sum_{i=1}^{N_p} ||\boldsymbol{r}_i^{L+1,K} - \boldsymbol{r}_i^{L+1,K-1}||}{\sum_{i=1}^{N_p} ||\boldsymbol{r}_i^{L+1,K} - \boldsymbol{r}_i^{L}||}\\
\quad \varpi_{\theta K} \stackrel{\text{def}}{=} \dfrac{\sum_{i=1}^{N_p} ||\theta_i^{L+1,K} - \theta_i^{L+1,K-1}||}{\sum_{i=1}^{N_p} ||\theta_i^{L+1,K} - \theta_i^{L}||}\\
\quad (b) Z_K \stackrel{\text{def}}{=} \mathbf{MAX}(z_{rK}, z_{\theta K})\ \ where\ \ z_{rK} \stackrel{\text{def}}{=} \dfrac{\varpi_{rK}}{TOL_r},\quad z_{\theta K} \stackrel{\text{def}}{=} \dfrac{\varpi_{\theta K}}{TOL_\theta}\\
\quad (c) \Lambda_K \stackrel{\text{def}}{=} \mathbf{MIN}(\Lambda_{rK}, \Lambda_{\theta K})\ \ where\\
\quad \Lambda_{rK} \stackrel{\text{def}}{=} \left(\dfrac{(\frac{TOL_r}{\varpi_{r0}})^{\frac{1}{pK_d}}}{(\frac{\varpi_{rK}}{\varpi_{r0}})^{\frac{1}{pK}}} \right),\\
\quad \Lambda_{\theta K} \stackrel{\text{def}}{=} \left(\dfrac{(\frac{TOL_\theta}{\varpi_{\theta 0}})^{\frac{1}{pK_d}}}{(\frac{\varpi_{\theta K}}{\varpi_{\theta 0}})^{\frac{1}{pK}}} \right)\\
(5)\ \text{IF TOL. NOT MET } (Z_K > 1) \text{ AND } K < K_d \text{ REPEAT ITERATION (K} = \text{K} + 1)\\
(6)\ \text{IF TOL. MET } (Z_K \leq 1) \text{ AND } K < K_d \text{ THEN}:\\
\quad (a)\ \text{INCREMENT TIME}: t = t + \Delta t\\
\quad (b)\ \text{CONSTRUCT NEW TIME STEP}: \Delta t = \Lambda_K \Delta t,\\
\quad (c)\ \text{SELECT MINIMUM}: \Delta t = \mathbf{MIN}(\Delta t^{lim}, \Delta t)\\
\quad (d)\ \text{UPDATE LASER FIELD I}_\text{i} \text{ (FOR ALL PARTICLES ITERATIVELY)}\\
\quad (e)\ \text{AND GO TO (1)}\\
(7)\ \text{IF TOL. NOT MET } (Z_K > 1) \text{ AND } K = K_d \text{ THEN}:\\
\quad (a)\ \text{CONSTRUCT NEW TIME STEP}: \Delta t = \Lambda_K \Delta t\\
\quad (b)\ \text{UPDATE LASER FIELD I}_\text{i} \text{ (FOR ALL PARTICLES ITERATIVELY)}\\
\quad (c)\ \text{RESTART AT TIME} = \text{t AND GO TO (1)}
\end{array}
\tag{6.51}
$$

The overall goal is to deliver solutions where staggering (incomplete coupling) error is controlled and the temporal discretization accuracy dictates the upper limits on the time-step size (Δt^{lim}) (Fig. 6.6).

6.4.3 Interaction Lists

There are a variety of options for simulation speed up involving particle calculations. For example, one can construct a so-called *interaction list* of neighboring particles within a radius of influence with which a specific particle interacts for a few time-steps. As the overall system configuration changes, one can periodically update the

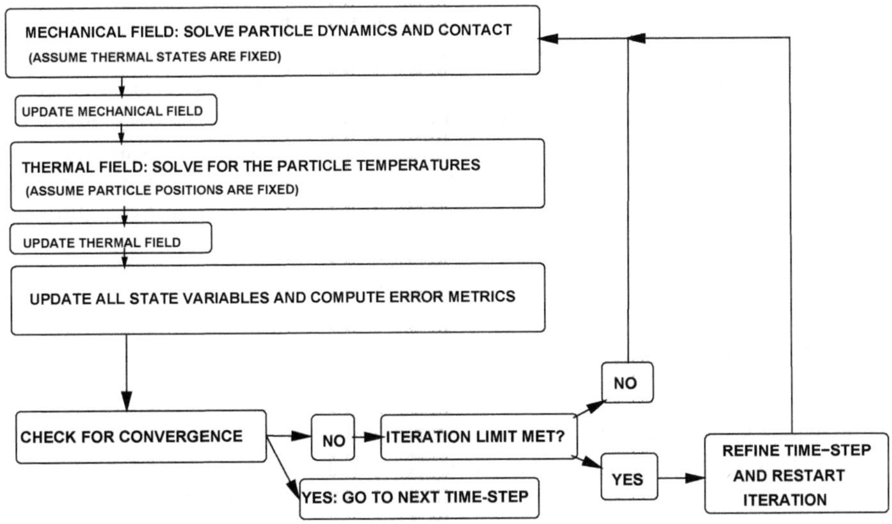

Fig. 6.6 A flowchart for the modular, staggered computation

lists (see Pöschel and Schwager [45]). For example, this dramatically reduces the computation time used in contact search, which is an N^2 operation. In this chapter, we adopt the following process:

- An initial full-blown contact search is initiated at the beginning of the simulation to generate an interaction list of nearest neighbors for each particle.
- The interaction lists are used for all inter-particle calculations for a select subinterval of the total simulation time, $0 \leq \Delta T \leq T$.
- After a select subinterval of the total simulation time (ΔT), the interaction lists are updated for each particle.

Interaction lists were used throughout the upcoming simulations, with the results being tested against direct computation (no interaction truncation) in order to determine appropriate parameter settings for the truncation radius and the frequency of list updating. The upcoming simulation results were insensitive to the truncations and are on the order of 50–100 times faster than direct computation. Finally, we remark that one can also employ domain decomposition techniques whereby the domain is partitioned into subdomains, the particles within each subdomain are sent to a processor and stepped forward in time, but with the positions of the particles outside of the subdomain fixed (relative to the particles in that subdomain). This is done for all of the subdomains separately; then, the position of all of the particles are updated and this information is shared between processors, with the procedure being repeated as needed. It is important to note that a significant acceleration in the computation can be achieved via sorting and binning methods, which proceed by partitioning the whole domain into bins. The particles are sorted by the bins in which they reside. The particle interaction proceeds, bin by bin, where the particles within a bin potentially only interact with particles in other nearest neighbor bins. Essentially, for a given particle in a bin, contact searches are conducted with particles in the neighboring bins only. The approach is relatively straightforward to implement and can speed up the computation significantly (see Zohdi [1–43]). There are a variety of related techniques to further accelerate computations. For example, one can assume that particles stay in the bins for a few time-steps, and that one does not need to re-sort immediately. This can be also used in conjunction with interaction lists, although this was not done in this work.

6.5 Numerical Examples

We consider a model problem of a droplet of equally sized particles (Fig. 6.7) to illustrate how to assemble the system. The absolute dimensions are unimportant for the model problem and have been normalized so that the initial droplet diameter was on the order of unity.[3] In order to generate an initial particle configuration, we randomly dispersed $N_p = 2000$ nonoverlapping particles within a spherical domain, and then the domain is dropped onto a large surface. The configuration of the sample, before it was dropped, was generated using a classical Random Sequential Addition (RSA) algorithm (Widom [139]), which places nonoverlapping particles randomly into the domain of interest. As alluded to before, one could start with a denser starting configuration by using equilibrium-driven Metropolis algorithm or alternative methods based on simultaneous particle flow and growth (see Kansaal et al. [72], Donev et al. [73–75], and Torquato [71]), although this was not necessary for this example. As an example, the relevant simulation parameters chosen were (in SI units if not explicitly stated):

- The particle radii, $R_i = 0.05\,\text{m}$,
- The normal contact parameter was $K_{po} = 10^7\,\text{N/m}^2$, with thermal softening $K_p = \mathbf{MAX}(K_{po}\left(e^{-a\frac{\theta}{\theta^*}-1}\right), K_p^{lim})$, where $\theta^* = 500°\,\text{K}$, $K_p^{lim} = 10^6\,\text{N/m}^2$, and the exponent in the contact law was set to $p_p = 2$,
- The stiffness of the wall in contact law, $K_{wo} = 10^9\,\text{N/m}^2$, with thermal softening $K_w = \mathbf{MAX}(K_{wo}\left(e^{-a\frac{\theta}{\theta^*}-1}\right), K_w^{lim})$, where $\theta^* = 500°\,\text{K}$, $K_w^{lim} = 10^8\text{N/m}^2$, and the exponent in the contact law was set to $p_w = 2$,
- The contact damping parameter, $c^{cd} = 10^5$,
- The friction contact parameter, $K^f = 10^7$,
- The coefficient of static friction, $\mu_s = 0.4$,
- The coefficient of dynamic friction, $\mu_d = 0.3$,
- The normal bond parameter, $K^{nb} = 10^6\,\text{N/m}^2$, and the exponent in the binding law was set to $p_b = 2$,
- The rotational bond parameter, $K^{rb} = 10^3$,

[3] All system parameters can be scaled to describe any specific system of interest. They were selected simply for illustration purposes.

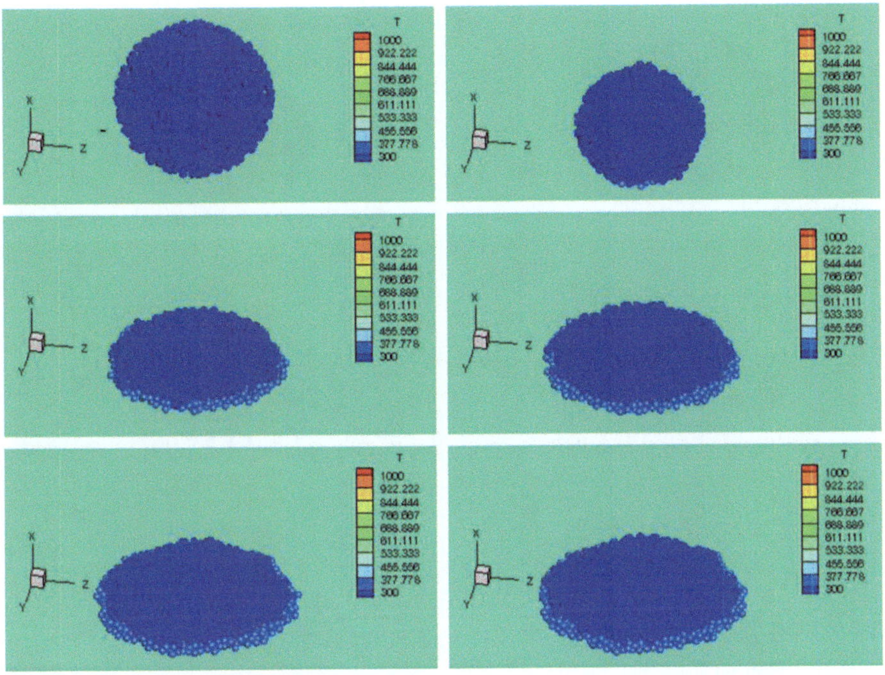

Fig. 6.7 **Scenario 1**: left to right and top to bottom, a sequence of frames for deposition of particles *without* laser input

- The near-field parameters, $\bar{\alpha}_1 = 0.5$, $\beta_1 = 1$, $\bar{\alpha}_2 = 0.01$, $\beta_2 = 2$, where the $\bar{\alpha}$ is per unit mass,
- The effective charge of the particles was $\bar{q} = 1$ (per unit mass), the electric field was $\boldsymbol{E}^{ext} = (-10, 0, 0)$ coulombs, the magnetic field was $\boldsymbol{B}^{ext} = (0, 0, 0)$ Tesla, and the magnetization parameter was $\gamma = 0$,
- The interstitial damping coefficient, $c^e = 1$ (assumed Stokesian-like),
- The density of the particles, $\rho = 2000\,\mathrm{kg/m^3}$,
- The heating absorption coefficient, $a = 0.5$,
- The conductivity, $\mathbb{K} = 100\,\mathrm{W/m\text{-}K}$,
- The initial material temperature, $\theta_i(t = 0) = 300°$ K,
- The wall temperature, $\theta_w(t = 0) = 500°$ K,
- The heat capacity, $C = 100\,\mathrm{J/kg\text{-}K}$,
- The target number of fixed-point iterations, $K_d = 10$,
- The trapezoidal time-stepping parameter, $\phi = 0.5$,
- The simulation duration, 2 s,
- The initial time-step size, 0.00025 s,
- The time-step upper bound, 0.00025 s, and
- The tolerance for the fixed-point iteration, 5×10^{-4}.

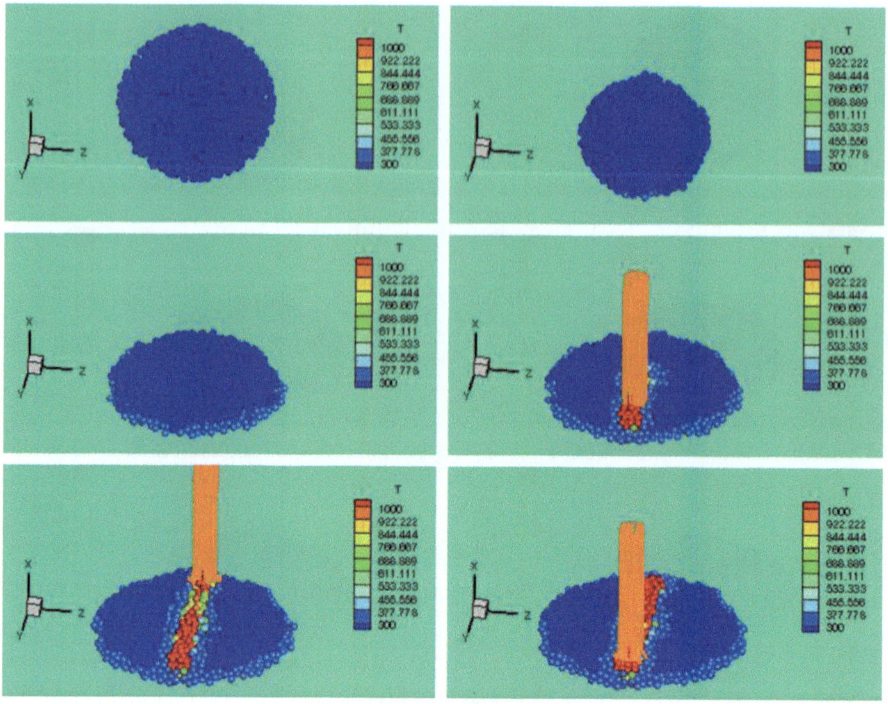

Fig. 6.8 **Scenario 2**: left to right and top to bottom, a sequence of frames for deposition of particles *with* laser input

To illustrate the results of the computational framework, two examples were considered:

- **Scenario 1**: The droplet is dropped onto a slightly heated lower surface (by gravity with an initial uniform downward velocity) and allowed to freely spread/deform according to its interaction with the surface; as shown in Fig. 6.7, the temperature values are depicted by the colors. The overall temperature is shown in Fig. 6.9. The temperatures rise very slightly, since the contact duration time is a fraction of a second, which does not allow enough time for significant conduction to occur (as well as because of the small individual contact areas between the particles and the surface).
- **Scenario 2**: The droplet is dropped onto a slightly heated lower surface (as in scenario 1, by gravity with an initial uniform downward velocity) and allowed to freely spread/deform according to its interaction with the surface. However, after waiting a period for the droplet to settle, a laser source is initiated (at one-quarter of the simulation time), moving back and forth in a linear manner; as shown in Fig. 6.8, the temperature values are depicted by the colors. The overall temperature is shown in Fig. 6.9. The temperatures rise significantly, due to the input of the laser.

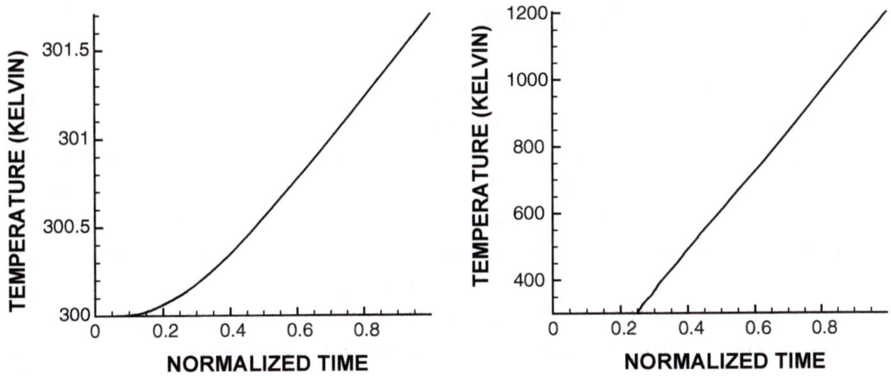

Fig. 6.9 Temperatures for scenario 1 (no laser input) and scenario 2 (with selective laser input)

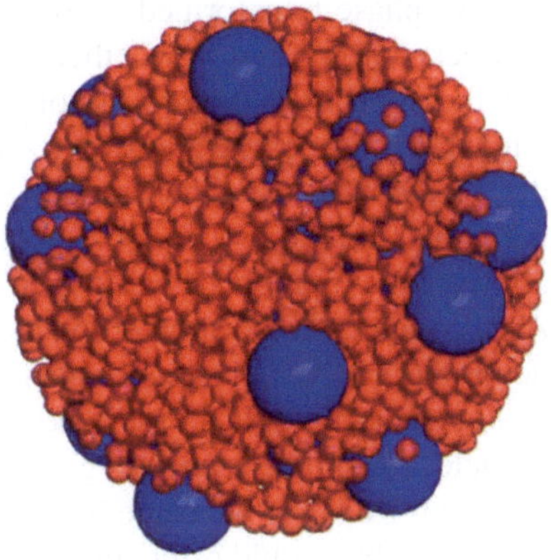

Fig. 6.10 A multiphase droplet representation which can be used for depositions that are comprised of multiple particle types, as well as particles with a fluid (solvent) binder

It is important to note that this framework can also handle depositions that are comprised of multiple particle types, as well as particles with a fluid (solvent) binder (Fig. 6.10). This simply requires a state variable, per particle, which indicates what type of particle it is, and the types of interaction rules with various other types of particles. Otherwise, the framework is identical. We remark that there are several applications where the deposition is particle-functionalized ink (a collection of particles with a solvent binder). This is common in industrial "additive" processes such as inkjet and electrohydrodynamic printing, which do not allow precise control over the structure of the printed lines, and often results in lines with scalloped edges or nonuniform width, thus offering limited control over the height of the printed features (Ridley et al. [140], Huang et al. [141], Sirringhaus et al. [142], Ahmad et al. [143], Samarasinghe et al. [144], and Wang et al. [145]).

6.6 Summary for DEM Approaches

This section illustrated a modular framework for the deposition of particulate materials and selective laser heating, using a direct particle representation in conjunction with a staggering scheme to couple submodels of each type of physics together. The physics of this system is strongly coupled since the dynamics controls which particles are in contact, which also dictates the contacts, which in turn controls the conductive heating and the induced thermal fields, which softens and binds the material. The strongly coupled system was solved iteratively within each time-step using a staggering scheme which employs temporal adaptivity to control the error. The submodular approach allows for easy replacement of models if so desired. Numerical examples were provided. There are a number of enhancements that can be made, for example, a description of the complete melting and vaporization of the material. However, this requires modifications of the existing framework. Specifically, the state of phase-transformed particles (e.g., from a solid to a liquid) would need to be tracked according to whether they are currently solid, melted, or vaporized, and whether they are in the process of transitioning from one state to the other, since the latent heats of melting and vaporization affect the heat transfer (see Chapter Appendix 1). Applications where melting is important include densification of deposited particulate materials (Fig. 6.11), whereby the top layer of the material is melted, allowing the liquid material to seep into the voids between the particles underneath. Applications where vaporization is important include cutting groves, trenches in the deposited material. There are many possibilities for modeling the melted or vaporized state of a particle into a liquid. For example, one approach is to break the original particle into smaller particles governed by different interaction rules (Fig. 6.10).The approach would be the same for the vaporization into a gas. This is computationally expensive, but would offer simulations that are quite realistic. A simpler approach is to assign a solid particle, which has become melted, a different set of properties, such

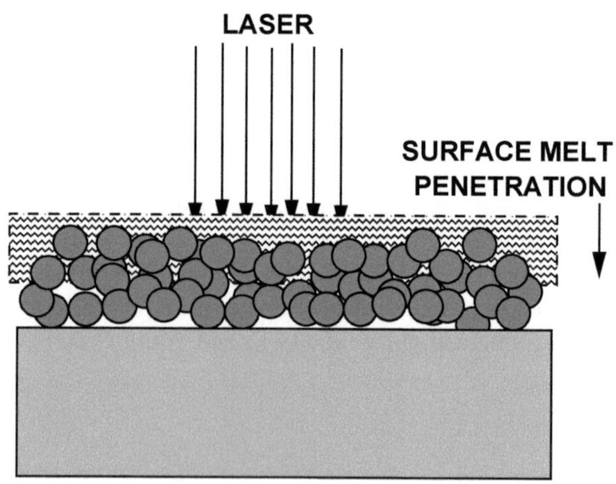

Fig. 6.11 Melting of a deposited material top layer and subsequent densification of the lower layer by allowing the liquid to seep into the voids in between the particles below

as a reduced stiffness, and a loss of bonds with surrounding particles. An alternative approach is to treat the phase-transformed material as a continuum and to then simulate a hybrid discrete particle surrounding continuum model. These techniques are currently under investigation by the author. In Chapter Appendix 1, these types of hybrid methods are discussed at length, as well as phase transformations in Chapter Appendix 2.

Generally, the deposition of particles to build or enhance components is a subclass of additive manufacturing, which in turn is a subfield of advanced manufacturing. Broadly speaking, advanced manufacturing of materials is frequently cited as a blend of techniques from various fields. The "greenness" of an overall process is often considered as a criterion for the approach to be judged as competitive (Dornfeld and Wright [146], Allwood [147], Reich-Weiser et al. [148] and Rosen, Dincer and Kanoglu [149]). The true greenness of a combination of techniques which, individually, may be safe and sustainable, is unclear. Take, for example, laser ablation of rare earth materials, functionalized inks laden with small-scale particles, which, if done in mass, can produce a highly toxic environment in a nonvacuum processing area. Thus, ablation is of particular interest, due to the potential for dangerous ejecta for large-scale processes. For example, thermal or photothermal ablation usually refers to laser light conversion to lattice vibration before breaking of bonds to liberate atomic-scale material, while photochemical/electronic ablation is a direct electronically induced vibration. Hydrodynamical ablation refers to micrometer droplets following from the molten phase, whereas exfoliation is an erosive mechanism by which material is removed as flakes. Note that all of the mechanisms can occur simultaneously. We refer the reader to the state of the art in Grigoropoulos [150], which categorizes various mechanisms for ablation. While discrete element-based methods are useful to describe the dynamics of the deposition of particles and the flow of laser-input energy, they are deficient in the ability to described inherently continuum-based fields such as stresses, relative to CM approaches.

Remark There are applications where Discrete Element Method (DEM) modeling alone is not adequate (because too many very small particles would be needed to describe the interstitial solvent) nor are pure continuum formulations (CF)-based descriptions alone sufficient (because the mesh sizes have to be extremely fine to describe the particles). In these cases, hybrid methods, combining aspects of discrete and continuum methods are advantageous, in particular for describing flowing particle-laden fluids. This is discussed in the Monograph Appendices 4 and 5.

6.7 Chapter Appendix 1: Contact Area Parameter and Alternative Models

Following Zohdi [1–43], and referring to Fig. 6.12, one can solve for an approximation of the common contact radius a_{ij} (and the contact area, $A^c_{ij} = \pi a^2_{ij}$) by solving the following three equations,

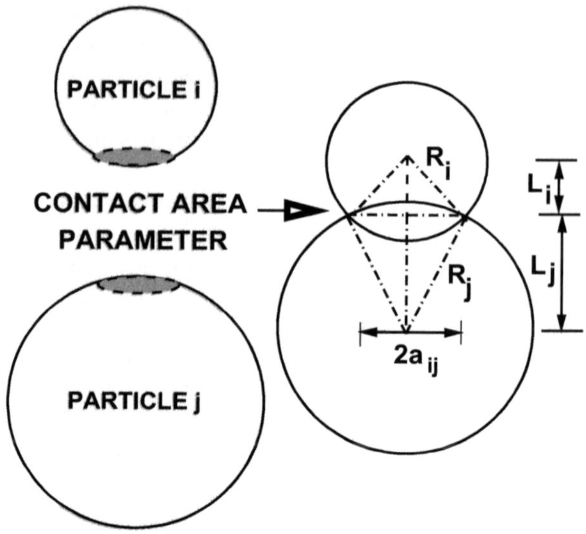

Fig. 6.12 An approximation of the contact area parameter for two particles in contact

$$a_{ij}^2 + L_i^2 = R_i^2, \tag{6.52}$$

and

$$a_{ij}^2 + L_j^2 = R_j^2, \tag{6.53}$$

and

$$L_i + L_j = ||\boldsymbol{r}_i - \boldsymbol{r}_j||, \tag{6.54}$$

where R_i is the radius of particle i, R_j is the radius of particle j, L_i is the distance from the center of particle i and the common contact interpenetration line, and L_j is the distance from the center of particle j and the common contact interpenetration line, where the extent of interpenetration is

$$\delta_{ij} = R_i + R_j - ||\boldsymbol{r}_i - \boldsymbol{r}_j||. \tag{6.55}$$

The above equations yield an expression a_{ij}, which yields an expression for the contact area parameter

$$A_{ij}^c = \pi a_{ij}^2 = \pi(R_i^2 - L_i^2), \tag{6.56}$$

where

$$L_i = \frac{1}{2}\left(||\boldsymbol{r}_i - \boldsymbol{r}_j|| - \frac{R_j^2 - R_i^2}{||\boldsymbol{r}_i - \boldsymbol{r}_j||}\right). \tag{6.57}$$

One could easily construct more elaborate relations connecting the relative proximity of the particles and other metrics to the contact force, $\boldsymbol{\Psi}_{ij}^{con,n} \propto \mathcal{F}(\boldsymbol{r}_i, \boldsymbol{r}_j, \boldsymbol{n}_{ij}, R_i, R_j, ...)$, building on, for example, Hertzian contact models. This poses no difficulty in the direct numerical method developed. For the remainder of the analysis, we shall use the deformation metric in Eq. 6.8. For detailed treatments, see Wellman et al. [50, 151–154] and Avci and Wriggers [121]. We note that with the appropriate definition of parameters, one can recover Hertz, Bradley, Johnson-Kendel-Roberts, Derjaguin-Muller-Toporov contact models. For example, Hertzian contact is widely used, with the assumptions being

- Frictionless, continuous, surfaces,
- Each of contacting bodies are elastic half-spaces, whereby the contact area dimensions are smaller radii of the bodies, and
- The bodies remain elastic (infinitesimal strains),

resulting in the following contact force:

$$\boldsymbol{\Psi}_{ij}^{con,n} = \frac{4}{3}(R^*)^{1/2} E^* \delta_{ij}^{3/2}, \tag{6.58}$$

which has the general form of $\boldsymbol{\Psi}^{con,n} = K_{ij}^* \delta_{ij}^p$, where

- $R^* = \left(\frac{1}{R_i} + \frac{1}{R_j}\right)^{-1}$ and
- $E^* = \left(\frac{1-\nu_i^2}{E_i} + \frac{1-\nu_j^2}{E_j}\right)^{-1}$,

where E is the Young's modulus and ν is the Poisson ratio. The contact area with such a model has already been incorporated in the relation above and is equal to $A_{ij}^c = \pi a^2$ where $a = \sqrt{R^* \delta_{ij}}$. For more details, we refer the reader to Johnson [155]. Furthermore, we remark that the normal contact between a particle and a wall, with a Hertzian model, is given by

$$\boldsymbol{\Psi}_i^{wall,n} = \frac{4}{3}(R^*)^{1/2} E^* \delta_{iw}^{3/2} = K_{iw}^* \delta_{iw}^p, \tag{6.59}$$

where $R_j = R_w = \infty$ (see Eq. 6.58)

- $R^* = R_i$ and
- $E^* = \left(\frac{1-\nu_i^2}{E_i} + \frac{1-\nu_j^2}{E_j}\right)^{-1}$.

It is obvious that for a deeper understanding of the deformation within a particle, it must be treated as a deformable continuum, which would require a highly resolved spatial discretization, for example, using the Finite Element Method for the contacting bodies. This requires a large computational effort. For the state of the art in Finite Element Methods and Contact Mechanics, see the books of Wriggers [156, 157]. For work specifically focusing on the continuum mechanics of particles, see Zohdi and Wriggers [1–43].

6.8 Chapter Appendix 2: Phase Transformations

To include phase transformations, we consider seven cases, which are implemented in a predictor-corrector manner by first solving the governing equations to obtain predicted temperature, and then checking the following:

- *Solid → solid-no melting with $C_i = C_S$:* If $\theta(t) < \theta_m$ and $\theta(t + \Delta t) < \theta_m$, then retain $C(\theta) = C_S$,
- *Solid → liquid-melting with $C_i = C_S$:* If $\theta(t) < \theta_m$ and $\theta(t + \Delta t) \geq \theta_m$, then re-solve the governing equations with $C(\theta) = C_S + \frac{\delta \mathcal{P}^{S \to L}}{\delta \theta}$,
- *Liquid → liquid-melted with $C_i = C_L$:* If $\theta(t) \geq \theta_m$ and $\theta(t + \Delta t) \geq \theta_m$, then retain $C(\theta) = C_L$,
- *Liquid → solid-solidification with $C_i = C_L$:* If $\theta(t) \geq \theta_m$ and $\theta(t + \Delta t) < \theta_m$ then re-solve the governing equations with $C(\theta) = C_L + \frac{\delta \mathcal{P}^{L \to S}}{\delta \theta}$,
- *Liquid → vapor-vaporization with $C_i = C_L$:* If $\theta(t) < \theta_v$ and $\theta(t + \Delta t) \geq \theta_v$, then re-solve governing equations with $C(\theta) = C_L + \frac{\delta \mathcal{P}^{L \to V}}{\delta \theta}$,
- *Vapor → vapor-remains a vapor with $C_i = C_V$:* If $\theta(t) \geq \theta_v$ and $\theta(t + \Delta t) \geq \theta_v$, then retain $C(\theta) = C_V$,
- *Vapor → liquid-condensation with $C_i = C_v$:* If $\theta(t) \geq \theta_v$ and $\theta(t + \Delta t) < \theta_v$ then re-solve the governing equations with $C(\theta) = C_V + \frac{\delta \mathcal{P}^{V \to L}}{\delta \theta}$,

where C_S is the heat capacity of the solid, C_L is the heat capacity of the liquid, and C_V is the heat capacity of the vapor and

- $0 < \delta \mathcal{P}^{S \to L}$ is the latent heat of melting.
- $0 < \delta \mathcal{P}^{L \to S}$ is the latent heat of solidification.
- $0 < \delta \mathcal{P}^{L \to V}$ is the latent heat of vaporization.
- $0 < \delta \mathcal{P}^{V \to L}$ is the latent heat of condensation.
- $0 < \delta \theta$ is small and can be thought of as a "bandwidth" for a phase transformation. For more details on melting processes, see Davis [158].

We note that latent heats have a tendency to resist the phase transformations, achieved by adding the positive terms to the instantaneous heat capacity, thus enforcing reduced temperature (during the phase transformation).[4] This approach is relatively straightforward to include within the staggering framework. We note that because of changes in the heat capacity (due to phase transformations) that the thermal storage term is no longer simply $m_i \frac{d}{dt}(C_i(\theta_i - \theta_o)) = m_i C_i \dot{\theta}_i$ but is

$$m_i \frac{d}{dt}(C_i(\theta_i - \theta_o)) = m_i \dot{C}_i(\theta_i - \theta_o) + m_i C_i \dot{\theta}_i. \tag{6.60}$$

As a consequence,

$$m_i \dot{C}_i(\theta_i - \theta_o) + m_i C_i \dot{\theta}_i = \mathcal{Q}_i + \mathcal{H}_i. \tag{6.61}$$

[4] In the idealized limit, the temperature would be constant.

Rearranging yields

$$m_i C_i \dot{\theta}_i = \mathcal{Q}_i + \mathcal{H}_i - m_i \dot{C}_i (\theta_i - \theta_o). \tag{6.62}$$

Evaluating each term at $t + \phi \Delta t$ yields

$$\begin{aligned} m_i(\phi C_i(t+\Delta t) + (1-\phi) C_i(t)) \left(\frac{\theta(t+\Delta t) - \theta_i(t)}{\Delta t} \right) &= \phi(\mathcal{Q}_i(t+\Delta t) \\ + \mathcal{H}_i(t+\Delta t)) + \phi(\mathcal{Q}_i(t) + \mathcal{H}_i(t)) \\ - m_i \left(\frac{C_i(t+\Delta t) - C_i(t)}{\Delta t} \right) (\phi \theta_i(t+\Delta t) + (1-\phi)\theta_i(t) - \theta_o) \end{aligned} \tag{6.63}$$

This can be written as an recursive iteration:

$$\begin{aligned} \theta^{K+1}(t+\Delta t) = \theta_i(t) + \Delta t \frac{\left(\phi(\mathcal{Q}_i^K(t+\Delta t) + \mathcal{H}_i^{\dot{K}}(t+\Delta t)) + \phi(\mathcal{Q}_i(t) + \mathcal{H}_i(t)) \right)}{m_i(\phi C_i^K(t+\Delta t) + (1-\phi) C_i(t))} \\ - \Delta t \frac{m_i \left(\frac{C_i^K(t+\Delta t) - C_i(t)}{\Delta t} \right) (\phi \theta_i^K(t+\Delta t) + (1-\phi)\theta_i(t) - \theta_o)}{m_i(\phi C_i^K(t+\Delta t) + (1-\phi) C_i(t))} \end{aligned} \tag{6.64}$$

The subsequent convergence of the thermal calculation is rather quick, since the time-steps are extremely small. For more details on convergence on iterative time-stepping schemes, see Zohdi [1–43].

Remark A number of extensions can be made to the DEM models introduced in the chapter, depending on the specific manufacturing process to be characterized. In the following chapters, we illustrate some of these extensions.

References

1. Zohdi, T.I.: Genetic design of solids possessing a random-particulate microstructure. Philosoph. Trans. R. Soc. Math. Phys. Eng. Sci. **361**(1806), 1021–1043 (2003)
2. Zohdi, T.I.: On the compaction of cohesive hyperelastic granules at finite strains. Proc. R. Soc. **454**(2034), 1395–1401 (2003)
3. Zohdi, T.I.: Computational design of swarms. Int. J. Numer. Methods Eng. **57**, 2205–2219 (2003)
4. Zohdi, T.I.: Constrained inverse formulations in random material design. Comput. Methods Appl. Mech. Eng. 1–20 **192**, 28–30, 18, 3179–3194 (2003)
5. Zohdi, T.I.: Staggering error control for a class of inelastic processes in random microheterogeneous solids. Int. J. Nonlinear Mech. **39**, 281–297 (2004)
6. Zohdi, T.I.: Modeling and simulation of a class of coupled thermo-chemo-mechanical processes in multiphase solids. Comput. Methods Appl. Mech. Eng. **193**(6–8), 679–699 (2004)
7. Zohdi, T.I.: Modeling and direct simulation of near-field granular flows. Int. J. Solids Struct. **42**(2), 539–564 (2004)
8. Zohdi, T.I.: A computational framework for agglomeration in thermo-chemically reacting granular flows. Proc. R. Soc. **460**(2052), 3421–3445 (2004)

9. Zohdi, T.I.: Statistical ensemble error bounds for homogenized microheterogeneous solids. J. Appl. Math. Phys. (Zeitschrift für Angewandte Mathematik und Physik) **56**(3), 497–515 (2005)
10. Zohdi, T.I.: Charge-induced clustering in multifield particulate flow. Int. J. Numer. Methods Eng. **62**(7), 870–898 (2005)
11. Zohdi, T.I.: On the optical thickness of disordered particulate media. Mech. Mater. **38**, 969–981 (2006)
12. Zohdi, T.I., Kuypers, F.A.: Modeling and rapid simulation of multiple red blood cell light scattering. Proc. R. Soc. Interface **3**(11), 823–831 (2006)
13. Zohdi, T.I.: Computation of the coupled thermo-optical scattering properties of random particulate systems. Comput. Methods Appl. Mech. Eng. **195**, 5813–5830 (2006)
14. Zohdi, T.I.: Computation of strongly coupled multifield interaction in particle-fluid systems. Comput. Methods Appl. Mech. Eng. **196**, 3927–3950 (2007)
15. Zohdi, T.I.: Particle collision and adhesion under the influence of near-fields. J. Mech. Mater. Struct. **2**(6), 1011–1018 (2007)
16. Zohdi, T.I.: On the computation of the coupled thermo-electromagnetic response of continua with particulate microstructure. Int. J. Numer. Methods Eng. **76**, 1250–1279 (2008)
17. Zohdi, T.I.: Mechanistic modeling of swarms. Comput. Methods Appl. Mech. Eng. **198**(21–26), 2039–2051 (2009)
18. Zohdi, T.I.: Charged wall-growth in channel-flow. Int. J. Eng. Sci. **48**, 1520 (2010)
19. Zohdi, T.I.: On the dynamics of charged electromagnetic particulate jets. Arch. Comput. Methods Eng. **17**(2), 109–135 (2010)
20. Zohdi, T.I., Kuypers, F.A., Lee, W.C.: Estimation of Red Blood Cell volume fraction from overall permittivity measurement. Int. J. Eng. Sci. **48**, 1681–1691 (2010)
21. Zohdi, T.I.: Simulation of coupled microscale multiphysical-fields in particulate-doped dielectrics with staggered adaptive FDTD. Comput. Methods Appl. Mech. Eng. **199**, 79–101 (2010)
22. Zohdi, T.I.: Dynamics of clusters of charged particulates in electromagnetic fields. Int. J. Numer. Methods Eng. **85**, 1140–1159 (2011)
23. Zohdi, T.I.: Joule-heating field phase-amplification in particulate-doped dielectrics. Int. J. Eng. Sci. **49**, 30–40 (2011)
24. Zohdi, T.I.: Estimation of electrical-heating load-shares for sintering of powder mixtures. Proc. R. Soci. **468**, 2174–2190 (2012)
25. Zohdi, T.I.: Modeling and simulation of the optical response rod-functionalized reflective surfaces. Comput. Mech. **50**(2), 257–268 (2012)
26. Zohdi, T.I.: On the reduction of heat generation in lubricants using microscale additives. Int. J. Eng. Sci. **62**, 84–89 (2013)
27. Zohdi, T.I.: Electromagnetically-induced vibration in particulate-doped materials. ASME J. Vib. Acoust. **135**(3) (2013). https://doi.org/10.1115/1.4023251
28. Zohdi, T.I.: Numerical simulation of charged particulate cluster-droplet impact on electrified surfaces. J. Comput. Phys. **233**, 509–526 (2013)
29. Zohdi, T.I.: On inducing compressive residual stress in microscale print-lines for flexible electronics. Int. J. Eng. Sci. **62**, 157–164 (2013)
30. Zohdi, T.I.: Rapid simulation of laser processing of discrete particulate materials. Arch. Comput. Methods Eng. **20**, 309–325 (2013)
31. Zohdi, T.I.: A direct particle-based computational framework for electrically-enhanced thermo-mechanical sintering of powdered materials. Math. Mech. Solids **19**(1), 93–113 (2014)
32. Zohdi, T.I.: On cross-correlation between thermal gradients and electric fields. Int. J. Eng. Sci. **74**, 143–150 (2014)
33. Zohdi, T.I.: Mechanically-driven accumulation of microscale material at coupled solid-fluid interfaces in biological channels. Proc. R. Soc. Interface **11**, 20130922 (2014)
34. Zohdi, T.I.: A computational modeling framework for heat transfer processes in laser-induced dermal tissue removal. Comput. Mech. Eng. Sci. **98**(3), 261–277 (2014)

35. Zohdi, T.I.: Additive particle deposition and selective laser processing-a computational manufacturing framework. Comput. Mech. **54**, 171–191 (2014)
36. Zohdi, T.I.: Embedded electromagnetically sensitive particle motion in functionalized fluids. Comput. Part. Mech. **1**, 27–45 (2014)
37. Zohdi, T.I.: Rapid computation of statistically-stable particle/feature ratios for consistent substrate stresses in printed flexible electronics. J. Manuf. Sci. Eng. ASME (2015). MANU-14-1476. https://doi.org/10.1115/1.4029327
38. Zohdi, T.I.: A computational modelling framework for high-frequency particulate obscurant cloud performance. Int. J. Eng. Sci. **89**, 75–85 (2015)
39. Zohdi, T.I.: On necessary pumping pressures for industrial process-driven particle-laden fluid flows. J. Manuf. Sci. Eng. ASME (2015). https://doi.org/10.1115/1.4030620
40. Zohdi, T.I.: On the thermal response of a laser-irradiated powder particle in additive manufacturing. CIRP J. Manuf. Sci. Technol. **10**, 7783, Aug 2015
41. Zohdi, T.I.: Modeling and simulation of cooling-induced residual stresses in heated particulate mixture depositions. Comput. Mech. **56**, 613–630 (2015)
42. Zohdi, T.I.: Modeling and efficient simulation of the deposition of particulate flows onto compliant substrates. Int. J. Eng. Sci. **99**, 74–91 (2015). https://doi.org/10.1016/j.ijengsci.2015.10.012
43. Zohdi, T.I.: Modeling and simulation of laser processing of particulate-functionalized materials. Arch. Comput. Methods Eng. 1–25 (2015). https://doi.org/10.1007/s11831-015-9160-1
44. Duran, J.: Sands, powders and grains. An introduction to the physics of Granular Matter. Springer (1997)
45. Pöschel, T., Schwager, T.: Computational Granular Dynamics. Springer (2004)
46. Onate, E., Idelsohn, S.R., Celigueta, M.A., Rossi, R.: Advances in the particle finite element method for the analysis of fluid-multibody interaction and bed erosion in free surface flows. Comput. Methods Appl. Mech. Eng. **197**(19–20), 1777–1800 (2008)
47. Onate, E., Celigueta, M.A., Idelsohn, S.R., Salazar, F., Surez, B.: Possibilities of the particle finite element method for fluid-soil-structure interaction problems. Comput. Mech. **48**, 307–318 (2011)
48. Rojek, J., Labra, C., Su, O., Onate, E.: Comparative study of different discrete element models and evaluation of equivalent micromechanical parameters. Int. J. Solids Struct. **49**, 1497–1517 (2012). https://doi.org/10.1016/j.ijsolstr.2012.02.032
49. Carbonell, J.M., Onate, E., Suarez, B.: Modeling of ground excavation with the particle finite element method. J. Eng. Mech. ASCE **136**, 455–463 (2010)
50. Labra, C., Onate, E.: High-density sphere packing for discrete element method simulations. Commun. Numer. Methods Eng. **25**(7), 837–849 (2009)
51. Mukherjee, D., Zohdi, T.I.: Electromagnetic control of charged particulate spray systems—Models for planning the spray gun operations. Comput.-Aided Des. **46**, 211–215 (2014)
52. Mukherjee, D., Zaky, Z., Zohdi, T.I., Salama, A., Sun, S.: Investigation of guided particle transport for noninvasive healing of damaged piping system using electro-magneto-mechanical methods. J. Soc. Pet. Eng. J. SPE **169639**, 1–12 (2015)
53. Mukherjee, D., Zohdi, T.I.: A discrete element based simulation framework to investigate particulate spray deposition processes. J. Comput. Phys. **290**, 298–317 (2015)
54. Mukherjee, D., Zohdi, T.I.: Computational modeling of the dynamics and interference effects of an erosive granular jet impacting a porous, compliant surface. Granul. Matter **17**, 231–252 (2015)
55. Akisanya, A.R., Cocks, A.C.F., Fleck, N.A.: The yield behavior of metal powders. Int. J. Mech. Sci. **39**, 1315–1324 (1997)
56. Anand, L., Gu, C.: Granular materials: constitutive equations and shear localization. J. Mech. Phys. Solids **48**, 1701–1733 (2000)
57. Brown, S., Abou-Chedid, G.: Yield behavior of metal powder assemblages. J. Mech. Phys. Solids **42**, 383–398 (1994)

58. Domas, F.: Eigenschaft profile und Anwendungsübersicht von EPE und EPP. Technical report of the BASF Company (1997)
59. Fleck, N.A.: On the cold compaction of powders. J. Mech. Phys. Solids **43**, 1409–1431 (1995)
60. Gethin, D.T., Lewis, R.W., Ransing, R.S.: A discrete deformable element approach for the compaction of powder systems. Model. Simul. Mater. Sci. Eng. **11**(1), 101–114 (2003)
61. Gu, C., Kim, M., Anand, L.: Constitutive equations for metal powders: application to powder forming processes. Int. J. Plast. **17**, 147–209 (2001)
62. Lewis, R.W., Gethin, D.T., Yang, X.S.S., Rowe, R.C.: A combined finite-discrete element method for simulating pharmaceutical powder tableting. Int. J. Numer. Methods Eng. **62**, 853869 (2005)
63. Ransing, R.S., Lewis, R.W., Gethin, D.T.: Using a deformable discrete-element technique to model the compaction behaviour of mixed ductile and brittle particulate systems. Philos. Trans. R. Soc.—Ser. A: Math. Phys. Eng. Sci. **362**(1822), 1867–1884 (2004)
64. Tatzel, H.: Grundlagen der Verarbeitungstechnik von EPP-Bewährte und neue Verfahren. Technical report of the BASF Company (1996)
65. Martin, P.: (2009). Handbook of Deposition Technologies for Films and Coatings, 3rd edn. Elsevier
66. Martin, P. (2011). Introduction to Surface Engineering and Functionally Engineered Materials. Scrivener and Elsevier
67. Choi, S., Park, I., Hao, Z., Holman, H.Y., Pisano, A.P., Zohdi, T.I.: Ultra-fast self-assembly of micro-scale particles by open channel flow. Langmuir **26**(7), 4661–4667 (2010)
68. Choi, S., Jamshidi, A., Seok, T.J., Zohdi, T.I., Wu, M.C., Pisano, A.P.: Fast, High-throughput creation of size-tunable micro, nanoparticle clusters via evaporative self-assembly in picoliter-scale droplets of particle suspension. Langmuir **28**(6), 3102–11 (2012)
69. Choi, S., Pisano, A.P., Zohdi, T.I.: An analysis of evaporative self-assembly of micro particles in printed picoliter suspension droplets. J. Thin Solid Films **537**(30), 180–189 (2013)
70. Demko, M., Choi, S., Zohdi, T.I., Pisano, A.P.: High resolution patterning of nanoparticles by evaporative self-assembly enabled by in-situ creation and mechanical lift-off of a polymer template. Appl. Phys. Lett. **99**, 253102-1–253102-3 (2012)
71. Torquato, S.: Random Heterogeneous Materials: Microstructure and Macroscopic Properties. Springer, New York (2002)
72. Kansaal, A., Torquato, S., Stillinger, F.: Diversity of order & densities in jammed hard-particle packings. Phys. Rev. E **66**, 041109 (2002)
73. Donev, A., Cisse, I., Sachs, D., Variano, E. A., Stillinger, F., Connelly, R., Torquato, S., Chaikin, P.: Improving the density of jammed disordered packings using ellipsoids. Science 13, **303**, 990–993, Feb 2004
74. Donev, A., Stillinger, F.H., Chaikin, P.M., Torquato, S.: Unusually dense crystal ellipsoid packings. Phys. Rev. Lett. **92**, 255506 (2004b)
75. Donev, A., Torquato, S., Stillinger, F.: Neighbor list collision-driven molecular dynamics simulation for nonspherical hard particles-I. Algorithmic details. J. Comput. Phys. **202**, 737 (2005a)
76. Donev, A., Torquato, S., Stillinger, F.: Neighbor list collision-driven molecular dynamics simulation for nonspherical hard particles-II. Application to ellipses and ellipsoids. J. Comput. Phys. **202**, 765 (2005b)
77. Donev, A., Torquato, S., Stillinger, F.H.: Pair correlation function characteristics of nearly jammed disordered and ordered hard-sphere packings. Phys. Rev. E **71**, 011105 (2005c)
78. Jaeger, H.M., Nagel, S.R.: La Physique de l'Etat Granulaire. La Recherche **249**, 1380 (1992)
79. Jaeger, H.M., Nagel, S.R.: Physics of the Granular State. Science **255**, 1523 (1992)
80. Nagel, S.R.: Instabilities in a Sandpile. Rev. Mod. Phys. **64**, 321 (1992)
81. Liu, Y., Nakamura, T., Dwivedi, G., Valarezo, A., Sampath, S.: Anelastic behavior of plasma sprayed zirconia coatings. J. Am. Ceram. Soc. **91**, 4036–4043 (2008)
82. Liu, Y., Nakamura, T., Srinivasan, V., Vaidya, A., Gouldstone, A., Sampath, S.: Nonlinear elastic properties of plasma sprayed zirconia coatings and associated relationships to processing conditions. Acta Materialia **55**, 4667–4678 (2007)

83. Jaeger, H.M., Nagel, S.R.: La Fisica del Estado Granular. Mundo Cient. **132**, 108 (1993)
84. Jaeger, H.M., Knight, J.B., Liu, C.H., Nagel, S.R.: What is shaking in the sand box? Mat. Res. Soc. Bull. **19**, 25 (1994)
85. Jaeger, H.M., Nagel, S.R., Behringer, R.P.: The Physics of Granular Materials. Phys. Today **4**, 32 (1996)
86. Jaeger, H.M., Nagel, S.R., Behringer, R.P.: Granular Solids, Liquids & Gases. Rev. Mod. Phys. **68**, 1259 (1996)
87. Jaeger, H.M., Nagel, S.R.: Dynamics of Granular material. Am. Sci. **85**, 540 (1997)
88. Tai, Y.-C., Noelle, S., Gray, J.M.N.T., Hutter, K.: Shock capturing & front tracking methods for granular avalanches. J. Comput. Phy. **175**, 269–301 (2002)
89. Tai, Y.-C., Noelle, S., Gray, J.M.N.T., Hutter, K.: An accurate shock-capturing finite-difference method to solve the Savage-Hutter equations in avalanche dynamics. Ann. Glaciol. **32**, 263–267 (2001)
90. Jenkins, J.T., La Ragione, L.: Particle spin in anisotropic granular materials. Int. J. Solids Struct. **38**, 1063–1069 (1999)
91. Gray, J.M.N.T., Wieland, M., Hutter, K.: Gravity-driven free surface flow of granular avalanches over complex basal topography. Proc. R. Soc. Lond. A **455**, 1841–1874 (1999)
92. Wieland, M., Gray, J.M.N.T., Hutter, K.: Channelized free-surface flow of cohesionless granular avalanches in a chute with shallow lateral curvature. J. Fluid Mech. **392**, 73–100 (1999)
93. Berezin, Y.A., Hutter, K., Spodareva, L.A.: Stability properties of shallow granular flows. Int. J. Nonlinear Mech. **33**(4), 647–658 (1998)
94. Gray, J.M.N.T., Hutter, K.: Pattern formation in granular avalanches. Contin. Mech. Thermodyn. **9**, 341–345 (1997)
95. Gray, J.M.N.T.: Granular flow in partially filled slowly rotating drums. J. Fluid Mech. **441**, 1–29 (2001)
96. Hutter, K.: Avalanche dynamics. In: Singh, V.P. (ed.) Hydrology of Disasters, pp. 317–394. Kluwer Academic Publishers, Dordrecht etc (1996)
97. Hutter, K., Koch, T., Plüss, C., Savage, S.B.: The dynamics of avalanches of granular materials from initiation to runout. Part II. Experiments. Acta Mechanica **109**, 127–165 (1995)
98. Hutter, K., Rajagopal, K.R.: On flows of granular materials. Contin. Mech. Thermodyn. **6**, 81–139 (1994)
99. Koch, T., Greve, R., Hutter, K.: Unconfined flow of granular avalanches along a partly curved surface. II. Experiments & numerical computations. Proc. R. Soc. Lond. A **445**, 415–435 (1994)
100. Greve, R., Hutter, K.: Motion of a granular avalanche in a convex & concave curved chute: experiments & theoretical predictions. Philos. Trans. R. Soc. Lond. A **342**, 573–600 (1993)
101. Hutter, K., Siegel, M., Savage, S.B., Nohguchi, Y.: Two-dimensional spreading of a granular avalanche down an inclined plane. Part I: Theory. Acta Mechanica **100**, 37–68 (1993)
102. Behringer, R.P.: The dynamics of flowing sand. Nonlinear Sci Today **3**, 1 (1993)
103. Behringer, R.P., Baxter, G.W.: Pattern formation, complexity & time-dependence in granular flows. In: Mehta, A. (ed.) Granular Matter—An Interdisciplinary Approach, pp. 85–119. Springer, New York (1993)
104. Behringer, R.P., Miller, B.J.: Stress fluctuations for sheared 3D granular materials. In: Behringer, R., Jenkins, J. (eds.) Proceedings, Powders & Grains, vol. 97, pp. 333–336. Balkema (1997)
105. Behringer, R.P., Howell, D., Veje, C.: Fluctuations in Granular flows. Chaos **9**, 559–572 (1999)
106. Jenkins, J.T., Strack, O.D.L.: Mean-field inelastic behavior of random arrays of identical spheres. Mech. Mater. **16**, 25–33 (1993)
107. Jenkins, J.T., Koenders, M.A.: The incremental response of random aggregates of identical round particles. Eur. Phys. J. E-Soft Matter. **13**, 113–123 (2004)
108. Jenkins, J.T., Johnson, D., La Ragione, L., Makse, H.: Fluctuations and the effective moduli of an isotropic, random aggregate of identical, frictionless spheres. J. Mech. Phys. Solids **53**, 197–225 (2005)
109. Jackson, J.D.: Classical Electrodynamics, 3rd edn. Wiley (1998)

110. Frenklach, M., Carmer, C.S.: Molecular dynamics using combined quantum & empirical forces: application to surface reactions. Adv. Class. Trajectory Methods **4**, 27–63 (1999)
111. Haile, J.M.: Molecular Dynamics Simulations: Elementary Methods. Wiley (1992)
112. Hase, W.L.: Molecular dynamics of clusters, surfaces, liquids, & interfaces. Advances in classical trajectory methods, vol. 4. JAI Press (1999)
113. Schlick, T.: Molecular Modeling & Simulation. An Interdisciplinary Guide. Springer, New York (2000)
114. Rapaport, D.C.: The Art of Molecular Dynamics Simulation. Cambridge University Press (1995)
115. Moelwyn-Hughes, E.A.: Physical Chemistry. Pergamon (1961)
116. Stillinger, F.H., Weber, T.A.: Computer simulation of local order in condensed phases of silicon. Phys. Rev. B **31**, 5262–5271 (1985)
117. Tersoff, J.: Empirical interatomic potential for carbon, with applications to amorphous carbon. Phys. Rev. Lett. **61**, 2879–2882 (1988)
118. Feynman, R.P., Leighton, R.B., Sands, M.: Feynman Lect. Phys. **2** (2006). ISBN 0-8053-9045-6
119. Cullity, B.D., Graham, C.D.: Introduction to Magnetic Materials, 2nd edn. p. 103. Wiley-IEEE Press (2008). ISBN 0-471-47741-9
120. Boyer, T.H.: The force on a magnetic dipole. Am. J. Phys. **56**(8), 688692 (1988). Bibcode:1988AmJPh.56.688B. https://doi.org/10.1119/1.15501
121. Avci, B., Wriggers, P.: A DEM-FEM coupling approach for the direct numerical simulation of 3D particulate flows. J. Appl. Mech. **79**, 010901-1–7 (2012)
122. Zienkiewicz, O.C.: Coupled problems & their numerical solution. In: Lewis, R.W., Bettes, P., Hinton, E. (eds.) Numerical Methods in Coupled Systems, pp. 35–58. Wiley, Chichester (1984)
123. Zienkiewicz, O.C., Paul, D.K., Chan, A.H.C.: Unconditionally stable staggered solution procedure for soil-pore fluid interaction problems. Int. J. Numer. Methods Eng. **26**, 1039–1055 (1988)
124. Lewis, R.W., Schrefler, B.A., Simoni, L.: Coupling versus uncoupling in soil consolidation. Int. J. Numer. Anal. Methods Geomech. **15**, 533–548 (1992)
125. Lewis, R.W., Schrefler, B.A.: The Finite Element Method in the Static and Dynamic Deformation and Consolidation of Porous Media. 2nd edn. Wiley Press (1998)
126. Park, K.C., Felippa, C.A.: Partitioned analysis of coupled systems. In: Belytschko, T., Hughes, T.J.R. (eds.) Computational Methods for Transient Analysis (1983)
127. Farhat, C., Lesoinne, M., Maman, N.: Mixed explicit/implicit time integration of coupled aeroelastic problems: three-field formulation, geometric conservation and distributed solution. Int. J. Numer. Methods Fluids **21**, 807–835 (1995)
128. Farhat, C., Lesoinne, M.: Two efficient staggered procedures for the serial and parallel solution of three-dimensional nonlinear transient aeroelastic problems. Comput. Methods Appl. Mech. Eng. **182**, 499–516 (2000)
129. Farhat, C., van der Zee, G., Geuzaine, P.: Provably second-order time-accurate loosely-coupled solution algorithms for transient nonlinear computational aeroelasticity. Comput. Methods Appl. Mech. Eng. **195**, 1973–2001 (2006)
130. Piperno, S.: Explicit/implicit fluid/structure staggered procedures with a structural predictor & fluid subcycling for 2D inviscid aeroelastic simulations. Int. J. Numer. Meth. Fluids **25**, 1207–1226 (1997)
131. Piperno, S., Farhat, C., Larrouturou, B.: Partitioned procedures for the transient solution of coupled aeroelastic problems—Part I: model problem, theory, and two-dimensional application. Comput. Methods Appl. Mech. Eng. **124**(1–2), 79–112 (1995)
132. Piperno, S., Farhat, C.: Partitioned procedures for the transient solution of coupled aeroelastic problems—Part II: energy transfer analysis and three-dimensional applications. Comput. Methods Appl. Mech. Eng. **190**, 3147–3170 (2001)
133. Michopoulos, J., Farhat, C., Fish, J.: Modeling and simulation of multiphysics systems. J. Comput. Inf. Sci. Eng. **5**(3), 198–213 (2005)

134. Steuben, J.C., Iliopoulos, A.P., Michopoulos, J.G.: Discrete element modeling of particle-based additive manufacturing processes. Comput Methods Appl M **305**(0045-7825), 537–561 (2016). https://doi.org/10.1016/j.cma.2016.02.023
135. Steuben, J.C., Iliopoulos, A.P., Michopoulos, J.G.: On Multiphysics Discrete Element Modeling of Powder-Based Additive Manufacturing Processes. V01AT02A032 (2016). http://dx.doi.org/10.1115/DETC2016-59634
136. Steuben, J.C., Iliopoulos, A.P., Michopoulos, J.G.: Recent developments of the multiphysics discrete element method for additive manufacturing modeling and simulation. In: ASME 2017 International Design Engineering Technical Conferences and Computers and Information in Engineering Conference, DETC/CIE2017-67597, DETC2017-DVD, ASME (2017)
137. Lesoinne, M., Farhat, C.: Free staggered algorithm for nonlinear transient aeroelastic problems. AIAA J. **36**(9), 1754–1756 (1998)
138. Le Tallec, P., Mouro, J.: Fluid structure interaction with large structural displacements. Comput. Methods Appl. Mech. Eng. **190**(24–25), 3039–3067 (2001)
139. Widom, B.: Random sequential addition of hard spheres to a volume. J. Chem. Phys. **44**, 3888–3894 (1966)
140. Ridley, B.A., Nivi, B., Jacobson, J.M.: All-inorganic field effect transistors fabricated by printing. Science **286**, 746–749 (1999)
141. Huang, D., Liao, F., Molesa, S., Redinger, D., Subramanian, V.: Plastic-compatible low-resistance printable gold nanoparticle conductors for flexible electronics. J. Electrochem. Soc. **150**(7), G412–417 (2003)
142. Sirringhaus, H., Kawase, T., Friend, R.H., Shimoda, T., Inbasekaran, M., Wu, W., Woo, E.P.: High-resolution inkjet printing of all-polymer transistor circuits. Science **290**, 2123–2126 (2000)
143. Ahmad, Z., Rasekh, M., Edirisinghe, M.: Electrohydrodynamic direct writing of biomedical polymers and composites. Macromol. Mater. Eng. **295**, 315–319 (2010)
144. Samarasinghe, S.R., Pastoriza-Santos, I., Edirisinghe, M.J., Reece, M.J., Liz-Marzan, L.M.: Printing gold nanoparticles with an electrohydrodynamic direct write device. Gold Bull. **39**, 48–53 (2006)
145. Wang, J.Z., Zheng, Z.H., Li, H.W., Huck, W.T.S., Sirringhaus, H.: Dewetting of conducting polymer inkjet droplets on patterned surfaces. Nat. Mater. **3**, 171–176 (2004)
146. Dornfeld, D., Wright, P.: Technology wedges for implementing green manufacturing. Trans. N. Am. Manuf. Res. Inst. **35**, 193–200 (2007)
147. Allwood, J.: What is sustainable manufacturing? Lecture. Cambridge University (2005)
148. Reich-Weiser, C., Vijayaraghavan, A., Dornfeld, D.A.: Metrics for manufacturing sustainability. In: Proceedings of the IMSEC. ASME, Evanston, IL, 7–10 Oct 2008
149. Rosen, M., Dincer, I., Kanoglu, M.: Role of exergy in increasing efficiency and sustainability and reducing environmental impact. Energy Policy **36**, 128–137 (2008)
150. Grigoropoulos, C.P.: Transport in Laser Microfabrication. Cambridge University Press (2009)
151. Wellmann, C., Lillie, C., Wriggers, P.: A contact detection algorithm for superellipsoids based on the common-normal concept. Eng. Comput. **25**, 432–442 (2008)
152. Lagaros, N., Papadrakakis, M., Kokossalakis, G.: Structural optimization using evolutionary algorithms. Comput. Struct. **80**, 571–589 (2002)
153. Leonardi, A., Wittel, F.K., Mendoza, M., Herrmann, H.J.: Coupled DEM-LBM method for the free-surface simulation of heterogeneous suspensions. Comput. Part. Mech. **1**(1), 3–13 (2014)
154. Wellmann, C., Wriggers, P.: A two-scale model of granular materials. Comput. Methods Appl. Mech. Eng. **205–208**, 46–58 (2012)
155. Johnson, K.: Contact Mechanics. Cambridge University Press (1985)
156. Wriggers, P.: Computational Contact Mechanics. Wiley (2002)
157. Wriggers, P.: Nonlinear Finite Element Analysis. Springer (2008)
158. Davis, S.H.: Theory of Solidification. Cambridge University Press (2001)

Chapter 7
DEM Extensions: Electrically Driven Deposition of Polydisperse Particulate Powder Mixtures

A key part of emerging advanced additive manufacturing methods is the deposition of specialized particulate mixtures of materials onto substrates. For example, in many cases these materials are polydisperse powder mixtures whereby one set of particles is chosen with the objective to electrically, thermally, or mechanically functionalize the overall mixture material and another set of finer-scale particles serves as an interstitial filler/binder. Often, achieving controllable, precise deposition is difficult or impossible using mechanical means alone. It is for this reason that electromagnetically driven methods are being pursued in industry, whereby the particles are ionized and an electromagnetic field is used to guide them into place. The goal of this chapter is to further explore this issue, which has been discussed in previous chapters.

7.1 Introduction

As mentioned, achieving precisely controlled deposition of such dry particulate-based materials is difficult or impossible by solely mechanical means. It is for this reason that electrically driven methods are being pursued in industry, whereby the particles are ionized and an electric field is used to guide them into place (Fig. 7.1). The goal of this work is to develop a model and simulation method to investigate the behavior of such depositions as a function of the applied electric field. There is a direct correlation between the ionization strength between particles in the powder and the more fluid-like behavior. Effectively, with no ionization, the system behaves as a loose powder, which is difficult to control. As the ionization is ramped up, the balance between mutual attraction and repulsion leads to surface-tension-like effects. Thus, the expectation is that at high external fields and high ionization, the deposition will yield coherent aggregate "droplets" of the powder-mixture material (Fig. 7.2).

T. I. Zohdi, *Modeling and Simulation of Functionalized Materials for Additive Manufacturing and 3D Printing: Continuous and Discrete Media*, Lecture Notes in Applied and Computational Mechanics 60, https://doi.org/10.1007/978-3-319-70079-3_7

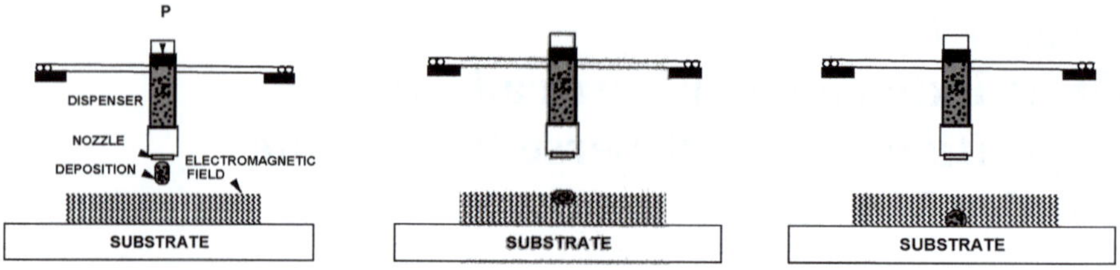

Fig. 7.1 A schematic of a 3D printer (deposition) mechanism

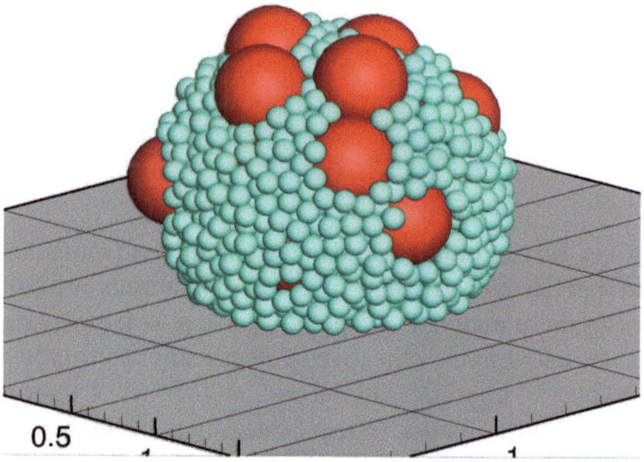

Fig. 7.2 A multiphase droplet in contact with a substrate modeled using the Discrete Element Method

7.2 Algorithm

As in previous sections, within a time-step, the solution STEPS are based on a global fixed-point iteration:

- **(1): Set $i = 1$ (particle counter) and $K = 0$ (iteration counter):**
- **(2): If $i > N_p$, then go to (4)**
- **(3): If $i \leq N_p$, then (a) compute position $\boldsymbol{r}_i^{L+1,K}$ and (b) go to (2) for next particle ($i = i + 1$)**
- **(4): Compute iterative error metrics:**

$$Z_K \stackrel{\text{def}}{=} \frac{\varpi_K}{TOL_r} \text{ and } \Lambda_K \stackrel{\text{def}}{=} \left(\frac{(\frac{TOL}{\varpi_0})^{\frac{1}{pK_d}}}{(\frac{\varpi_K}{\varpi_0})^{\frac{1}{pK}}} \right), \tag{7.1}$$

where $\varpi_K \stackrel{\text{def}}{=} \frac{\sum_{i=1}^{N_p} ||\boldsymbol{r}_i^{L+1,K} - \boldsymbol{r}_i^{L+1,K-1}||}{\sum_{i=1}^{N_p} ||\boldsymbol{r}_i^{L+1,K} - \boldsymbol{r}_i^{L}||}$.

- **(5): If $Z_K \leq 1$ (met tolerance) and $K < K_d$ (below preset number of iterations), then (a) increment time: $t = t + \Delta t$, (b) construct the next time-step: $\Delta t^{new} = \Lambda_K \Delta t^{old}$, (c) select the minimum size: $\Delta t = \text{MIN}(\Delta t^{lim}, \Delta t^{new})$, and (d) update the particle positions and go to (1)**

- **(6): If $Z_K > 1$ (tolerance not met) and $K < K_d$ (still iterating), then (a) update the iteration counter: $K = K + 1$, (b) reset the particle counter: $i = 1$, and (c) go to (2)**
- **(7): If $Z_K > 1$ (tolerance not met) and $K = K_d$ (at the iteration counter limit), then (a) construct a new time-step: $\Delta t^{new} = \Lambda_K \Delta t^{old}$ and (b) restart at time t and go to (1)**

We observe that in STEP (5), Λ_K may enlarge the time-step if the error is lower than the preset tolerance. Thus, because the scheme can also enlarge the time-steps if the iterative system converges quickly within an existing time-step, to ensure the accuracy of the time-stepping scheme, we add an upper bound to control temporal discretization error, i.e., $\Delta t \leq \Delta t^{lim}$.

7.3 Numerical Examples of Involving Polydisperse Depositions

In order to characterize the physical process, we select a model problem, where we consider a group of N_p spherical particles, of two sizes, randomly dispersed, and initially generated within in a cylindrical domain of normalized radius $R = 1$ (diameter $D = 2R = 2$) and length $L = 8$ (Fig. 7.3). The ratio of smaller particle diameter, d_s, to total domain diameter, D, was $d_s/D = 0.05$ for smaller particles and $d_l/D = 0.2$ for the larger particles. A Random Sequential Addition (RSA) algorithm (Widom [1]) was employed to initially to place particles in the domain. Thereafter, the

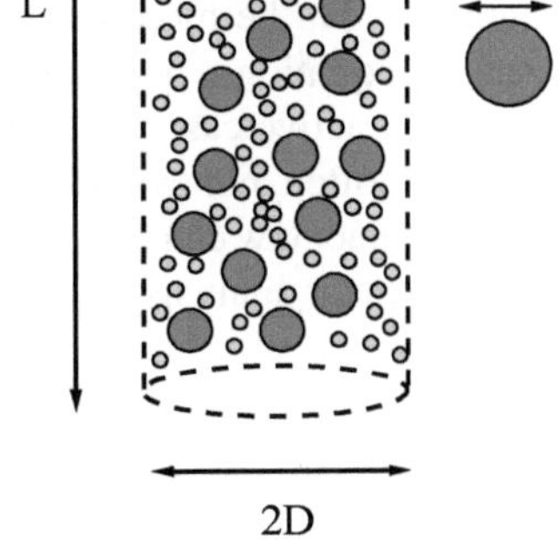

Fig. 7.3 Schematic of the initial configuration for the model problem

dynamics of the particles drove them to a more compacted state.[1] Although we have formulated the system with both electric and magnetic (electromagnetic) fields, for the examples of interest, consistent with most mainstream industrial processes, we will only include the electric field in the simulations. There are of course applications where an applied magnetic field would be of interest; however, this is beyond the scope of the current paper. The following simulation parameters were chosen[2]:

- Magnetic field, $\boldsymbol{B}^{ext} = (0, 0, 0)$ Tesla,
- Electric field, $\boldsymbol{E}^{ext} = (-100, 0, 0)$ N/C,
- Charge per unit particle surface area, $q = 100$ C/kg,
- Density of air, $\rho_g = 1.225\ \mathrm{kg/m^3}$,
- Particles were distributed randomly in a parallelepiped domain: $(8 \times 2 \times 2)$ m,
- Total number of particles, $N = 2000$: 1900 small particles and 100 large particles,
- Radius of small particles, $R_s = 0.05$ m,
- Radius of large particles, $R_l = 0.2$ m,
- Density of materials, $\rho_1 = 2000\ \mathrm{kg/m^3}$ (binder particles), $\rho_2 = 5000\ \mathrm{kg/m^3}$ (functionalizing particles).
- Contact damping parameter, $c^{cd} = 10^5$,
- Friction contact parameter $K^f = 10^7$,
- Coefficient of static friction, $\mu_s = 0.4$,
- Coefficient of dynamic friction, $\mu_d = 0.3$,
- Normal bond parameter, $K^{nb} = 10^6\ \mathrm{N/m^2}$ and the exponent in the binding law was set to $p_b = 2$,
- Rotational/tangential bond parameter, $K^{rb} = 10^3$,
- Near-field parameters, $\bar{\alpha}_1 = 0.5$, $\beta_1 = 1$, $\bar{\alpha}_2 = 0.01$, $\beta_2 = 2$, where the $\bar{\alpha}$ is per unit mass,
- Total simulation event duration, 1.0 s,
- Desired number of fixed-point iterations, $K_d = 10$,
- Trapezoidal-like time-stepping parameter, $\phi = 0.5$,
- Initial time-step size, $\Delta t = 0.0000025$ s,
- Time-step upper bound, $\Delta t^{l-m} = 0.00025$ s, and
- Tolerance for the fixed-point iteration, 10^{-6}.

Presently in this example, we did not consider thermal effects. However, the algorithm can be modified to account for coupled thermal effects by solving heat transfer equations, in addition to the linear and angular momentum balances. In thermally active cases, the parameters such as K_{pij} may be temperature-dependent and may soften. For example, the normal stiffness constant for the ith and jth particles in the contact law can be written as follows (here Θ is the temperature, which is fixed in the present analysis, and Θ^* is a thermal constant):

---

[1]For more details on packing of particles, see Torquato [2], Kansaal et al. [3], and Donev et al. [4–8].

[2]This parameter set is not intended to simulate any specific scenario. The units are SI with properties being the same for the small and large particles, unless explicitly stated otherwise

$$K_{pi} = \mathbf{MAX}(K_{pio}\left(e^{-a_i(\frac{\Theta_i}{\Theta_i^*}-1)}\right), K_{pi}^{lim}), \tag{7.2}$$

and for particle j

$$K_{pj} = \mathbf{MAX}(K_{pjo}\left(e^{-a_j(\frac{\Theta_j}{\Theta_j^*}-1)}\right), K_{pj}^{lim}), \tag{7.3}$$

and the average taken at the interface, providing the parameter needed in the contact law, $K_{pij} = \frac{1}{2}\left(K_{pi} + K_{pj}\right)$. In the current example, $K_{po} = 10^7\,\mathrm{N/m^2}$, where $K_p = \mathbf{MAX}(K_{po}\left(e^{-(a\frac{\Theta}{\Theta^*}-1)}\right), K_p^{lim})$, where $\Theta^* = 500\,^\circ\mathrm{K}$, $K_p^{lim} = 10^6\,\mathrm{N/m^2}$, and exponent in the contact law was set to $p_p = 2$ (the temperature was fixed to be $\Theta = 300\,^\circ\mathrm{K}$, and the thermal sensitivity parameter was set to $a = 1$). There are many possible representations for temperature-dependency. The overall model has a modular structure which allows one to replace models easily. Although a fully coupled thermal model was not considered here, we refer the reader to Zohdi [9–15] for more details in the general area of thermal multiphysics (as well as the previous chapter).

Proceeding, for all particles, we applied an initial velocity of $\boldsymbol{v}(t=0) = (-1, 0, 0)$, projecting them directly toward the substrate, in conjunction with gravity ($\boldsymbol{g} = -(9.81, 0, 0)\,\mathrm{m/s^2}$). The electric field starts below a y–z parallel flat plane at $(2, 0, 0)$. The center of the starting configuration is at $(3.5, 0, 0)$, and the substrate is at $(-2, 0, 0)$. In the upcoming simulations, the smaller blue particles are "binder" particles and the larger red particles are "functionalizing" particles. We considered four different cases:

- **Example #1: no external electric field and no particle-to-particle ionization interaction effects**—In this case, the particle system simply falls due to initial starting velocity. The impact with the substrate is relatively mild, but not controlled (Fig. 7.4).
- **Example #2: external electric field and no particle-to-particle ionization interaction effects**—In this case, the particle system falls due to initial starting velocity and the attraction of the electric field. The impact with the substrate is intense and not controlled. The strong electric field causes more intense impact than before (Fig. 7.5). To qualitatively consider the added effects of the electric field in the impact velocity, consider an isolated charged mass with position vector denoted by $\boldsymbol{r}$, governed by ($\dot{\boldsymbol{r}} = \boldsymbol{v}, \ddot{\boldsymbol{r}} = \dot{\boldsymbol{v}}$)

$$m\dot{\boldsymbol{v}} = q\boldsymbol{E}^{ext}, \tag{7.4}$$

with $\boldsymbol{r}(t=0) = r_o\boldsymbol{e}_1$, $\boldsymbol{v}(t=0) = v_o\boldsymbol{e}_1$, and $\boldsymbol{E}^{ext} = E_1^{ext}\boldsymbol{e}_1$, consequently, for a single particle

$$v_1(t) = v_o + \frac{q}{m}E_1^{ext}t \Rightarrow r_1(t) = r_o + v_o t + \frac{q}{2m}E_1^{ext}t^2. \tag{7.5}$$

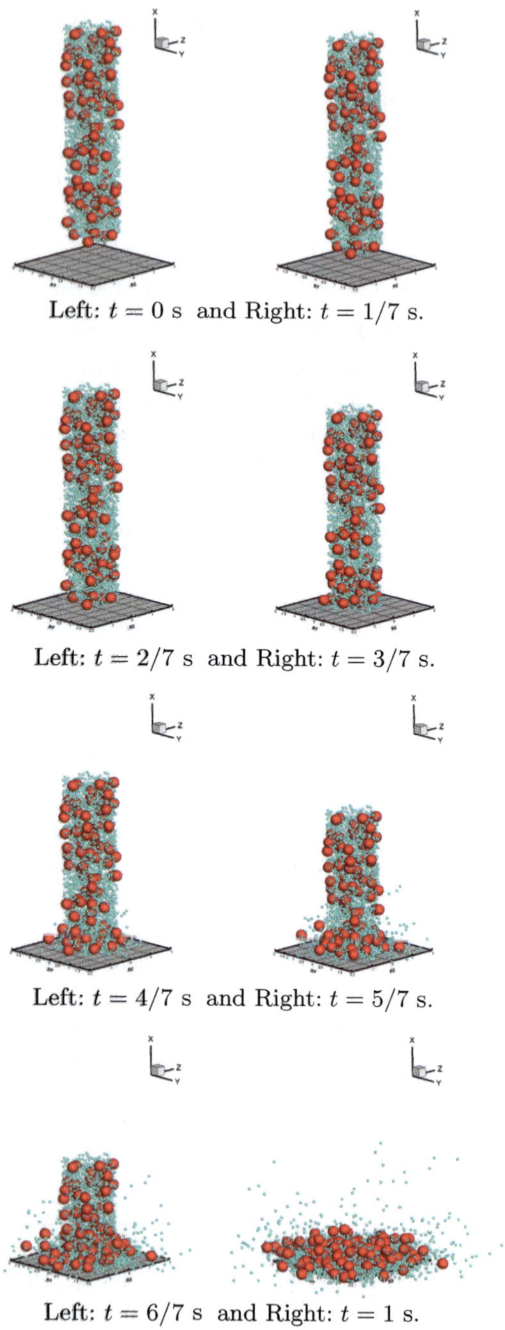

Fig. 7.4 Example #1: no external electric field and no particle-to-particle ionization interaction effects

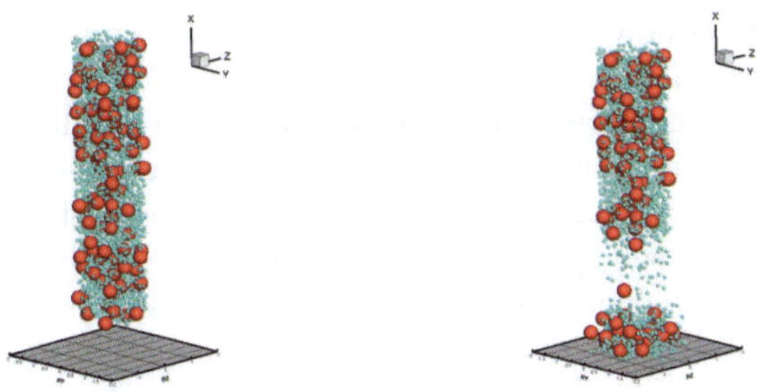

Left: $t = 0$ s and Right: $t = 1/7$ s.

Left: $t = 1/7$ s and Right: $t = 3/7$ s.

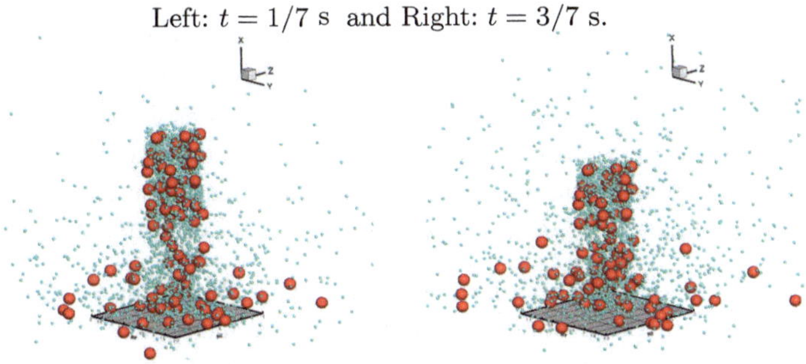

Left: $t = 4/7$ s and Right: $t = 5/7$ s.

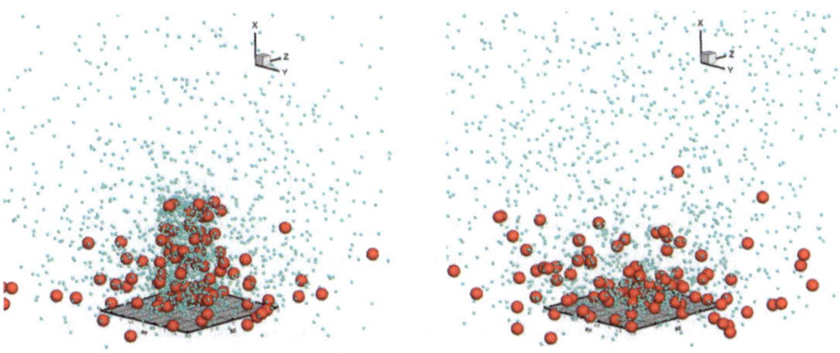

Left: $t = 6/7$ s and Right: $t = 1$ s.

Fig. 7.5 Example #2: external electric field and no particle-to-particle ionization interaction effects

This directly shows how the intensity of the impact is correlated with the mass, charge, and electric field.

- **Example #3: no external electric field and particle-to-particle ionization interaction effects**—In this case, the particle system falls due to initial starting velocity. The mutual ionization forces the stream to collapse upon itself and form three distinct "droplets." The two upper droplets reconnect. The two connected drops then continue to fall and connect with the lower droplet, and the entire system then progresses to make contact with the substrate. The impact on the substrate is mild (Fig. 7.6).
- **Example #4: external electric field and particle-to-particle ionization interaction effects**—In this case, the particle system falls due to initial starting velocity and the electric field. The mutual ionization forces the stream to collapse upon itself and form three distinct "droplets." The two upper droplets reconnect, while the lower one attaches to the substrate. The two connected drops then continue to fall and connect with the droplet on the substrate. The impact on the substrate is more intense, but controlled by the electric field, which pins the material to the substrate (Figs. 7.7 and 7.8). For comparison purposes, we also ran the simulation with only binder particles as well (Fig. 7.9). The breakup is less immediate, primarily because of less Plateau–Rayleigh perturbations (described further momentarily).

Thus, it is observed that there is a direct correlation with the ionization strength between particles in the powder and the more fluid-like behavior, which the electric field can control. With no ionization, the system behaves as a loose powder, which is nearly impossible to control as a deposition. As the ionization is increased, the balance between mutual attraction and repulsion leads to surface-tension-like effects yielding coherent aggregate "droplets" of the powder-mixture material.

Remark 1 In the analysis of standard (nonparticulate) fluids, the breakup of a long column of fluid with perturbations (longitudinal waviness) was first investigated experimentally by Plateau in 1873, who found that a vertically falling stream of water will break up into drops if its wavelength is greater than approximately 3.13–3.18 times its diameter. Subsequently, Rayleigh analytically proved that a wavy falling column of nonviscous liquid (with circular cross section) should break up into drops if its wavelength exceeded its circumference. This type of instability is driven by surface tension, which forces fluids to minimize their surface area. See Papageorgiou [16] and Eggers [17] for more details. In the case of a charged particulate medium, the degree of near-field strength plays the role of surface tension. An uncharged particulate medium will not exhibit this phenomenon.

Remark 2 As we have mentioned previously, because the scheme can also enlarge the time-steps if the iterative system converges quickly within an existing time-step, to ensure the accuracy of the time-stepping scheme, we add an upper bound to control temporal discretization error, i.e., $\Delta t \leq \Delta t^{lim}$. We started all simulations with extremely small time-step sizes and allowed the error estimation and time-step adaptation to auto-correct the proper size. Figure 7.10 shows a typical scenario, taken

Fig. 7.6 Example #3: no external electric field and particle-to-particle ionization interaction effects. We note the presence of a pseudo Plateau–Rayleigh instability that leads to a breakup of the stream into droplets

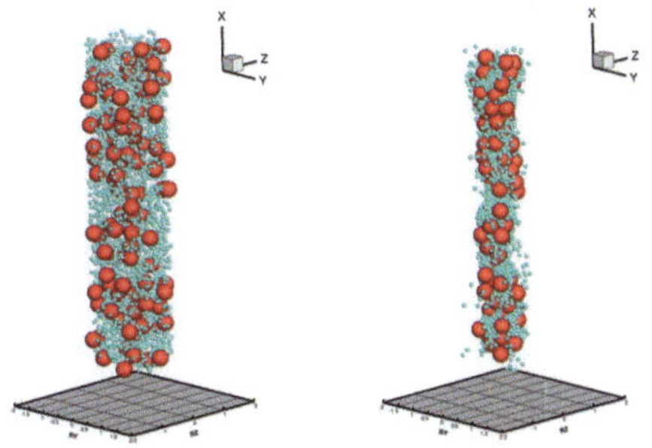

Left: $t = 0$ s and Right: $t = 1/7$ s.

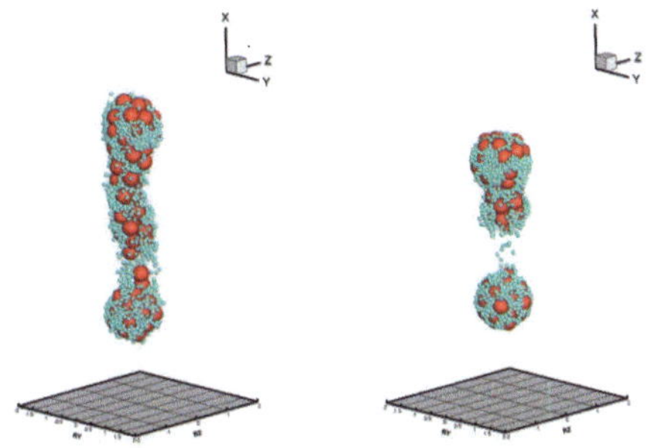

Left: $t = 2/7$s and Right: $t = 3/7$ s.

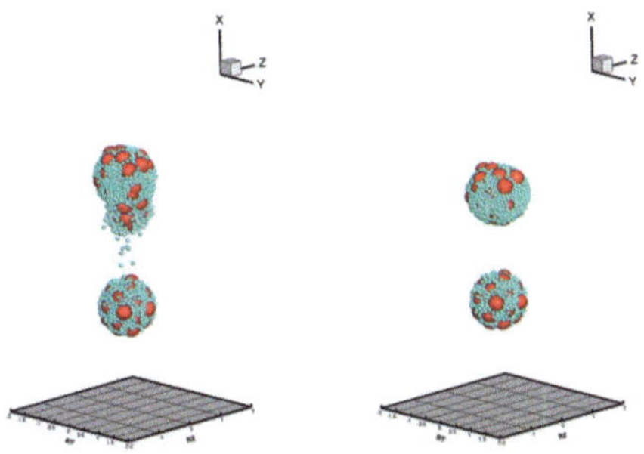

Left: $t = 4/7$ s and Right: $t = 5/7$ s.

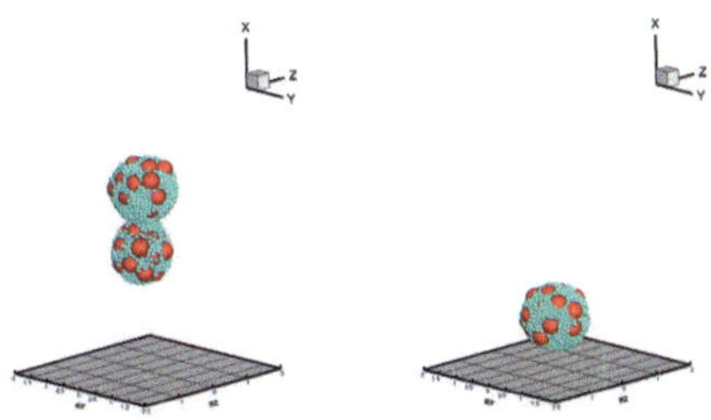

Left: $t = 6/7$ s and Right: $t = 1$ s.

Fig. 7.7 Example #4: external electric field and particle-to-particle ionization interaction

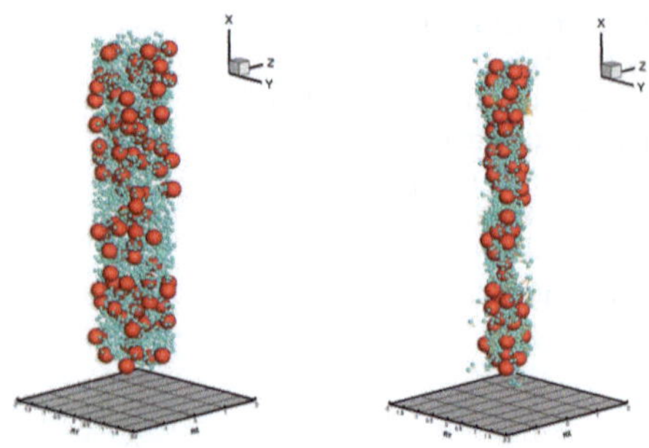

Left: $t = 0$ s and Right: $t = 1/7$ s.

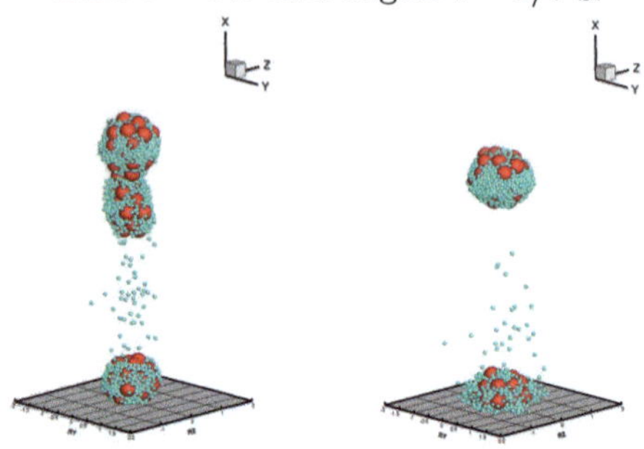

Left: $t = 2/7$ s and Right: $t = 3/7$ s.

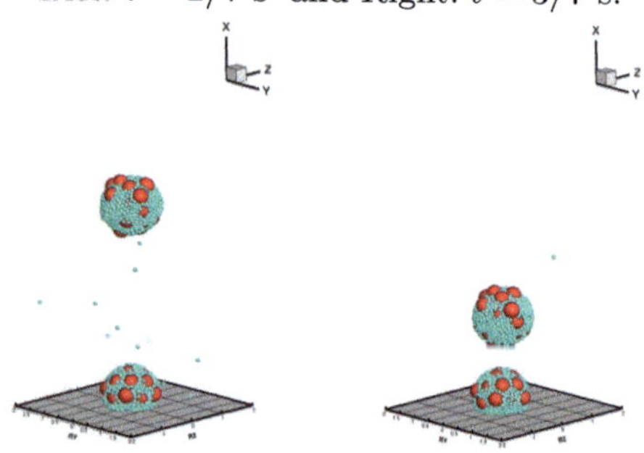

Left: $t = 4/7$ s and Right: $t = 5/7$ s.

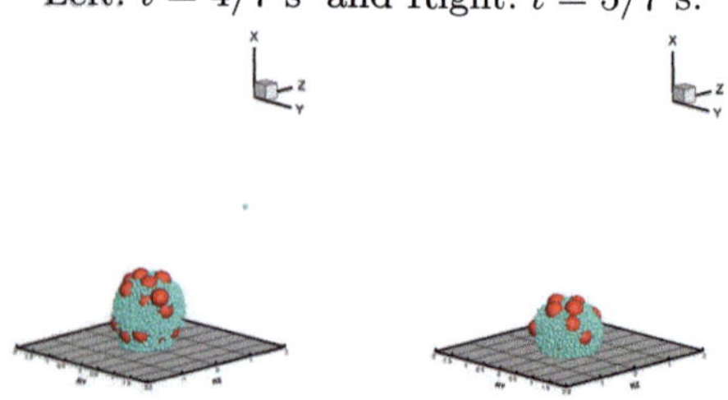

Left: $t = 6/7$ s and Right: $t = 1$ s.

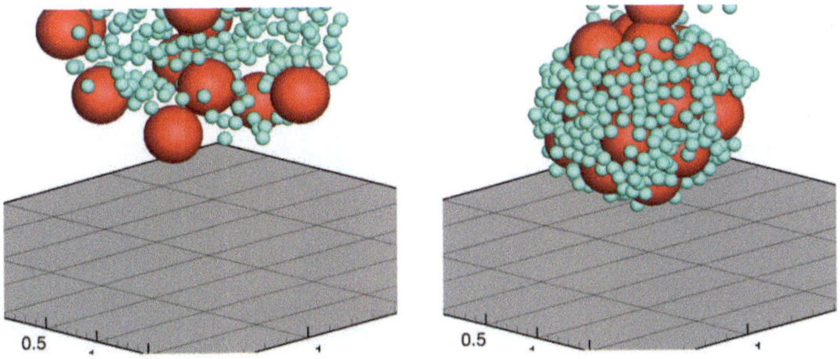

Left: $t = 0$ s and Right: $t = 1/7$ s.

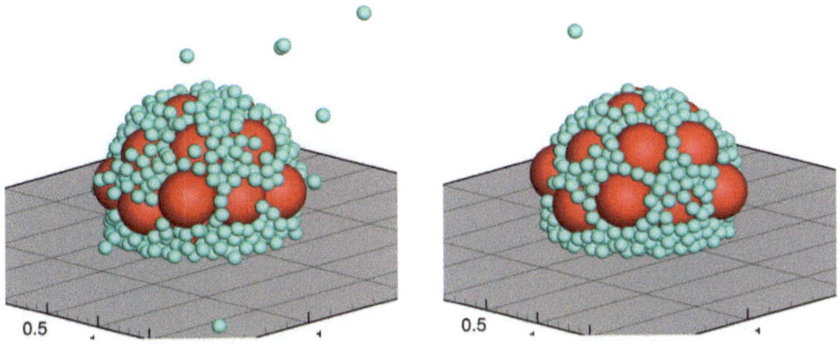

Left: $t = 2/7$ s and Right: $t = 3/7$ s.

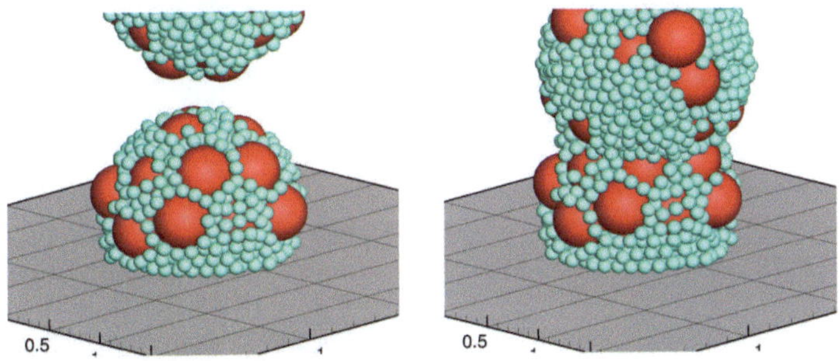

Left: $t = 4/7$ s and Right: $t = 5/7$ s.

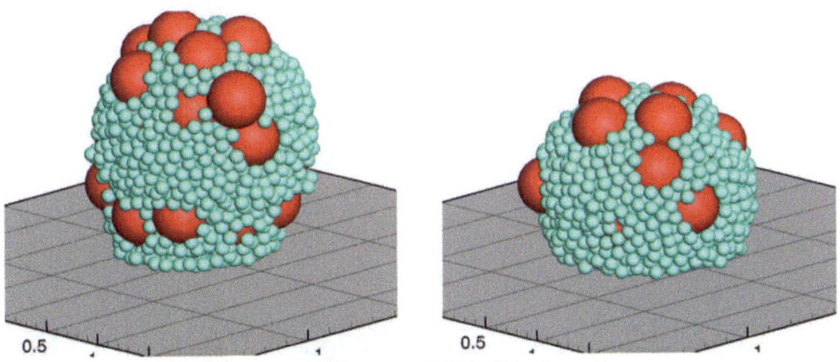

Left: $t = 6/7$ s and Right: $t = 1$ s.

Fig. 7.8 A zoom for example #4: external electric field and particle-to-particle ionization interaction

Fig. 7.9 Baseline "reference" monodisperse example: external electric field and particle-to-particle ionization interaction

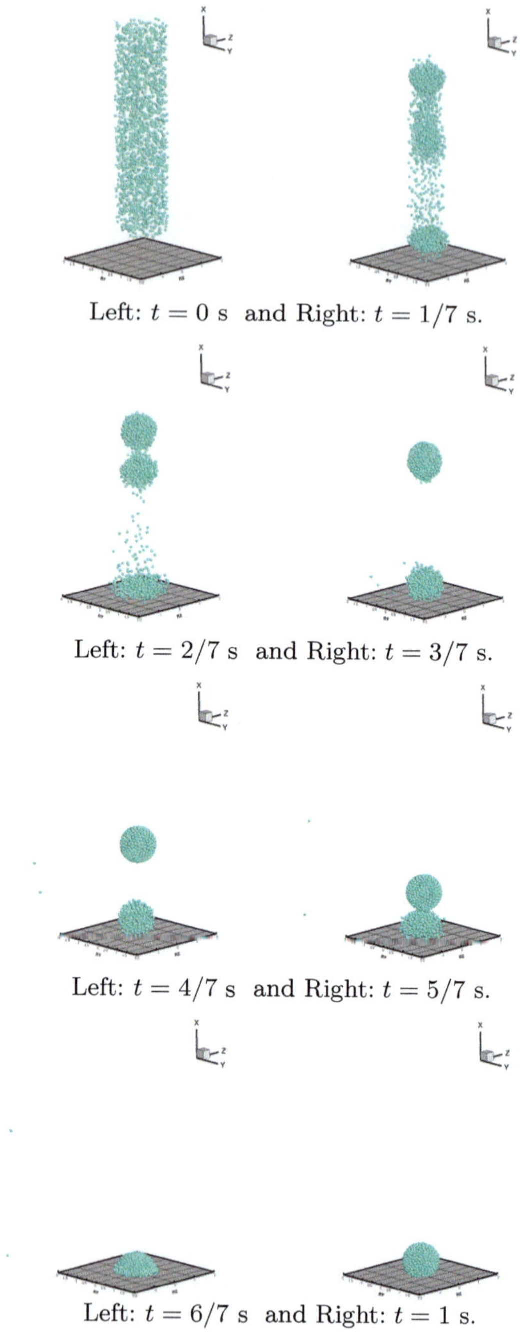

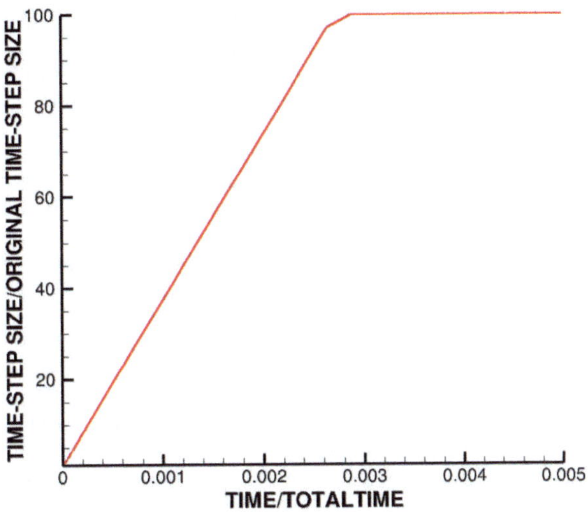

Fig. 7.10 Time-step adaptation for Example #4: external electric field and particle-to-particle ionization interaction

from Example # 4, where the sizes were initially small and adapted until they met the limit set by the algorithm (user specified).

Remark 3 The use of so-called interaction lists is advantageous to speed up calculations and to extend such simulations to very large particle systems. These lists are constructed, for each particle, by taking neighboring particles within a radius of influence. The list is then updated periodically during the simulations. This significantly reduces the computation time used in contact search and other intra-particle calculations, which are N^2 operations. In the simulations that were presented: (a) For each particle, a nearest neighbor list was constructed at the beginning of the simulation (b) For a subinterval of time, the interaction for each particle was restricted to these neighbors, and (c) the lists were updated after that interval expired and the process repeated. See Zohdi [9–15, 18, 19] for details.

References

1. Widom, B.: Random sequential addition of hard spheres to a volume. J. Chem. Phys. **44**, 3888–3894 (1966)
2. Torquato, S.: Random Heterogeneous Materials: Microstructure and Macroscopic Properties. Springer, New York (2002)
3. Kansaal, A., Torquato, S., Stillinger, F.: Diversity of order & densities in jammed hard-particle packings. Phys. Rev. E. **66**, 041109 (2002)
4. Donev, A., Cisse, I., Sachs, D., Variano, E.A., Stillinger, F., Connelly, R., Torquato, S., Chaikin, P.: Improving the density of jammed disordered packings using ellipsoids. Science **13**(303), 990–993 (2004)
5. Donev, A., Stillinger, F.H., Chaikin, P.M., Torquato, S.: Unusually dense crystal ellipsoid packings. Phys. Rev. Lett. **92**, 255506 (2004b)

6. Donev, A., Torquato, S., Stillinger, F.: Neighbor list collision-driven molecular dynamics simulation for nonspherical hard particles-I algorithmic details. J. Comput. Phys. **202**, 737 (2005a)
7. Donev, A., Torquato, S., Stillinger, F.: Neighbor list collision-driven molecular dynamics simulation for nonspherical hard particles-II. Application to ellipses and ellipsoids. J. Comput. Phys. **202**, 765 (2005b)
8. Donev, A., Torquato, S., Stillinger, F.H.: Pair correlation function characteristics of nearly jammed disordered and ordered hard-sphere packings. Phys. Rev. E **71**, 011105 (2005c)
9. Zohdi, T.I.: Rapid simulation of laser processing of discrete particulate materials. Arch. Comput. Methods Eng. **20**, 309–325 (2013)
10. Zohdi, T.I.: A direct particle-based computational framework for electrically-enhanced thermo-mechanical sintering of powdered materials. Math. Mech. Solids **19**(1), 93–113 (2014)
11. Zohdi, T.I.: Additive particle deposition and selective laser processing-a computational manufacturing framework. Comput. Mech. **54**, 171–191 (2014)
12. Zohdi, T.I.: Embedded electromagnetically sensitive particle motion in functionalized fluids. Comput. Part. Mech. **1**, 27–45 (2014)
13. Zohdi, T.I.: Modeling and simulation of cooling-induced residual stresses in heated particulate mixture depositions. Comput. Mech. **56**, 613–630 (2015)
14. Zohdi, T.I.: Modeling and efficient simulation of the deposition of particulate flows onto compliant substrates. Int. J. Eng. Sci. **99**, 74–91 (2015). https://doi.org/10.1016/j.ijengsci.2015.10.012
15. Zohdi, T.I.: Modeling and simulation of laser processing of particulate-functionalized materials. Arch. Comput. Methods Eng., 1–25 (2015). https://doi.org/10.1007/s11831-015-9160-1
16. Papageorgiou, D.T.: On the breakup of viscous liquid threads. Phys. Fluids. **7**(7), 15291521 (1995)
17. Eggers, J.: Nonlinear dynamics and breakup of free-surface flows. Rev. Mod. Phys. **69**(3), 865 (1997)
18. Zohdi, T.I.: Numerical simulation of charged particulate cluster-droplet impact on electrified surfaces. J. Comput. Phys. **233**, 509–526 (2013)
19. Zohdi, T.I.: On inducing compressive residual stress in microscale print-lines for flexible electronics. Int. J. Eng. Sci. **62**, 157–164 (2013)

Chapter 8
DEM Extensions: Electrically Aided Compaction and Sintering

8.1 Introduction

One commonly used approach in processing powdered materials is sintering. Generally, sintering refers to processing a powdered material, comprised of fine-scale particles, by compacting it in a press (Fig. 8.1) and utilizing heat to bind the material. Sintering has distinct advantages over other methods, for example, (1) high purity of processed materials, (2) relatively few STEPS in fabrication (thus retaining the purity), and (3) the production of near net shape of the desired product. Most importantly, it is a method that can be utilized to produce products with complex shapes that cannot be easily made with other methods. Innovative methods for processing compacted microscale powders are rapidly being developed in industry in order to meet the specifications demanded by new products.

The delivery of heat can be achieved in a variety of ways, for example, by heating the walls of the press, with the heat transfer taking place primarily by thermal conduction. In an effort to enhance the heat transfer process, in particular within the interior of the material to be processed, electrically aided heat generation, drawing upon the material's inherent resistance, via Joule-heating, is one method and is considered in this work.[1]

8.1.1 *Objectives*

The objective of this chapter is to develop a direct particle-based model which captures three main physical events:

- *Particle dynamics*, which primarily entails: (a) the movement of the particles induced by contact with the compressing walls and (b) particle-to-particle contact forces,
- *Particle electrical current flow*, which primarily entails: (a) current flow through the particles and (b) current exchange between the particles and walls,

[1] Monograph Appendix 2 gives a detailed analysis of Joule-heating phenomena.

T. I. Zohdi, *Modeling and Simulation of Functionalized Materials for Additive Manufacturing and 3D Printing: Continuous and Discrete Media*, Lecture Notes in Applied and Computational Mechanics 60, https://doi.org/10.1007/978-3-319-70079-3_8

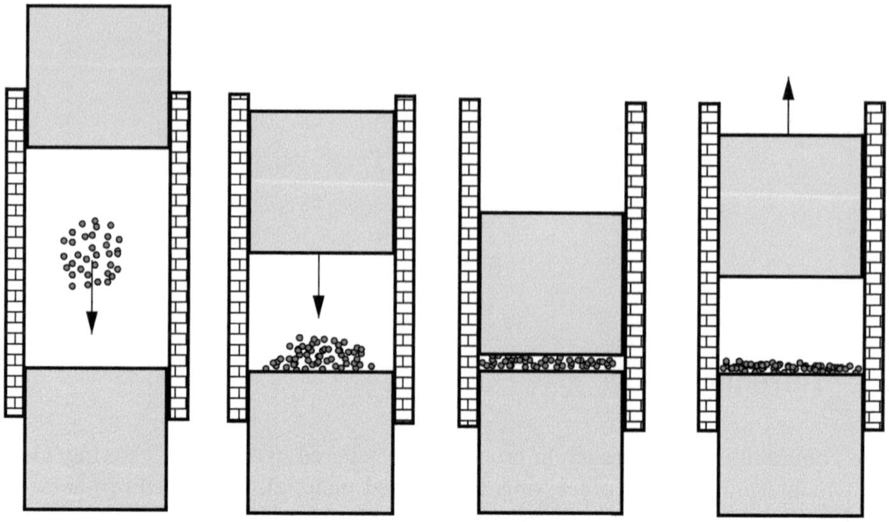

Fig. 8.1 *A model problem*: The sequence of events: **a** particles dropped onto a surface, **b** an electrified head compresses and electrifies the particles, **c** the particles heat up and fuse together, and **d** the electrified head is removed

- *Particle thermodynamics,* which primarily entails: (a) heat generation via Joule-heating, (b) heat transfer between particles in contact by conduction, and (c) thermal softening of the particles.

This is a strongly coupled multiphysical system, since the dynamics controls which particles are in mechanical contact, thus dictating the electrical contacts, which in turn controls the Joule-heating and the induced thermal fields, thus softening and binding the material. As in the previous chapters, the approach taken in the present work is to construct a submodel for each primary physical process. These submodels are coupled to one another. In order to resolve the coupling, a recursive multiphysical staggering scheme is constructed. As before, the general methodology is as follows (at a given time increment): (1) Each field equation is solved individually, "freezing" the other (coupled) fields in the system, allowing only the primary field to be active and (2) after the solution of each field equation, the primary field variable is updated, and the next field equation is treated in a similar manner. As the physics changes, the field that is most sensitive (exhibiting the largest amount of relative nondimensional change) dictates the time-step size. This approach can be classified as an implicit, staggered, time-stepping scheme, in conjunction with an iterative solution method that automatically adapts the time-step sizes to control the rates of convergence within a time-step. If the process does not converge (below an error tolerance) within a preset number of iterations, the time-step is adapted (reduced) by utilizing an estimate of the spectral radius of the coupled system. The modular approach allows for easy replacement of submodels, if needed.

Remark 1 The approach outlined here can be thought of as a *middle ground* between semianalytical approaches which provide qualitative information (Zohdi [1]) and computationally expensive "brute-force" continuum approaches (Zohdi [2]) which attempt to resolve the evolution of fine-scale thermal gradients, transient electromagnetic fields, stress fields, and chemical/damage fields by solving a coupled system of PDEs associated with (1) Maxwell's equations, (2) the first law of thermodynamics, (3) the balance of linear momentum, and (4) reaction–diffusion laws. For the mentioned "brute-force" continuum approach to accurately resolve the coupled time-transient spatial electromagnetic, thermal, mechanical, and chemical fields, Zohdi [2] developed a staggered, temporally adaptive, FDTD (Finite Difference Time Domain) method. This is a computationally intensive approach, owing to the fact that one needs literally millions of numerical unknowns, due to the fine mesh sizes needed. We remark that there are other continuum-based methods, such as (a) *the Multiresolution Time Domain Method*, which is based on wavelet-based discretization, (b) *the Finite Element Method*, which is based on discretization of variational formulations and which is ideal for irregular geometries (see Demkowicz [3] and Demkowicz et al. [4] for the state of the art in adaptive finite element methods for Maxwell's equations), (c) *the Pseudospectral Time Domain Method*, which is based on Fourier and Chebyshev transforms, followed by a lattice or grid discretization of the transformed domain, (d) *the Discrete Dipole Approximation*, which is based on an array of dipoles solved iteratively with the Conjugate Gradient method and a Fast Fourier Transform to multiply matrices, (e) *the Method of Moments*, which is based on integral formulations employing Boundary Element Method discretization, often accompanied by the Fast Multipole Method to accelerate summations needed during the calculations, and (f) *the Partial Element Equivalent Circuit Method*, which is based on integral equations that are interpreted as circuits in discretization cells. However, they are also primarily based on continuum models and are computationally intensive.

Remark 2 There has been considerable research activity in *nonelectrical* compaction of powders, for example, see Akisanya et al. [5], Anand and Gu [6], Brown and Abou-Chedid [7], Domas [8], Fleck [9], Gethin et al. [10], Gu et al. [11], Lewis et al. [12], Ransing et al. [13], Tatzel [14], and Zohdi [1, 2, 15–40].

8.2 Direct Particle Representation

As before, we consider a group of nonintersecting particles ($i = 1, 2, \ldots, N_p$). The equation of motion for the ith particle in system is

$$m_i \ddot{\boldsymbol{r}}_i = \boldsymbol{\Psi}_i^{tot}(\boldsymbol{r}_1, \boldsymbol{r}_2, \ldots, \boldsymbol{r}_{N_p}) = \boldsymbol{\Psi}_i^{con} + \boldsymbol{\Psi}_i^{fric} + \boldsymbol{\Psi}_i^{wall} + \boldsymbol{\Psi}_i^{bond} + \boldsymbol{\Psi}_i^{env}, \quad (8.1)$$

where $\boldsymbol{r}_i$ is the position vector of the ith particle and $\boldsymbol{\Psi}_i^{tot}$ represents all forces acting on particle i, which is decomposed into the sum of forces due to normal contact forces ($\boldsymbol{\Psi}_i^{con}$), sliding frictional forces ($\boldsymbol{\Psi}_i^{fric}$), wall forces ($\boldsymbol{\Psi}_i^{wall}$, having both contact and friction), adhesive bonding forces ($\boldsymbol{\Psi}_i^{bond}$) with other particles, and forces arising from the surrounding interstitial environment ($\boldsymbol{\Psi}_i^{env}$).[2] We utilize the same dynamics model as in previous chapters.

Remark: The primary additional feature in this type of model is the current that flows through the system and the subsequent Joule-heating. The previous effects have been described in the previous chapter.

8.3 Thermal Fields

8.3.1 *Governing Equations*

We assume that radiative, convective, and strain-rate effects are negligible for the particle's thermodynamics. Only Joule-heating and conduction are considered important. Thus, for each particle $i = 1, 2, \ldots, N_p$,

$$m_i C_i \dot{\theta}_i = \mathcal{Q}_i + \mathcal{H}_i, \tag{8.2}$$

where $\mathcal{Q}_i$ represents the conductive contribution from surrounding particles in contact (including walls) and $\mathcal{H}_i$ represents the Joule-heating term. It is assumed that the temperature fields are uniform within the (small) particles. This assumption is justified, i.e., a lumped thermal model, ignoring temperature gradients and assuming a uniform temperature within a particle, when the Biot number is small. The Biot number for spheres scales with the ratio of the particle volume (V) to the particle surface area (A_s) in the following manner (R being the particle radius)

$$\frac{V}{A_s} = \frac{\frac{4}{3}\pi R^3}{4\pi R^2} = \frac{R}{3}, \tag{8.3}$$

which indicates that a uniform temperature distribution is appropriate, since the particles, by definition, are small. Assuming that the fields are uniform in each particle allows for the following (for particle i)

$$\mathcal{Q} = -\int_{\partial\omega} \boldsymbol{Q} \cdot \boldsymbol{n}\, dA \approx \sum_{j=1}^{N_{pc}} I\!K_{ij} \frac{\theta_j - \theta_i}{||\boldsymbol{r}_i - \boldsymbol{r}_j||} A_{ij}^c, \tag{8.4}$$

where the summation extends over all particles $j = 1, 2, 3, \ldots, N_{pc}$ that are in contact with particle i (Fig. 8.2).[3] This yields

---

[2]Such forces can occur from viscous, surrounding, interstitial fluid.

[3]$I\!K_{ij}$ can be approximated by an average interfacial value of the $i - j$ pair, $I\!K_{ij} \approx \frac{I\!K_i + I\!K_j}{2}$. If the materials are the same, this collapses to simply $I\!K$. As for the mechanical contact, A_{ij}^c is the contact area associated with the particle pair (ij).

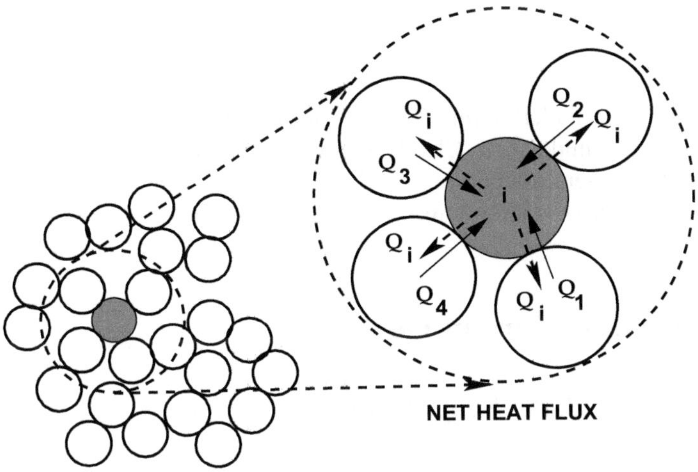

Fig. 8.2 Heat flux exchange between particles

$$m_i C_i \frac{d\theta_i}{dt} = \underbrace{\sum_{j=1}^{N_{pc}} \mathbb{K}_{ij} \frac{\theta_j - \theta_i}{||\boldsymbol{r}_i - \boldsymbol{r}_j||} A^c_{ij}}_{\mathcal{Q}_i} + \mathcal{H}_i, \tag{8.5}$$

where the specific form of Joule-heating is

$$\mathcal{H}_i \approx a_i \frac{J_i^2}{\sigma_i} V_i, \tag{8.6}$$

where J_i is the current magnitude in particle i, σ_i is the conductivity, and $0 \leq a_i \leq 1$ is an absorption constant. More remarks on Joule-heating will be given shortly.

8.3.2 *Numerical Integration*

Integrating the energy equation yields for each particle ($i = 1, 2, \ldots, N_p$)

$$\begin{aligned} \theta_i(t + \Delta t) &= \theta_i(t) + \frac{1}{m_i C_i} \left(\int_t^{t+\Delta t} \mathcal{Q}_i \, dt + \int_t^{t+\Delta t} \mathcal{H}_i \, dt \right) \\ &\approx \theta_i(t) + \frac{\Delta t}{m_i C_i} \left(\phi(\mathcal{Q}_i(t + \Delta t) + \mathcal{H}_i(t + \Delta t)) + (1 - \phi)(\mathcal{Q}_i(t) + \mathcal{H}_i(t)) \right). \end{aligned} \tag{8.7}$$

We note that Eq. 8.7 represents a coupled system of the general form (similar to the equation that arises for the particle dynamics)

$$\theta_i(t+\Delta t) = \mathcal{G}_i(\theta_i(t+\Delta t)) + \mathcal{R}_i, \tag{8.8}$$

where for the "remainder" term, $\mathcal{R}_i \neq \mathcal{R}_i(\theta_i(t+\Delta t))$, and $\mathcal{G}_i$'s behavior is controlled by the magnitude of Δt. Clearly, the temperature is coupled to the mechanical behavior of the system. As before, shortly, we introduce a multiphysical staggering scheme to solve the overall system.

8.4 Modeling of Current Flow

8.4.1 Particle Model Simplification

In order to describe the electrical flow in the system, we consider a conservation of charge in integral form (posed over an arbitrary, general, domain ω)

$$\int_{\partial\omega} \boldsymbol{J} \cdot \boldsymbol{n}\, dA + \int_{\omega} \frac{\partial \mathcal{P}}{\partial t}\, dV = 0, \tag{8.9}$$

where $\boldsymbol{J}$ is the current field and $\mathcal{P}$ is the charge. We assume that the electrical current has evolved to steady state on timescales that are much shorter than the dynamics of the particles. Therefore, $\frac{\partial \mathcal{P}}{\partial t} = 0$ is assumed throughout the remainder of this work; thus,

$$\int_{\partial\omega} \boldsymbol{J} \cdot \boldsymbol{n}\, dA = 0. \tag{8.10}$$

For the particle model, we re-express Eq. 8.10 in terms of electrical contact and current flow (through the particle interfaces), characterized by fluxes in and out of particles, as

$$\sum_{j=1}^{N_{pc}} (\boldsymbol{J}_i + \boldsymbol{J}_j) \cdot \boldsymbol{n}_{ij} A^c_{ij} = 0, \tag{8.11}$$

where A^c_{ij} is the contact area associated with the particle pair (ij), $\boldsymbol{n}_{ij} = \frac{\boldsymbol{r}_j - \boldsymbol{r}_i}{||\boldsymbol{r}_i - \boldsymbol{r}_j||}$, and N_{pc} is the number of particles in contact with particle i. Assuming that the current flows in and out in a radial manner, see Fig. 8.3.

$$(\boldsymbol{J}_i + \boldsymbol{J}_j) = (J_i - J_j)\boldsymbol{n}_{ij}; \tag{8.12}$$

thus,

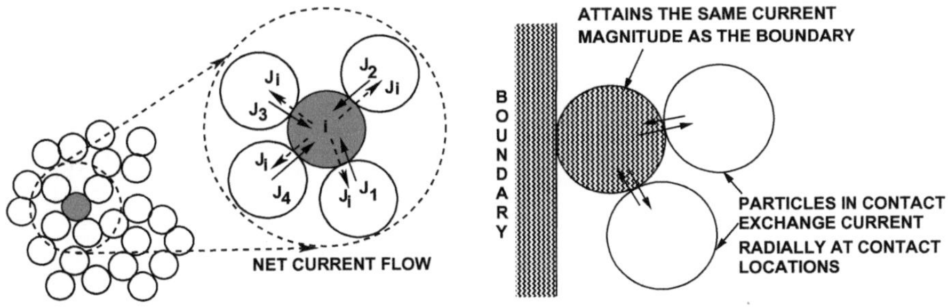

Fig. 8.3 Left: current exchange between particles. Right: The particles in contact with the walls are assumed to become fully electrified and to attain the magnitude of current in the wall as well as the temperature of the wall. From particles that contact the walls, the current flows radially to contact pairs

$$\sum_{j=1}^{N_{pc}}(J_i - J_j)A_{ij}^c = 0 \Rightarrow J_i = \frac{\sum_{j=1}^{N_{pc}} J_j A_{ij}^c}{\sum_{j=1}^{N_{pc}} A_{ij}^c}. \tag{8.13}$$

8.4.2 *Iterative Flux Summation/Solution Process*

The preceding relations lead to an implicit set of equations which are strongly coupled, as well as being coupled to the system dynamics. In order to deal with system, we employ a staggering scheme where, broadly speaking, the solution method sweeps through the system, *particle by particle*, updating the variables as it progresses. Specifically (where $K = 1, 2\ldots$ is an iteration counter), the current is solved iteratively:

$$J_i^{K+1} = \frac{\sum_{j=1}^{N_{pc}} J_j^K A_{ij}^{cK}}{\sum_{j=1}^{N_{pc}} A_{ij}^{cK}}. \tag{8.14}$$

This process is repeated for all particles. The dynamical and thermo-mechanical equations are then re-solved for the motion and temperature of the particles *freezing the electrical current variables*, and the entire procedure (all previous STEPS) is repeated.

Remark 1 Once the current in each particle is solved for, one can compute the Joule-heating via $\mathcal{H}_i \approx a\frac{J_i^2}{\sigma_i}V_i$.

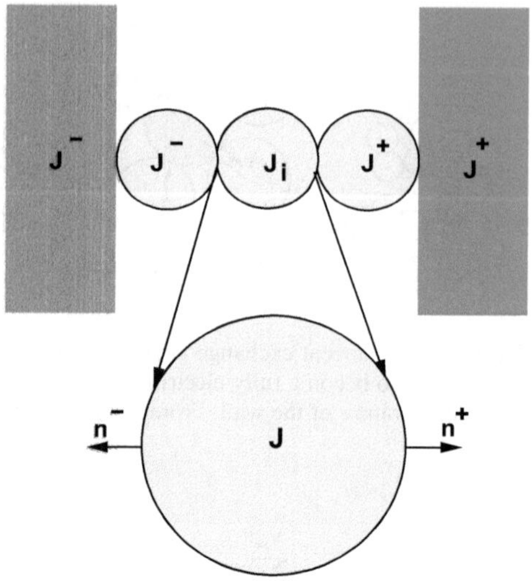

Fig. 8.4 A one-dimensional example of current flow

Remark 2 The particles in contact with the compacting walls/boundaries are assumed to become fully electrified and to attain the magnitude of current in the plate (Figs. 8.4, 8.5, 8.6, and 8.7). From that particle, the current flows radially to other particles in contact at their respective contact points (Fig. 8.8).

Remark 3 For an overview of the continuum modeling of current flow, Joule-heating, and so forth, see Zohdi [2].

Remark 4 As an example, consider the one-dimensional array of particles shown in Fig. 8.4. The two outer particles attain the values at the wall (due to the adopted approach for applying boundary conditions). According to the model, any current in a particle flows radially outward from the contact point into the neighboring particle. Thus, for particle i, we have two contact points

$$\begin{aligned}\sum_{j=1}^{2}(\boldsymbol{J}_i+\boldsymbol{J}_j)\cdot\boldsymbol{n}_{ij}A^c_{ij} &= (\boldsymbol{J}^+\cdot\boldsymbol{n}^+ + \boldsymbol{J}_i\cdot\boldsymbol{n}^+)A^{c,+} + (\boldsymbol{J}^-\cdot\boldsymbol{n}^- + \boldsymbol{J}_i\cdot\boldsymbol{n}^-)A^{c,-}\\ &= (-J^+ + J_i)A^{c,+} + (-J^- + J_i)A^{c,-}\\ &= 0. \end{aligned} \tag{8.15}$$

If the contact areas are the same, this collapses to the average, $J_i = \frac{1}{2}(J^+ + J^-)$.

8.4.3 Overall Solution Algorithm

The algorithm is as follows:

(1) GLOBAL FIXED – POINT ITERATION : (SET i = 1 AND K = 0) :
(2) IF $i > N_p$ THEN GO TO (4)
(3) IF $i \leq N_p$ THEN : (FOR PARTICLE i)
(a) COMPUTE POSITION :$\boldsymbol{r}_i^{L+1,K}$
(b) COMPUTE TEMPERATURE :$\theta_i^{L+1,K}$
(c) GO TO (2) AND NEXT PARTICLE (i = i + 1)
(4) ERROR MEASURES(NORMALIZED) :

$$(a)\varpi_{rK} \stackrel{def}{=} \frac{\sum_{i=1}^{N_p} ||\boldsymbol{r}_i^{L+1,K} - \boldsymbol{r}_i^{L+1,K-1}||}{\sum_{i=1}^{N_p} ||\boldsymbol{r}_i^{L+1,K} - \boldsymbol{r}_i^{L}||}$$

$$\varpi_{\theta K} \stackrel{def}{=} \frac{\sum_{i=1}^{N_p} ||\theta_i^{L+1,K} - \theta_i^{L+1,K-1}||}{\sum_{i=1}^{N_p} ||\theta_i^{L+1,K} - \theta_i^{L}||}$$

$$(b) Z_K \stackrel{def}{=} max(z_{rK}, z_{\theta K}) \quad \text{WHERE} \quad z_{rK} \stackrel{def}{=} \frac{\varpi_{rK}}{TOL_r}, \qquad z_{\theta K} \stackrel{def}{=} \frac{\varpi_{\theta K}}{TOL_\theta}$$

$$(c)\Lambda_K \stackrel{def}{=} min(\Lambda_{rK}, \Lambda_{\theta K}) \quad \text{WHERE}$$

$$\Lambda_{rK} \stackrel{def}{=} \left(\frac{(\frac{TOL_r}{\varpi_{r0}})^{\frac{1}{pK_d}}}{(\frac{\varpi_{rK}}{\varpi_{r0}})^{\frac{1}{pK}}} \right),$$

$$\Lambda_{\theta K} \stackrel{def}{=} \left(\frac{(\frac{TOL_\theta}{\varpi_{\theta 0}})^{\frac{1}{pK_d}}}{(\frac{\varpi_{\theta K}}{\varpi_{\theta 0}})^{\frac{1}{pK}}} \right)$$

(5)<u>IF TOL. NOT MET ($Z_K > 1$) AND $K < K_d$ REPEAT ITERATION (K = K + 1)</u>
(6) <u>IF TOL. MET ($Z_K \leq 1$) AND $K < K_d$ THEN :</u>
(a) INCREMENT TIME : $t = t + \Delta t$
(b) CONSTRUCT NEW TIME – STEP : $\Delta t = \Lambda_K \Delta t$,
(c) SELECT MINIMUM : $\Delta t = MIN(\Delta t^{lim}, \Delta t)$
(d) UPDATE CURRENT FIELD $\boldsymbol{J}_i$(FOR ALL PARTICLES ITERATIVELY)
(e) AND GO TO (1)
(7) <u>IF TOL. NOT MET ($Z_K > 1$) AND $K = K_d$ THEN :</u>
(a) CONSTRUCT NEW TIME – STEP :$\Delta t = \Lambda_K \Delta t$
(b) UPDATE CURRENT FIELD $\boldsymbol{J}_i$(FOR ALL PARTICLES ITERATIVELY)
(c) RESTART AT TIME = t AND GO TO (1)

(8.16)

The overall goal is to deliver solutions, where staggering (incomplete coupling) error is controlled and the temporal discretization accuracy dictates the upper limits on the time-step size (Δt^{lim}).

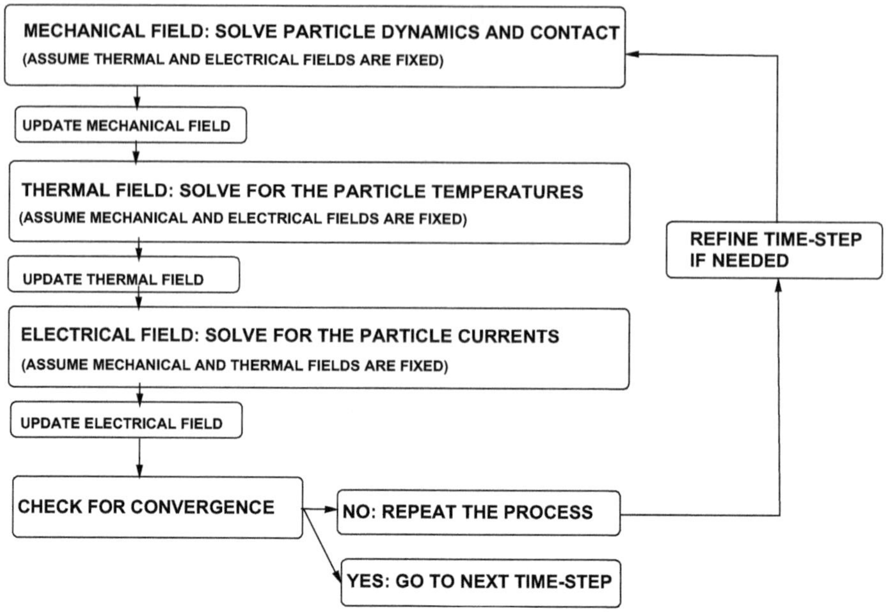

Fig. 8.5 A flowchart for the modular, staggered, computation

8.5 Numerical Examples

We consider a model problem of a group of equally sized particles placed between six walls (Fig. 8.1) to illustrate how to assemble the system. The absolute dimensions are unimportant for the model problem and have been normalized. Specifically, we considered a group of $N_p = 1000$ randomly positioned particles in initially cubical box domain with dimensions $(x = -1, x = 1)\times(y = -1, y = 1)\times(z = -1, z = 1)$ meters (the particles are then poured onto the lower wall). The particle radii were $R = 0.075$ m. In the x-direction (vertical), the top and bottom walls are electrified. The four other walls are neutral. The top is pushed downward at a constant rate (starting at $x = -1$), until it gets to $x = x^*$ at approximately 100% volume fraction, and then it is pulled away.[4] As an example, the relevant simulation parameters chosen were:

- The normal contact parameter was $K_{po} = 10^7\,\mathrm{N/m^2}$, with thermal softening $K_p = \mathbf{MAX}(K_{po}\left(e^{-a\frac{\theta}{\theta*}-1}\right), K_p^{lim})$, where $\theta* = 500\,°\mathrm{K}$, $K_p^{lim} = 10^6\,\mathrm{N/m^2}$, and the exponent in the contact law was set to $p_p = 2$,
- The stiffness of the wall in contact law, $K_{wo} = 10^9\,\mathrm{N/m^2}$, with thermal softening $K_w = \mathbf{MAX}(K_{wo}\left(e^{-a\frac{\theta}{\theta*}-1}\right), K_w^{lim})$, where $\theta* = 500\,°\mathrm{K}$, $K_w^{lim} = 10^8\,\mathrm{N/m^2}$, and the exponent in the contact law was set to $p_w = 2$,

[4]The transverse dimensions of the box were set to be approximately unity, initially. All system parameters can be scaled to describe any specific system of interest.

- The bond stiffness, $K_b = 10^6\,\mathrm{N/m^2}$, and the exponent in the binding law was set to $p_b = 2$,
- The damping coefficient, $c^{env} = c_o^{env} 6\pi R$, $c_o^{env} = 10$ (assumed Stokesian-like),
- The density of the particles, $\rho = 2000\,\mathrm{kg/m^3}$,
- The Joule-heating absorption coefficient, $a = 0.5$,
- The electrical conductivity, $\sigma = 0.1$,
- Coefficient of dynamic friction, $\mu_d = 0.1$,
- Conductivity (same for all particles in this example), $I\!K = 100\,\mathrm{Watts/m - Kelvin}$,
- Initial material temperature, $\theta_i(t = 0) = 300^\circ$ K,
- Wall temperatures, $\theta_w(t = 0) = 1000^\circ$ K,
- Heat capacity, $C = 100\,\mathrm{J/kg - Kelvin}$,
- Target number of fixed-point iterations, $K_d = 10$,
- The trapezoidal time-stepping parameter, $\phi = 0.5$,
- On the bottom and top wall (in the x-direction), $J^{ext} = 10^5$ amps,
- Simulation duration, 1 s,
- Initial time-step size, 0.005 s,
- Time-step upper bound, 0.005 s and
- Tolerance for the fixed-point iteration, 5×10^{-4}.

8.5.1 STEP 1: Pouring the Particles

In order to generate an initial particle configuration, we dropped a sample of nonoverlapping random particles onto the lower surface and allowed it to spread under the force of gravity, where it was constrained by the lateral walls. The volume fraction of monodisperse random particles should be in range of 50–60% at this stage, and the simulation reflects that this is achieved before compaction, under the force of gravity (Fig. 8.6).

Remark 1 The volume fraction was calculated based on the volume contained within the lowest compression point in Fig. 8.1.

Remark 2 The configuration of the sample, before it was dropped, was generated using a classical Random Sequential Addition (RSA) algorithm (Widom [41]), which places nonoverlapping particles randomly into the domain of interest. As mentioned previously, one could start with a denser starting configuration by using the equilibrium-driven Metropolis algorithm or alternative methods based on simultaneous particle flow and growth (see Kansaal et al. [42], Donev et al. [43–45], and Torquato [46]), although this was not necessary for this example.

8.5.2 STEP 2: Compacting the Particles

In Fig. 8.6, the temperature values are depicted by the colors. Initially, the particles are not touching the electrified walls in the vertical (x-direction), nor the lateral, unelectrified walls. As the electrified wall moves in the (minus) x-direction, the particles come into contact, become electrified (experience a current), and move

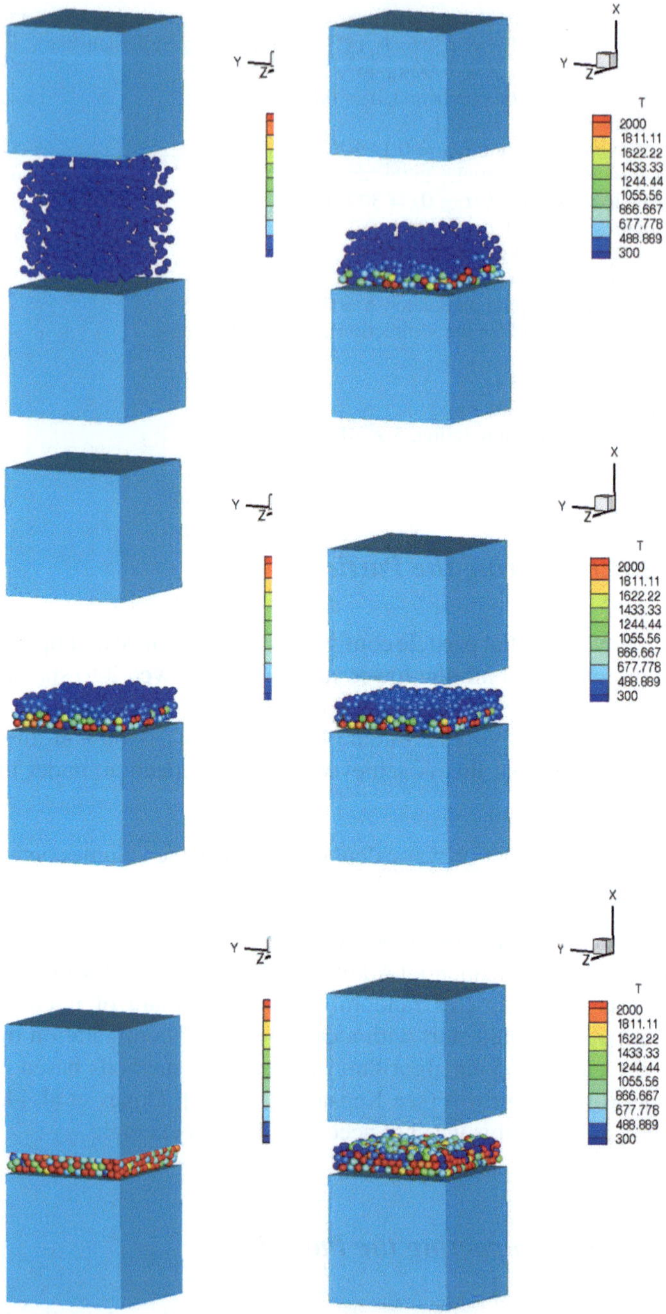

Fig. 8.6 A series of frames for compaction using the model (side walls not shown). Left to right and top to bottom: (1) Pouring of the particles, (2) Contact with the lower electrified wall, (3) Contact with the upper and lower electrified walls, and (4) Release of the upper electrified wall (loss of upper surface contact)

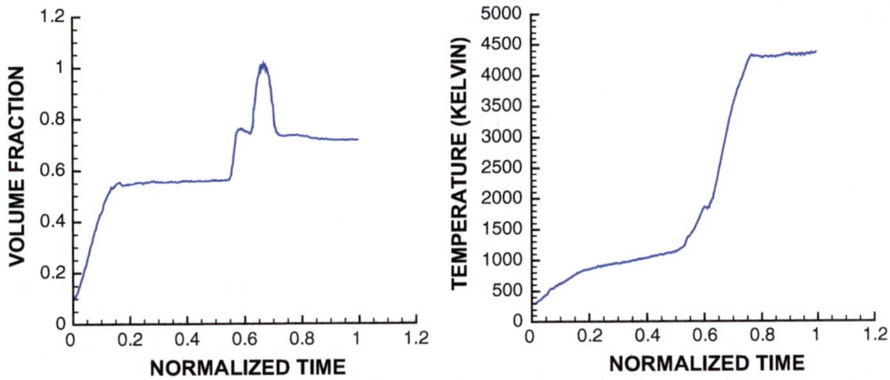

Fig. 8.7 *Densification and temperature:* The process can be described as having three phases: (1) Phase 1: pouring to roughly 50–60% volume fraction. The temperature rises due to contact with the lower electrified wall. (2) Phase 2: The compacting wall compresses the material to virtually 100% volume fraction. The temperature rises due to contact with the upper and lower electrified walls. (3) Phase 3: The compacting wall is released, and the material springs back somewhat, resulting in a lower volume fraction than the fully compressed state, but higher than the purely poured state. The temperature starts to drop due to conduction with the surrounding walls. The basic trends are: (1) The longer time the fully compressed state, the better the Joule-heating induced bonding and less springback. (2) The higher densification (higher volume fraction) in the fully compressed state, the better the Joule-heating induced bonding and less springback. The volume fraction was calculated based on the volume contained within the lowest compression point in Fig. 8.6

in the (minus) x-direction, as well as experience lateral movement (in the $y - z$) direction. The process can be described as having three phases:

- (1) Phase 1: pouring to roughly 50–60% volume fraction. The temperature rises due to contact with the lower electrified wall.
- (2) Phase 2: The compacting wall compresses the material to virtually 100% volume fraction. The temperature rises due to contact with the upper and lower electrified walls (Fig. 8.7).
- (3) Phase 3: The compacting wall is released, and the material springs back somewhat, resulting in a lower volume fraction than the fully compressed state, but higher than the purely poured state. For example, in the specific example chosen, the compaction process decreased the porosity by approximately 15% (Fig. 8.7).

The basic trends are:

- (1) The longer time the fully compressed state, the better the Joule-heating induced bonding and less springback and
- (2) The higher densification (higher volume fraction) in the fully compressed state, the better the Joule-heating induced bonding and less springback.

8.6 Extensions and Conclusions

This chapter developed a modular framework for the electrically enhanced sintering of powdered materials using a direct particle representation in conjunction with a staggering scheme to couple submodels of each type of physics together. The physics of this system are strongly coupled since the dynamics controls which particles are in contact, which also dictates the electrical contacts, which in turn controls the Joule-heating and the induced thermal fields, which softens and binds the material. The strongly multiphysics-coupled submodels were solved iteratively within each time-step using a staggering scheme which employs temporal adaptivity to control the error. The submodular approach allows for easy replacement of models, if so desired. Numerical examples were provided. There are a number of enhancements that can be made. For example, the properties of most electrically active materials are quite sensitive to the temperature; for the conductivity, one can use the following decomposition, employing thermoelectric saturation conditions (using sigmoid functions):

$$\sigma(\theta, \boldsymbol{E}) = \sigma(\theta_R, \boldsymbol{E}_R)\left(1 + \mathcal{K}_1(1 + e^{-\gamma_1(\theta-\theta_R)})^{-1} + \mathcal{K}_2(1 + e^{-\gamma_2||\boldsymbol{E}-\boldsymbol{E}_R||})^{-1}\right), \tag{8.17}$$

where the γs and $\mathcal{K}$s are material parameters, θ_R is a reference temperature, $\boldsymbol{E}$ is the electric field, and $\boldsymbol{E}_R$ is the reference electrical state. Generally speaking, for many materials, until a saturation threshold is met, $\sigma(\theta)$ decreases with θ. See the treatise of Jackson [47] for reviews of the rich variety of possible dielectrical responses of materials, including atomistic-level discussions to motivate nonlinear dielectric behavior.

In closing, it is important to note that a significant acceleration in the computation can be achieved via sorting and binning methods, which proceed by partitioning the whole domain into bins (Fig. 8.8). The particles are sorted by the bins in which they reside. The particle interaction proceeds, bin by bin, where the particles within a bin potentially only interact with particles in other nearest neighbor bins. Essentially, for a given particle in a bin, contact searches are conducted with particles in the

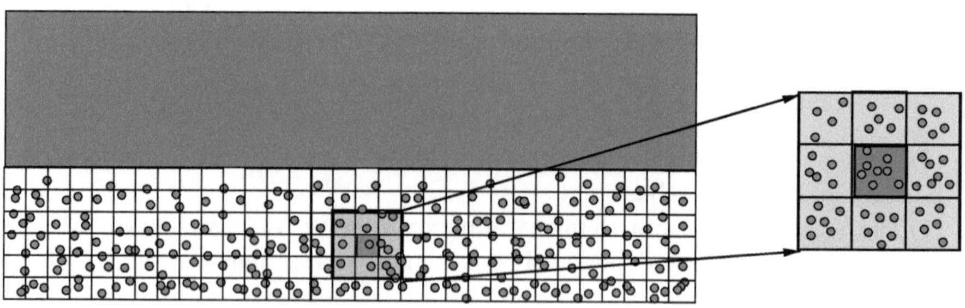

Fig. 8.8 An example of binning of the particles in the sample domain

neighboring bins only (Fig. 8.8). The approach is relatively straightforward to implement and can speed up the computation dramatically (see Zohdi [1, 2, 15–40]). There are a variety of related techniques to further accelerate computations. For example, one can assume that particles stay in the bins for a few time-steps and that one does not need to re-sort immediately. One can construct so-called interaction or Verlet lists of neighboring particles which a particle interacts with. One then updates the interaction list every few time-steps (see Pöschel and Schwager [48]). One can also employ domain decomposition techniques, whereby the domain is partitioned into subdomains and the particles within each subdomain are sent to a processor then stepped forward in time, but with the positions of the particles outside of the subdomain fixed (relative to the particles in that subdomain). This is done for all of the subdomains separately, then the positions of all of the particles are updated, and this information is shared between processors, with the process being repeated as needed.

8.7 Chapter Appendix 1: Joule-Heating

8.7.1 *Characterizing Electrical Losses*

The flow of current through materials usually leads to the phenomena of Joule-heating. To understand the phenomena, we consider a general electromagnetic system characterized by Faraday's law

$$\nabla \times \boldsymbol{E} = -\frac{\partial \boldsymbol{B}}{\partial t} \tag{8.18}$$

and Ampere's law

$$\nabla \times \boldsymbol{H} = \frac{\partial \boldsymbol{D}}{\partial t} + \boldsymbol{J} \tag{8.19}$$

where we recall that $\boldsymbol{E}$ is the electric field, $\boldsymbol{D}$ is the electric field flux, $\boldsymbol{J}$ is the electric current, $\boldsymbol{H}$ is the magnetic field, and $\boldsymbol{B}$ is the magnetic field flux. Joule-heating can be motivated by forming the inner product of the magnetic field with Faraday's law and the inner product of the electric field with Ampere's law and forming the difference to yield

$$\underbrace{\boldsymbol{E} \cdot (\nabla \times \boldsymbol{H}) - \boldsymbol{H} \cdot (\nabla \times \boldsymbol{E})}_{-\nabla \cdot (\boldsymbol{E} \times \boldsymbol{H}) = -\nabla \cdot \boldsymbol{S}} = \boldsymbol{E} \cdot \boldsymbol{J} + \underbrace{\boldsymbol{E} \cdot \frac{\partial \boldsymbol{D}}{\partial t} + \boldsymbol{H} \cdot \frac{\partial \boldsymbol{B}}{\partial t}}_{= \frac{\partial \mathcal{W}}{\partial t}}, \tag{8.20}$$

where $\mathcal{W} = \frac{1}{2}(\boldsymbol{E} \cdot \boldsymbol{D} + \boldsymbol{H} \cdot \boldsymbol{B}) = \frac{1}{2}(\boldsymbol{E} \cdot \boldsymbol{\epsilon} \cdot \boldsymbol{E} + \boldsymbol{H} \cdot \boldsymbol{\mu} \cdot \boldsymbol{H})$ is the electromagnetic energy and $\boldsymbol{S} = \boldsymbol{E} \times \boldsymbol{H}$ is the Poynting vector. Thus,

$$\frac{\partial \mathcal{W}}{\partial t} + \nabla \cdot \boldsymbol{S} = -\boldsymbol{J} \cdot \boldsymbol{E} \tag{8.21}$$

Equation 8.21 is usually referred to as Poynting's theorem and can be interpreted, *for simple material laws, where the previous representation for $\mathcal{W}$ holds*, as stating that the rate of change of electromagnetic energy within a volume, plus the energy flowing out through a boundary, is equal to the negative of the total work done by the fields on the sources and electrical conduction.

8.7.2 Joule-Heating

The interconversions of various forms of energy (electromagnetic, thermal, etc.) in a system are governed by the first law of thermodynamics, for example, case in the current configuration

$$\rho \dot{w} - \boldsymbol{T} : \nabla \boldsymbol{v} + \nabla \cdot \boldsymbol{q} - Z = 0, \tag{8.22}$$

where ρ is the mass density in current configuration, w is the stored energy per unit mass, $\boldsymbol{T}$ is Cauchy stress, $\boldsymbol{v}$ is the material velocity, and $\boldsymbol{q}$ is heat flux. We consider the absorbed energy that is available for heating to be proportional to the energy associated with electrical conduction, namely, from Eq. 8.21, $\boldsymbol{J} \cdot \boldsymbol{E}$, and account for it via $Z \stackrel{def}{=} \rho z = a\boldsymbol{J} \cdot \boldsymbol{E}$, where a is an absorption constant, $0 \le a \le 1$. *The systems considered in the body of the chapter are special cases of this general continuum formulation, which was treated in Zohdi* [2].

8.8 Chapter Appendix 2: Time-Scaling Arguments for $\frac{\partial \mathcal{P}}{\partial t} \approx 0$

Consider the following

$$\int_{\partial\Omega} \boldsymbol{J} \cdot \boldsymbol{n}\, dA + \int_{\Omega} \frac{\partial \mathcal{P}}{\partial t}\, dV = 0 \Rightarrow \nabla \cdot \boldsymbol{J} = 0, \tag{8.23}$$

where $\frac{\partial \mathcal{P}}{\partial t} = 0$ is assumed the current propagates through the particles at a much faster timescale than the deformation of the system. The velocity of the deformation of the system is far slower than the relative movement of charge (propagation of electricity) through the system (which is considered instantaneous). Changes in $\mathcal{P}$ are determined by the Gauss' law:

$$\int_{\partial\Omega} \boldsymbol{D} \cdot \boldsymbol{n}\, dA = \int_{\Omega} \mathcal{P}\, dV, \tag{8.24}$$

where $\boldsymbol{D}$ is the electric field flux. As an illustrative example, in order to appreciate the fast timescales that justify $\frac{\partial \mathcal{P}}{\partial t} \approx 0$, consider an arbitrary piece of continuum (undergoing no deformation) governed by

$$\nabla \cdot \boldsymbol{J} = \sigma \nabla \cdot \boldsymbol{E} = \frac{\sigma}{\epsilon} \nabla \cdot \boldsymbol{D} = \frac{\sigma}{\epsilon} \mathcal{P} = -\frac{\partial \mathcal{P}}{\partial t}, \tag{8.25}$$

where the following simple constitutive laws were used for illustration purposes: $\boldsymbol{J} = \sigma \boldsymbol{E}$, $\boldsymbol{D} = \epsilon \boldsymbol{E}$, and $\nabla \cdot \boldsymbol{D} = \mathcal{P}$. Solving for $\mathcal{P}$ yields

$$\mathcal{P}(\boldsymbol{x}, t) = \mathcal{P}(\boldsymbol{x}, t = 0) e^{-\frac{\sigma}{\epsilon} t}; \tag{8.26}$$

thus,

$$\frac{\partial \mathcal{P}}{\partial t} = -\frac{\sigma}{\epsilon} e^{-\frac{\sigma}{\epsilon} t}. \tag{8.27}$$

The term $e^{-\frac{\sigma}{\epsilon} t}$ is extremely small since the ratio $\frac{\sigma}{\epsilon}$ is huge for the materials of interest and $-\frac{\sigma}{\epsilon} e^{-\frac{\sigma}{\epsilon} t} \approx 0$ for virtually any timescales of interest, thus justifying $\int_{\partial \Omega} \boldsymbol{J} \cdot \boldsymbol{n} \, dA = 0$. In summary, any changes in $\mathcal{P}$ can be considered instantaneous, relative to mechanically induced deformations.

References

1. Zohdi, T.I.: Estimation of electrical-heating load-shares for sintering of powder mixtures. Proc. R. Soc. **468**, 2174–2190 (2012)
2. Zohdi, T.I.: Simulation of coupled microscale multiphysical-fields in particulate-doped dielectrics with staggered adaptive FDTD. Comput. Methods Appl. Mech. Eng. **199**, 79–101 (2010)
3. Demkowicz, L.: Computing with hp-Adaptive Finite Elements. I. One- and Two-dimensional Elliptic and Maxwell Problems. Chapman & Hall/CRC (2006)
4. Demkowicz, L., Kurtz, J., Pardo, D., Paszynski, M., Rachowicz, W. and Zdunek, A.: Computing with Hp-Adaptive Finite Elements, Vol. 2: Frontiers: Three Dimensional Elliptic and Maxwell Problems with Applications. CRC Press, Taylor and Francis (2007)
5. Akisanya, A.R., Cocks, A.C.F., Fleck, N.A.: The yield behavior of metal powders. Int. J. Mech. Sci. **39**, 1315–1324 (1997)
6. Anand, L., Gu, C.: Granular materials: constitutive equations and shear localization. J. Mech. Phys. Solids **48**, 1701–1733 (2000)
7. Brown, S., Abou-Chedid, G.: Yield behavior of metal powder assemblages. J. Mech. Phys. Solids **42**, 383–398 (1994)
8. Domas, F.: Eigenschaft profile und Anwendungsübersicht von EPE und EPP. Technical report of the BASF Company
9. Fleck, N.A.: On the cold compaction of powders. J. Mech. Phys. Solids **43**, 1409–1431 (1995)
10. Gethin, D.T., Lewis, R.W., Ransing, R.S.: A discrete deformable element approach for the compaction of powder systems. Modell. Simul. Mater. Sci. Eng. **11**(1), 101–114 (2003)
11. Gu, C., Kim, M., Anand, L.: Constitutive equations for metal powders: application to powder forming processes. Int. J. Plast. **17**, 147–209 (2001)

12. Lewis, R.W., Gethin, D.T., Yang, X.S.S., Rowe, R.C.: A combined finite-discrete element method for simulating pharmaceutical powder tableting. Int. J. Numer. Methods Eng. **62**, 853–869 (2005)
13. Ransing, R.S., Lewis, R.W., Gethin, D.T.: Using a deformable discrete-element technique to model the compaction behaviour of mixed ductile and brittle particulate systems. Philos. Trans. R. Soc.—Ser. A Math. Phys. Eng. Sci. **362**(1822), pp. 1867–1884 (2004)
14. Tatzel, H.: Grundlagen der Verarbeitungstechnik von EPP-Bewährte und neue Verfahren. Technical report of the BASF Company (1996)
15. Zohdi, T.I.: Genetic design of solids possessing a random-particulate microstructure. Philos. Trans. R. Soc. Math. Phys. Eng. Sci. **361**(1806), 1021–1043 (2003)
16. Zohdi, T.I.: On the compaction of cohesive hyperelastic granules at finite strains. Proc. R. Soc. **454**(2034), 1395–1401 (2003)
17. Zohdi, T.I.: Computational design of swarms. Int. J. Numer. Methods Eng. **57**, 2205–2219 (2003)
18. Zohdi, T.I.: Constrained inverse formulations in random material design. Comput. Methods Appl. Mech. Eng. 1–20. **192**, 28–30, 18, 3179–3194 (2003)
19. Zohdi, T.I.: Staggering error control for a class of inelastic processes in random microheterogeneous solids. Int. J. Nonlinear Mech. **39**, 281–297 (2004)
20. Zohdi, T.I.: Modeling and simulation of a class of coupled thermo-chemo-mechanical processes in multiphase solids. Comput. Methods Appl. Mech. Eng. **193**(6–8), 679–699 (2004)
21. Zohdi, T.I.: Modeling and direct simulation of near-field granular flows. Int. J. Solids Struct. **42**(2), 539–564 (2004)
22. Zohdi, T.I.: A computational framework for agglomeration in thermo-chemically reacting granular flows. Proc. R. Soc. **460**(2052), 3421–3445 (2004)
23. Zohdi, T.I.: Statistical ensemble error bounds for homogenized microheterogeneous solids. J. Appl. Math. Phys. (Zeitschrift für Angewandte Mathematik und Physik) **56**(3), 497–515 (2005)
24. Zohdi, T.I.: Charge-induced clustering in multifield particulate flow. Int. J. Numer. Methods Eng. **62**(7), 870–898 (2005)
25. Zohdi, T.I.: A simple model for shear stress mediated lumen reduction in blood vessels. Biomech. Model. Mechanobiol. **4**(1), 57–61 (2005)
26. Zohdi, T.I.: Computation of strongly coupled multifield interaction in particle-fluid systems. Comput. Methods Appl. Mech. Eng. **196**, 3927–3950 (2007)
27. Zohdi, T.I.: On the computation of the coupled thermo-electromagnetic response of continua with particulate microstructure. Int. J. Numer. Methods Eng. **76**, 1250–1279 (2008)
28. Zohdi, T.I.: Mechanistic modeling of swarms. Comput. Methods Appl. Mech. Eng. **198**(21–26), 2039–2051 (2009)
29. Zohdi, T.I.: On the dynamics of charged electromagnetic particulate jets. Arch. Comput. Methods Eng. **17**(2), 109–135 (2010)
30. Zohdi, T.I.: Dynamics of clusters of charged particulates in electromagnetic fields. Int. J. Numer. Methods Eng. **85**, 1140–1159 (2011)
31. Zohdi, T.I.: Joule-heating field phase-amplification in particulate-doped dielectrics. Int. J. Eng. Sci. **49**, 30–40 (2011)
32. Zohdi, T.I.: Numerical simulation of charged particulate cluster-droplet impact on electrified surfaces. J. Comput. Phys. **233**, 509–526 (2013)
33. Zohdi, T.I.: Rapid simulation of laser processing of discrete particulate materials. Arch. Comput. Methods Eng. **20**, 309–325 (2013)
34. Zohdi, T.I.: A direct particle-based computational framework for electrically-enhanced thermo-mechanical sintering of powdered materials. Math. Mech. Solids 19(1), 93–113 (2014)
35. Zohdi, T.I.: Additive particle deposition and selective laser processing-a computational manufacturing framework. Comput. Mech. **54**, 171–191 (2014)
36. Zohdi, T.I.: Embedded electromagnetically sensitive particle motion in functionalized fluids. Comput. Part. Mech. **1**, 27–45 (2014)
37. Zohdi, T.I.: Rapid computation of statistically-stable particle/feature ratios for consistent substrate stresses in printed flexible electronics. J. Manuf. Sci. Eng. ASME. MANU-14-1476 (2015). https://doi.org/10.1115/1.4029327

38. Zohdi, T.I.: Modeling and simulation of cooling-induced residual stresses in heated particulate mixture depositions. Comput. Mech. **56**, 613–630 (2015)
39. Zohdi, T.I.: Modeling and efficient simulation of the deposition of particulate flows onto compliant substrates. Int. J. Eng. Sci. **99**, 74–91 (2015). https://doi.org/10.1016/j.ijengsci.2015.10.012
40. Zohdi, T.I.: Modeling and simulation of laser processing of particulate-functionalized materials. Arch. Comput. Methods Eng., 1–25 (2015). https://doi.org/10.1007/s11831-015-9160-1
41. Widom, B.: Random sequential addition of hard spheres to a volume. J. Chem. Phys. **44**, 3888–3894 (1966)
42. Kansaal, A., Torquato, S., Stillinger, F.: Diversity of order and densities in jammed hard-particle packings. Phys. Rev. E **66**, 041109 (2002)
43. Donev, A., Cisse, I., Sachs, D., Variano, E. A., Stillinger, F., Connelly, R., Torquato, S., Chaikin, P.: Improving the density of jammed disordered packings using ellipsoids. Science **303**, 990–993 (2004)
44. Donev, A., Stillinger, F.H., Chaikin, P.M., Torquato, S.: Unusually dense crystal ellipsoid packings. Phys. Rev. Lett. **92**, 255506 (2004)
45. Donev, A., Torquato, S., Stillinger, F.: Neighbor list collision-driven molecular dynamics simulation for nonspherical hard particles-I Algorithmic details. J. Comput. Phys. **202**, 737 (2005)
46. Torquato, S.: Random Heterogeneous Materials: Microstructure and Macroscopic Properties. Springer, New York (2002)
47. Jackson, J.D.: Classical Electrodynamics, 3rd edn. Wiley (1998)
48. Pöschel, T., Schwager, T.: Computational Granular Dynamics. Springer (2004)

Chapter 9
DEM Extensions: Flexible Substrate Models

9.1 Introduction

In certain applications, because the substrate is fragile, knowledge of the induced stresses is important in order to properly control the process (Fig. 9.1). Such concerns have become increasingly important due to the rise of printed flexible electronics involving sensitive, potentially fragile dielectric and optical materials. As mentioned previously, applications include, optical coatings and photonics [1], MEMS applications [2, 3], and even biomedical devices [4]. There are a wide variety of additive-like processes, and we refer the reader to Gamota et al. [5], Sirringhaus et al. [6], Wang et al. [7], Huang et al. [8], Choi et al. [9–12], Demko et al. [13, 14], Fathi et al. [15], Martin [16, 17], and Zohdi [18–50] for details. These types are similar to those in the area spray coatings. We refer the reader to the extensive works of Sevostianov and Kachanov [51–53], Nakamura and coworkers: Dwivedi et al. [54], Liu et al. [55, 56], Nakamura and Liu [57], Nakamura et al. [58] and Qian et al. [59] and to Martin [16, 17] for the state of the art in deposition technologies. Oftentimes, the objective is to produce multilayer coatings on curved surfaces (see, e.g., [60]). The interested reader is referred to the recent overview article by Huang et al. [61] on the wide array of activities in additive manufacturing. This chapter develops a computational mechanics framework to investigate the behavior of such processes. Specifically, substrate stresses due to multiple, simultaneous, surface particle contact events are efficiently computed by superposing individual particle contact solutions, based on classical Boussinesq-like solutions, coupled to a multibody dynamics formulation for the interacting particles. Specifically, in the chapter:

- A multibody collision model (based on the previous chapters) is used to represent the interaction of the particles with each other, as well as with the substrate.
- Classical point-load solutions on a half-space are used to represent the contribution of each particle to the stresses on the substrate.

T. I. Zohdi, *Modeling and Simulation of Functionalized Materials for Additive Manufacturing and 3D Printing: Continuous and Discrete Media*, Lecture Notes in Applied and Computational Mechanics 60, https://doi.org/10.1007/978-3-319-70079-3_9

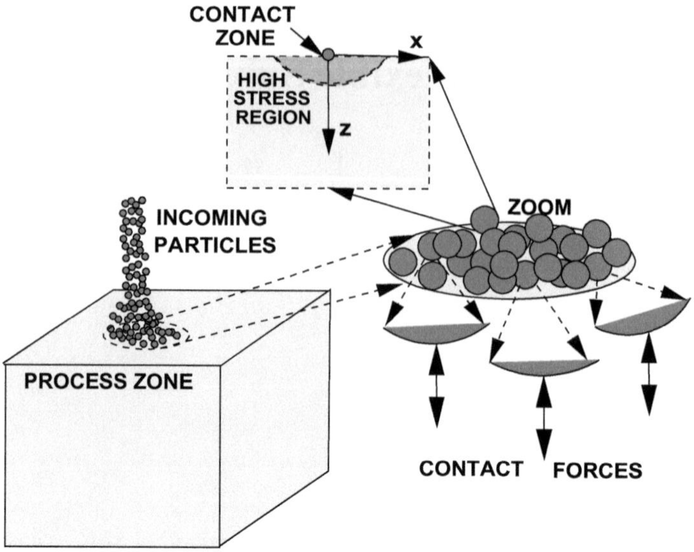

Fig. 9.1 Left: deposition of a stream of particles onto a substrate

- The response of the particles and substrate is coupled together with a recursive numerical scheme.
- Three-dimensional examples are provided to illustrate the technique.

Remark The modeling approach allows for rapid computation of deposition-induced stresses which allows one to conduct parameter studies, leaving more intensive finite element analyses, if warranted, for final process analysis stages. We note that the range of validity of this type of simulation is for relatively slow deposition where elastodynamic effects can be ignored in the substrate. Furthermore, we consider "dry" particle depositions where the interstitial fluid is of negligible importance. Particles in suspension are outside the scope of this chapter. We refer the reader to Kachanov and Abedian [62], Abedian and Kachanov [63], and Sevostianov and Kachanov [64] for details on the analysis of that class of particle-laden fluid materials. Such analyses can be useful for determining the rheology of so-called (particle) functionalized inks [18–50].

9.2 A Multibody Dynamics Model for the Particles

9.2.1 Overall Contributing Forces

As before, we consider a group of nonintersecting particles ($i = 1, 2, ..., N_p$). The objects in the system are assumed to be small enough to be considered (idealized) as particles, spherical in shape, and that the effects of their rotation with respect to their mass center are unimportant to their overall motion, although we will make further remarks on these effects shortly. The equation of motion for the ith particle in system is

$$m_i\ddot{\boldsymbol{r}}_i = \boldsymbol{\Psi}_i^{tot}(\boldsymbol{r}_1, \boldsymbol{r}_2, ..., \boldsymbol{r}_{N_p}) = \boldsymbol{\Psi}_i^{con} + \boldsymbol{\Psi}_i^{subs} + \boldsymbol{\Psi}_i^{bond} + \boldsymbol{\Psi}_i^{damp}, \tag{9.1}$$

where $\boldsymbol{r}_i$ is the position vector of the ith particle and $\boldsymbol{\Psi}_i^{tot}$ represents all forces acting on particle i, which is decomposed into the sum of forces due to:

- Inter-particle forces ($\boldsymbol{\Psi}_i^{con}$) generated by contact with other particles,
- Substrate forces ($\boldsymbol{\Psi}_i^{subs}$) generated by contact with constraining surfaces,
- Adhesive bonding forces ($\boldsymbol{\Psi}_i^{bond}$) with other particles and the substrate,
- Damping forces arising from the surrounding interstitial environment ($\boldsymbol{\Psi}_i^{damp}$) occurring from potentially viscous, surrounding, interstitial fluids and surfactants.

We refer the reader to the previous chapters.

9.3 Induced Substrate Stresses

Our basic approach is to represent the contribution of the contact induced by each particle as a point load on an infinite half-space. We break the point load into two components: (1) the normal load and the (2) tangential load. Afterward, we sum all of the contributions to obtain the total induced stress field in the substrate.

9.3.1 Individual Particle Contributions—Normal Load

The corresponding radially symmetric (θ-independent) solution (Fig. 9.1) for a normal load at $(x, y, z) = (0, 0, 0)$ in the z-direction is (in cylindrical coordinates, Boussinesq [76])[1]:

[1]See [77] for a complete, rigorous derivation.

$$
\begin{aligned}
\sigma_{rr} &= \tfrac{F_z}{2\pi}\left((1-2\nu)\left(\tfrac{1}{r^2}-\tfrac{z}{\gamma r^2}\right)-\tfrac{3zr^2}{\gamma^5}\right),\\
\sigma_{\theta\theta} &= -\tfrac{F_z}{2\pi}(1-2\nu)\left(\tfrac{1}{r^2}-\tfrac{z}{\gamma r^2}-\tfrac{z}{\gamma^3}\right),\\
\sigma_{zz} &= -\frac{3F_z}{2\pi}\frac{z^3}{\gamma^5},\\
\sigma_{rz} &= -\frac{3F_z}{2\pi}\frac{rz^2}{\gamma^5},\\
\sigma_{r\theta} &= 0\,(symmetry),\\
\sigma_{z\theta} &= 0\,(symmetry),
\end{aligned}
\tag{9.2}
$$

where $r \stackrel{\text{def}}{=} \sqrt{x^2+y^2}$ and $\gamma \stackrel{\text{def}}{=} \sqrt{x^2+y^2+z^2}$.

Remark Often, it is convenient to move back and forth from Cartesian and cylindrical bases, which can be achieved by simply rotating the system with

$$
\boldsymbol{\sigma}^{cart}(\theta) = \boldsymbol{R}^T(\theta)\cdot\boldsymbol{\sigma}^{cyl}\cdot\boldsymbol{R}(\theta), \tag{9.3}
$$

where $\boldsymbol{R}(\theta)$ is defined as

$$
\boldsymbol{R}(\theta) \stackrel{\text{def}}{=} \begin{bmatrix} cos\theta & sin\theta & 0 \\ -sin\theta & cos\theta & 0 \\ 0 & 0 & 1 \end{bmatrix}, \tag{9.4}
$$

and $\boldsymbol{R}^T(\theta) = \boldsymbol{R}^{-1}(\theta)$, because it is an orthonormal matrix.[2]

9.3.2 *Individual Particle Contributions—Tangential Load*

The effects of nonnormal (tangential) loadings can be included by utilizing the solutions for a tangential point load in the x−direction (see [77] for reviews):

[2]We note that the quantity $\boldsymbol{\sigma}' : \boldsymbol{\sigma}'$ is invariant under the rotational coordinate transformation, in other words, $\boldsymbol{\sigma}'^{,car} : \boldsymbol{\sigma}'^{,car} = (\boldsymbol{R}^T(\theta)\cdot\boldsymbol{\sigma}'^{,cyl}\cdot\boldsymbol{R}(\theta))^T : (\boldsymbol{R}^T(\theta)\cdot\boldsymbol{\sigma}'^{,cyl}\cdot\boldsymbol{R}(\theta)) = \boldsymbol{\sigma}'^{,cyl} : \boldsymbol{\sigma}'^{,cyl}$, and thus, this metric remains perfectly acceptable to use in the presence of nonnormal loading.

$$\sigma_{xx} = \frac{F_x}{2\pi}\left(-3\frac{x^3}{\gamma^5} + (1-2\nu)\left(\frac{x}{\gamma^3} - \frac{3x}{\gamma(\gamma+z)^2} + \frac{x^3}{\gamma^3(\gamma+z)^2} + \frac{2x^3}{\gamma^2(\gamma+z)^3}\right)\right)$$

$$\sigma_{yy} = \frac{F_x}{2\pi}\left(-3\frac{xy^2}{\gamma^5} + (1-2\nu)\left(\frac{x}{\gamma^3} - \frac{x}{\gamma(\gamma+z)^2} + \frac{xy^2}{\gamma^3(\gamma+z)^2} + \frac{2xy^2}{\gamma^2(\gamma+z)^3}\right)\right)$$

$$\sigma_{zz} = -\frac{3F_x}{2\pi}\frac{xz^2}{\gamma^5} \qquad (9.5)$$

$$\sigma_{xy} = \frac{F_x}{2\pi}\left(-3\frac{x^2y}{\gamma^5} + (1-2\nu)\left(-\frac{y}{\gamma(\gamma+z)^2} + \frac{x^2y}{\gamma^3(\gamma+z)^2} + \frac{2x^2y}{\gamma^2(\gamma+z)^3}\right)\right)$$

$$\sigma_{yz} = -\frac{3F_x}{2\pi}\frac{xyz}{\gamma^5}$$

$$\sigma_{zx} = -\frac{3F_x}{2\pi}\frac{x^2z}{\gamma^5}$$

and performing a coordinate transformation (to a tilde frame) $\tilde{x} = y$, $\tilde{y} = x$, $\tilde{z} = z$ to account for any F_y loading in the $y-$direction. The results from the loading in the $x-$, $y-$ and $z-$ directions can be superposed to produce the total loading.

9.3.3 *Superposition of Contributions for the Total Substrate Stresses*

The total stress at a point in the substrate is computed by summing all point-load (particle) contributions ($I = 1, 2, ..., N$, appropriately translated according to their position on the surface):

$$\boldsymbol{\sigma}_{tot-N}(\boldsymbol{x}) = \sum_{I=1}^{N} \boldsymbol{\sigma}_I(\boldsymbol{x}), \qquad (9.6)$$

where the contact forces will be determined from the solution of the multibody particle problem. We assume that there is only normal loading. The presence of tangential loading is discussed in the conclusions. From Eq. 9.2, one can determine the von Mises stress $(\boldsymbol{\sigma}_{tot-N})' = \sum_{I=1}^{N} \boldsymbol{\sigma}'_I$, where $\boldsymbol{\sigma}'_I = \boldsymbol{\sigma}_I - \frac{tr\boldsymbol{\sigma}_I}{3}\mathbf{1}$, which is usually important for failure assessment.

The computational algorithm is as follows:

- Initialization: Generate a starting configuration for the particles.
- STEP 1: Compute the forces from the particles in contact with the surface-this produces N loading sites.
- STEP 2: Compute the stress field contribution from each particle, $I = 1, 2, ...N$ on the surface.
- STEP 3: Sum the contributions of each particle $I = 1, 2, ...N$, to compute the total.
- STEP 4: Repeat STEPS 1-3 for each time-step.

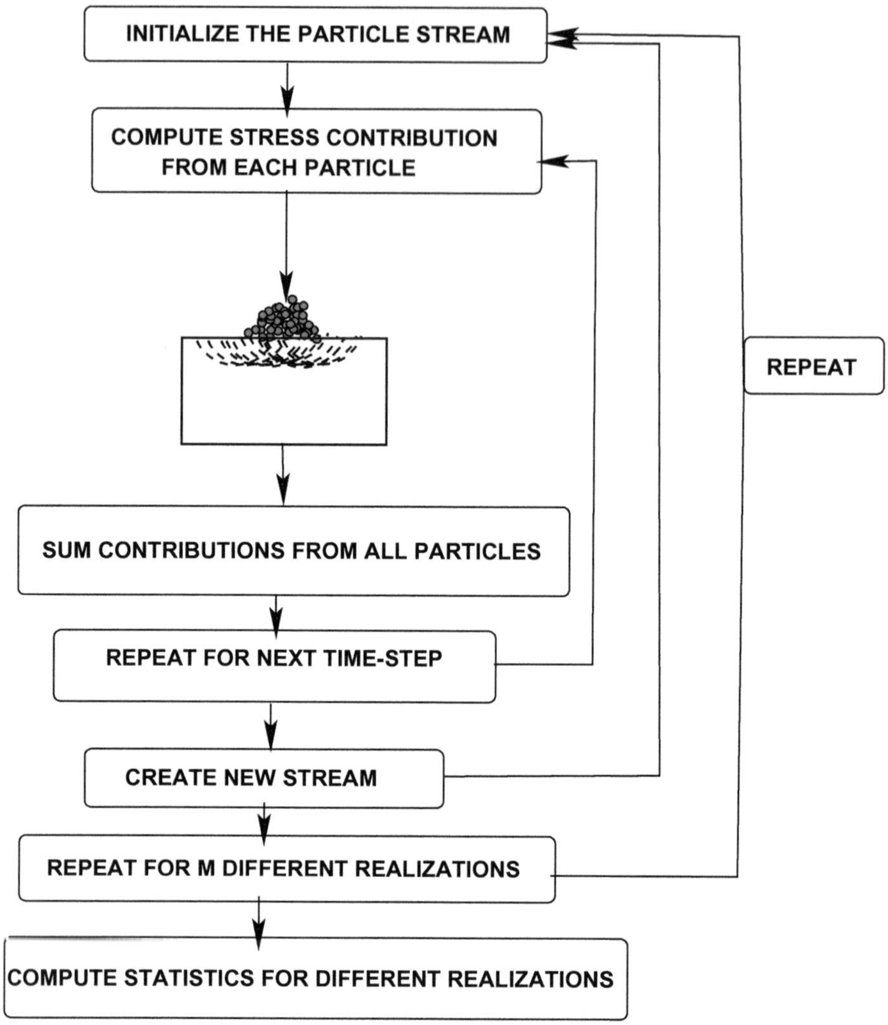

Fig. 9.2 Algorithm for computation of the loading of a surface

- STEP 5: Compute the response statistics in the target zone of interest as desired.

Figure 9.2 provides a corresponding flowchart for the process. The utility of the approach is that one can ascertain detailed spatial distribution of the stresses in the substrate. One can also post-process aggregate stresses. For example, one statistical metric is the volume average deviatoric stress metric in the volume with N surface load particles at any given moment in time:

$$\langle(\boldsymbol{\sigma}_{tot-N})' : (\boldsymbol{\sigma}_{tot-N})'\rangle_{\Omega} \stackrel{\text{def}}{=} \sqrt{\frac{1}{|\Omega|}\int_{\Omega}(\boldsymbol{\sigma}_{tot-N})' : (\boldsymbol{\sigma}_{tot-N})' \, dV}. \qquad (9.7)$$

9.4 Numerical Examples

We consider a model problem of an initially cylindrical stream of equally sized particles (Fig. 9.3) to illustrate the process. The absolute dimensions are unimportant for the model problem and have been normalized so that the initial droplet diameter was on the order of unity. The relevant simulation parameters chosen were (in SI units if not explicitly stated):

- The particle radii, $R_i = 0.05$ m,
- The normal contact parameter was $K_{po} = 10^7\,\mathrm{N/m^2}$, at a fixed temperature, $K_p = \mathbf{MAX}(K_{po}\left(e^{-a\frac{\Theta}{\Theta^*}-1}\right), K_p^{lim})$, where $\Theta^* = 500\,{}^\circ\mathrm{K}$, $K_p^{lim} = 10^6\,\mathrm{N/m^2}$, and the exponent in the contact law was set to $p_p = 2$,
- The stiffness of the wall in contact law, $K_w = 10^9\,\mathrm{N/m^2}$,
- The contact damping parameter, $c^{cd} = 10^5$,
- The friction contact parameter, $K^f = 10^7$,
- The coefficient of static friction, $\mu_s = 0.4$,
- The coefficient of dynamic friction, $\mu_d = 0.3$,
- The normal bond parameter, $K^{nb} = 10^6\,\mathrm{N/m^2}$, and the exponent in the binding law was set to $p_b = 2$,
- The rotational bond parameter, $K^{rb} = 10^3$,

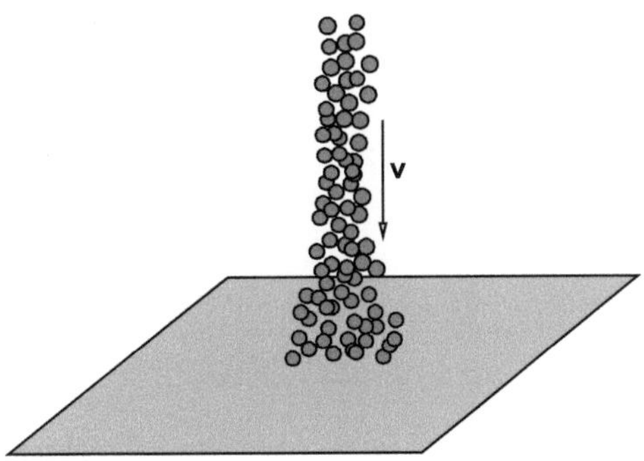

Fig. 9.3 Deposition scenario

- The interstitial damping coefficient, $c^e = 1$ (assumed Stokesian-like),
- The target number of fixed-point iterations, $K_d = 10$,
- The trapezoidal time-stepping parameter, $\phi = 0.5$,
- The simulation duration, 2 s,
- The initial time-step size, 0.00025 s,
- The time-step upper bound, 0.00025 s, and
- The tolerance for the fixed-point iteration, 5×10^{-4}.

All system parameters can be scaled to describe any specific system of interest. They were selected simply for illustration purposes. In order to generate an initial particle configuration, we randomly dispersed $N_p = 2000$ nonoverlapping particles within a cylindrical domain, then projected them onto the surface. The configuration of the sample, before it was dropped, was generated using a classical Random Sequential Addition (RSA) algorithm [78], which places nonoverlapping particles randomly into the domain of interest. As stated previously, one could start with a denser starting configuration by using the equilibrium-driven Metropolis algorithm or alternative methods based on simultaneous particle flow and growth (see [79–83]), although this was not necessary for this example. As an example, deposition of a stream of particles (Fig. 9.3) is dropped onto a lower surface (by gravity with an initial uniform downward velocity) and allowed to freely spread/deform according to its interaction with the surface (Figs. 9.4, 9.5, 9.6). The substrate colors indicate the norm of the total summed deviatoric stresses. Both the top (left) and bottom (right) are shown. The localized nature of the surface stresses is smeared out with increasing depth. All computations were run in a matter of a few minutes on a laptop.

Remark A rigorous, detailed analysis of the deformation and stress within an impacted substrate is an extremely complex process, requiring a finite element analysis of the deformation of contacting bodies. We refer the reader to Wriggers [84] for a thorough analysis of this topic, including thermo-mechanical heat generation. We also remark that the present analysis can be used to investigate shot-peening processes. We refer the reader to Afazov et al. [85], Bagherifard et al. [86], Elbella et al. [87], and Chen et al. [88] for the finite element analysis of the response of the substrate of a shot-peened solid and recently to Zohdi [45] for rapid computation of multiple contacting bodies on substrates for additive particle printing processes.

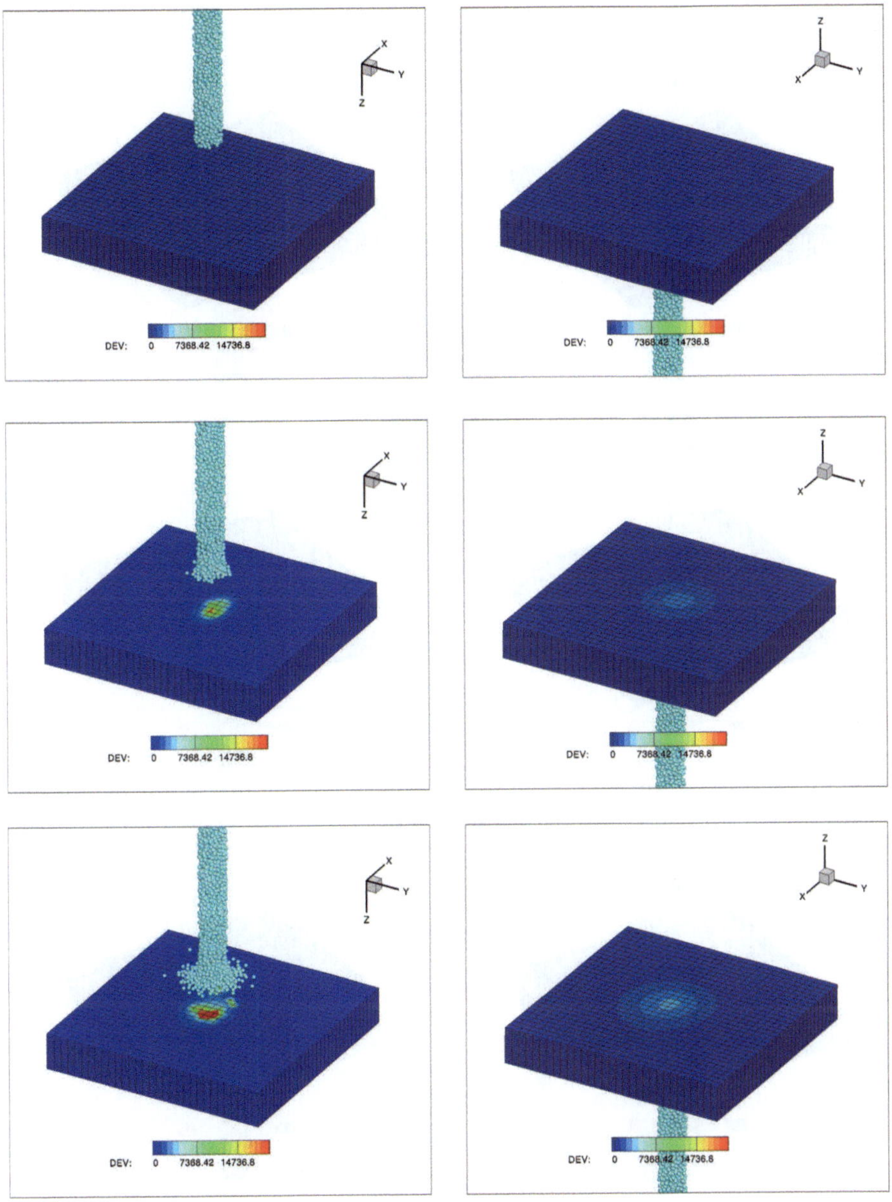

Fig. 9.4 Example: deposition onto a surface. The substrate colors indicate the norm of the total summed deviatoric stresses. Both the top (left) and bottom (right) are shown. The localized nature of the surface stresses is smeared out with increasing depth

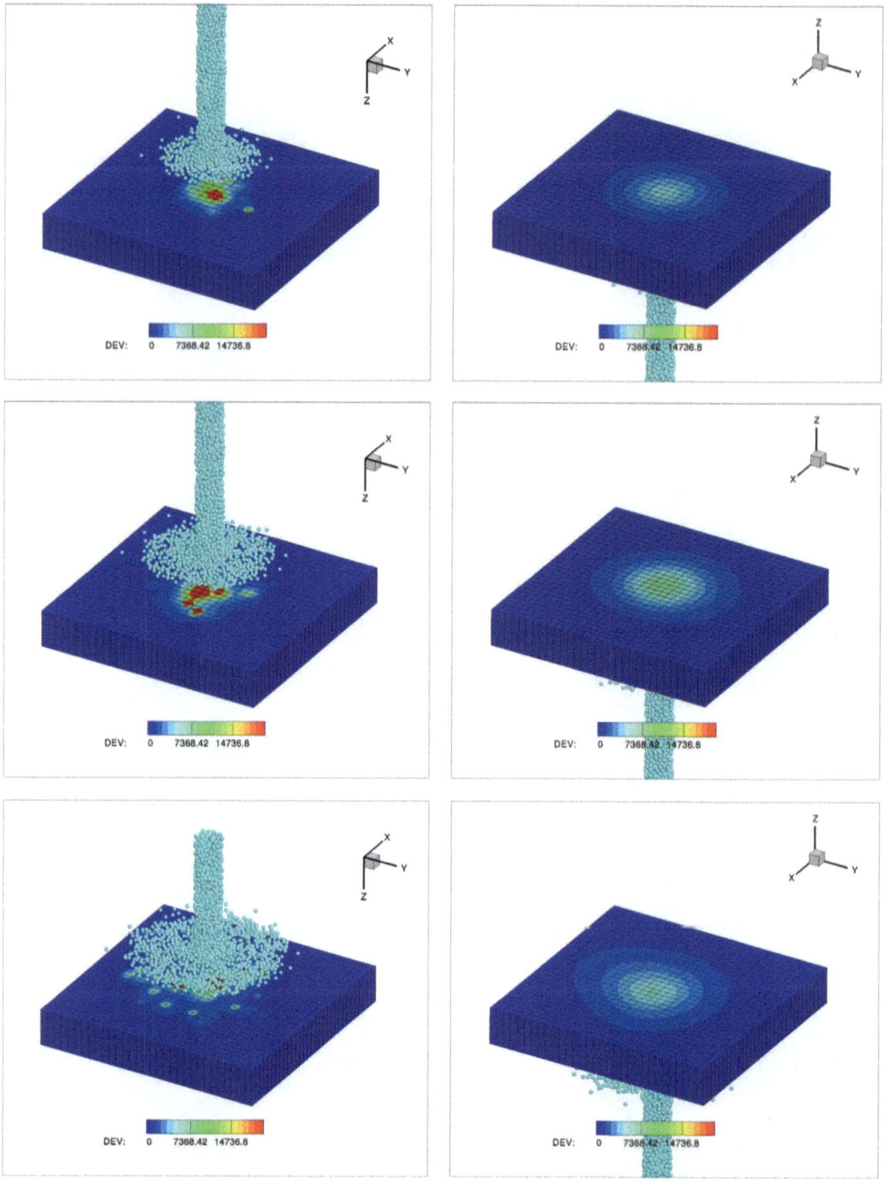

Fig. 9.5 Example: deposition onto a surface. The substrate colors indicate the norm of the total summed deviatoric stresses. Both the top (left) and bottom (right) are shown. The localized nature of the surface stresses is smeared out with increasing depth

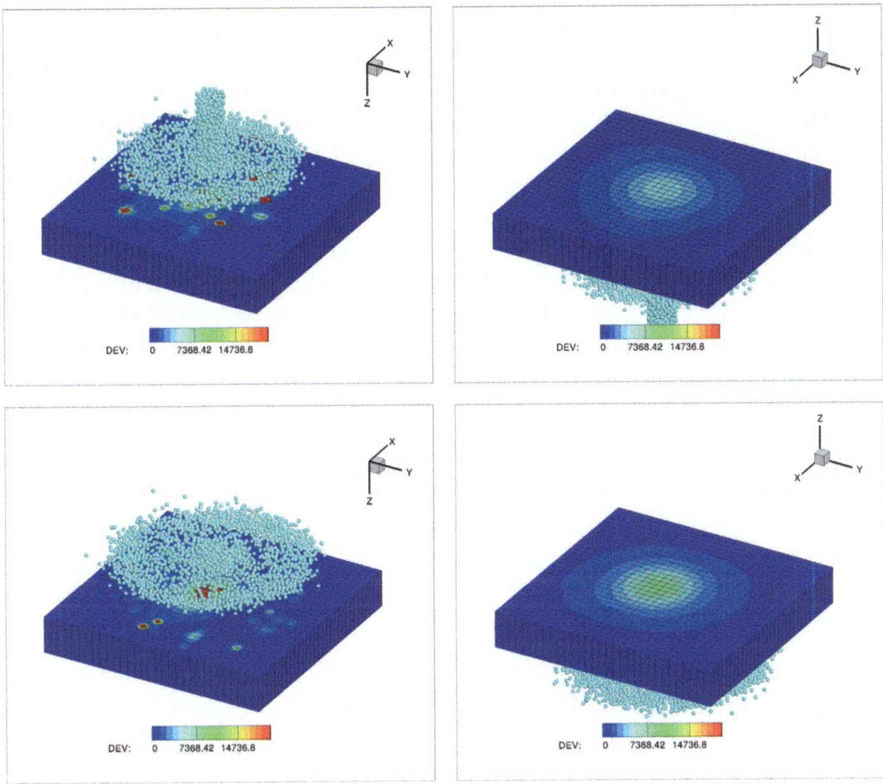

Fig. 9.6 Example: deposition onto a surface. The substrate colors indicate the norm of the total summed deviatoric stresses. Both the top (left) and bottom (right) are shown. The localized nature of the surface stresses is smeared out with increasing depth

9.5 Summary, Conclusions, and Extensions

In summary, this chapter developed a computational mechanics framework to investigate the deposition of streams of particles onto compliant substrates. Substrate stresses due to multiple surface particle contact events were efficiently computed by superposing individual particle contact solutions coupled to a multibody dynamics formulation for the interacting particles. In particular:

- A multibody collision model was developed to determine the interaction of the particles with each other, as well as with the substrate.
- Classical Boussinesq-like point-load solutions on a half-space were used to represent the contribution of each particle to the total stresses on the substrate.

Three-dimensional examples were provided to illustrate the technique. The utility of the approach is that one can very rapidly compute the results of the action of a given set of process parameters and thus run several "forward" problems for optimization studies. We remark that there are several manufacturing applications where particles are deposited onto substrates. However, there are variants which also utilize imprint lithography as a means of decreasing the feature size of patterned particles while allowing more precise control over the structure of the print [89–92]. In this fabrication method, the particle-laden inks are patterned by pressing with an elastomer mold and the particles are dried into their final configuration. While the resolution of imprint lithography is improved over inkjet printing, there exists a residual layer on the substrate that must be etched away after patterning. Control over the height of features can be corrupted by capillary action between the mold and the drying ink, in particular along the length of longer features. Thus, as a possible alternative to imprint lithography, particle self-assembly methods based on capillary filling of photoresist templates have been proposed [13, 14] and appear to be promising.

References

1. Nakanishi, H., Bishop, K.J.M., Kowalczyk, B., Nitzan, A., Weiss, E.A., Tretiakov, K.V., Apodaca, M.M., Klajn, R., Stoddart, J.F., Grzybowski, B.A.: Photoconductance and inverse photoconductance in thin films of functionalized metal nanoparticles. Nature **460**, 371–375 (2009)
2. Fuller, S.B., Wilhelm, E.J., Jacobson, J.M.: Ink-jet printed nanoparticle microelectromechanical systems. J. Microelectromech. Syst. **11**, 54–60 (2002)
3. Samarasinghe, S.R., Pastoriza-Santos, I., Edirisinghe, M.J., Reece, M.J., Liz-Marzan, L.M.: Printing gold nanoparticles with an electrohydrodynamic direct write device. Gold Bull. **39**, 48–53 (2006)
4. Ahmad, Z., Rasekh, M., Edirisinghe, M.: Electrohydrodynamic direct writing of biomedical polymers and composites. Macromol. Mater. Eng. **295**, 315–319 (2010)
5. Gamota, D., Brazis, P., Kalyanasundaram, K., Zhang, J.: Printed Organic and Molecular Electronics. Kluwer Academic Publishers, New York (2004)
6. Sirringhaus, H., Kawase, T., Friend, R.H., Shimoda, T., Inbasekaran, M., Wu, W., Woo, E.P.: High-resolution inkjet printing of all-polymer transistor circuits. Science **290**, 2123–2126 (2000)
7. Wang, J.Z., Zheng, Z.H., Li, H.W., Huck, W.T.S., Sirringhaus, H.: Dewetting of conducting polymer inkjet droplets on patterned surfaces. Nat. Mater. **3**, 171–176 (2004)
8. Huang, D., Liao, F., Molesa, S., Redinger, D., Subramanian, V.: Plastic-compatible low-resistance printable gold nanoparticle conductors for flexible electronics. J. Electrochem. Soc. **150**(7), G412–417 (2003)
9. Choi, S., Park, I., Hao, Z., Holman, H.Y., Pisano, A.P., Zohdi, T.I.: Ultra-fast self-assembly of micro-scale particles by open channel flow. Langmuir **26**(7), 4661–4667 (2010)
10. Choi, S., Stassi, S., Pisano, A.P., Zohdi, T.I.: Coffee-Ring Effect-Based Three Dimensional Patterning of Micro, Nanoparticle Assembly with a Single Droplet. Langmuir **26**(14), 11690–11698 (2010)
11. Choi, S., Jamshidi, A., Seok, T.J., Zohdi, T.I., Wu, M.C., Pisano, A.P.: Fast, High-throughput creation of size-tunable micro, nanoparticle clusters via evaporative self-assembly in picoliter-scale droplets of particle suspension. Langmuir **28**(6), 3102–11 (2012)
12. Choi, S., Pisano, A.P., Zohdi, T.I.: An analysis of evaporative self-assembly of micro particles in printed picoliter suspension droplets. J. Thin Solid Films **537**(30), 180–189 (2013)
13. Demko, M., Choi, S., Zohdi, T.I., Pisano, A.P.: High resolution patterning of nanoparticles by evaporative self-assembly enabled by in-situ creation and mechanical lift-off of a polymer template. Appl. Phy. Lett. **99**, 253102-1–253102-3 (2012)

14. Demko, M.T., Cheng, J.C., Pisano, A.P.: High-resolution direct patterning of gold nanoparticles by the microfluidic molding process. Langmuir 412–417 (2010)
15. Fathi, S., Dickens, P., Khodabakhshi, K., Gilbert, M.: Microcrystal particles behaviour in inkjet printing of reactive nylon materials. J. Manuf. Sci. Eng. **135**, 011009. https://doi.org/10.1115/1.4023272
16. Martin, P.: Handbook of Deposition Technologies for Films and Coatings. 3rd edn. Elsevier (2009)
17. Martin, P.: Introduction to surface engineering and functionally engineered materials. Scrivener and Elsevier (2011)
18. Zohdi, T.I.: Genetic design of solids possessing a random-particulate microstructure. Philos. Trans. R. Soc. Math. Phys. Eng. Sci. **361**(1806), 1021–1043 (2003)
19. Zohdi, T.I.: On the compaction of cohesive hyperelastic granules at finite strains. Proc. R. Soc. **454**(2034), 1395–1401 (2003)
20. Zohdi, T.I.: Computational design of swarms. Int. J. Numer. Method. Eng. **57**, 2205–2219 (2003)
21. Zohdi, T.I.: Constrained inverse formulations in random material design. Comput. Method. Appl. Mech. Eng. 1–20. **192**(28–30), 18, 3179–3194 (2003)
22. Zohdi, T.I.: Staggering error control for a class of inelastic processes in random microheterogeneous solids. Int. J. Nonlin. Mech. **39**, 281–297 (2004)
23. Zohdi, T.I.: Modeling and simulation of a class of coupled thermo-chemo-mechanical processes in multiphase solids. Comput. Method. Appl. Mech. Eng. **193**(6–8), 679–699 (2004)
24. Zohdi, T.I.: Modeling and direct simulation of near-field granular flows. Int. J. Solids Struct. **42**(2), 539–564 (2004)
25. Zohdi, T.I.: A computational framework for agglomeration in thermo-chemically reacting granular flows. Proc. R. Soc. **460**(2052), 3421–3445 (2004)
26. Zohdi, T.I.: Statistical ensemble error bounds for homogenized microheterogeneous solids. J. Appl. Math. Phys. (Zeitschrift für Angewandte Mathematik und Physik) **56**(3), 497–515 (2005)
27. Zohdi, T.I.: Charge-induced clustering in multifield particulate flow. Int. J. Numer. Method. Eng. **62**(7), 870–898 (2005)
28. Zohdi, T.I.: Computation of strongly coupled multifield interaction in particle-fluid systems. Comput. Method. Appl. Mech. Eng. **196**, 3927–3950 (2007)
29. Zohdi, T.I.: Particle collision and adhesion under the influence of near-fields. J. Mech. Mater. Struct. **2**(6), 1011–1018 (2007)
30. Zohdi, T.I.: On the computation of the coupled thermo-electromagnetic response of continua with particulate microstructure. Int. J. Numer. Method. Eng. **76**, 1250–1279 (2008)
31. Zohdi, T.I.: Mechanistic modeling of swarms. Comput. Method. Appl. Mech. Eng. **198**(21–26), 2039–2051 (2009)
32. Zohdi, T.I.: On the dynamics of charged electromagnetic particulate jets. Arch. Comput. Method. Eng. **17**(2), 109–135 (2010)
33. Zohdi, T.I.: Simulation of coupled microscale multiphysical-fields in particulate-doped dielectrics with staggered adaptive FDTD. Comput. Method. Appl. Mech. Eng. **199**, 79–101 (2010)
34. Zohdi, T.I.: Dynamics of clusters of charged particulates in electromagnetic fields. Int. J. Numer. Method. Eng. **85**, 1140–1159 (2011)
35. Zohdi, T.I.: Joule-heating field phase-amplification in particulate-doped dielectrics. Int. J. Eng. Sci. **49**, 30–40 (2011)
36. Zohdi, T.I.: Estimation of electrical-heating load-shares for sintering of powder mixtures. Proc. R. Soc. **468**, 2174–2190 (2012)
37. Zohdi, T.I.: Modeling and simulation of the optical response rod-functionalized reflective surfaces. Comput. Mech. **50**(2), 257–268 (2012)
38. Zohdi, T.I.: On the reduction of heat generation in lubricants using microscale additives. Int. J. Eng. Sci. **62**, 84–89 (2013)

39. Zohdi, T.I.: Numerical simulation of charged particulate cluster-droplet impact on electrified surfaces. J. Comput. Phys. **233**, 509–526 (2013)
40. Zohdi, T.I.: On inducing compressive residual stress in microscale print-lines for flexible electronics. Int. J. Eng. Sci. **62**, 157–164 (2013)
41. Zohdi, T.I.: Rapid simulation of laser processing of discrete particulate materials. Arch. Comput. Method. Eng. **20**, 309–325 (2013)
42. Zohdi, T.I.: A direct particle-based computational framework for electrically-enhanced thermo-mechanical sintering of powdered materials. Math. Mech. Solids **19**(1), 93–113 (2014)
43. Zohdi, T.I.: Additive particle deposition and selective laser processing-a computational manufacturing framework. Comput. Mech. **54**, 171–191 (2014)
44. Zohdi, T.I.: Embedded electromagnetically sensitive particle motion in functionalized fluids. Comput. Part. Mech. **1**, 27–45 (2014)
45. Zohdi, T.I.: Rapid computation of statistically-stable particle/feature ratios for consistent substrate stresses in printed flexible electronics. J. Manuf. Sci. Eng. ASME MANU-14-1476. https://doi.org/10.1115/1.4029327
46. Zohdi, T.I.: Modeling and simulation of the post-impact trajectories of particles in oblique precision shot-peening. Comput. Part. Mech. (2015). https://doi.org/10.1007/s40571-015-0048-5
47. Zohdi, T.I.: Modeling and simulation of cooling-induced residual stresses in heated particulate mixture depositions. Comput. Mech. **56**, 613–630 (2015)
48. Zohdi, T.I.: Modeling and efficient simulation of the deposition of particulate flows onto compliant substrates. Int. J. Eng. Sci. **99**, 74–91 (2015). https://doi.org/10.1016/j.ijengsci.2015.10.012
49. Zohdi, T.I.: Modeling and simulation of laser processing of particulate-functionalized materials. Arch. Comput. Method. Eng. 1–25 (2015). https://doi.org/10.1007/s11831-015-9160-1
50. Zohdi, T.I., Wriggers, P.: Introduction to Computational Micromechanics. Springer, Second Reprinting (Peer Reviewed) (2005, 2008)
51. Sevostianov, I., Kachanov, M.: Modeling of the anisotropic elastic properties of plasma-sprayed coatings in relation to their microstructure. Acta Mater. **48**(6), 1361–1370 (2000)
52. Sevostianov, I., Kachanov, M.: Thermal conductivity of plasma sprayed coatings in relation to their microstructure, with Sevostianov, I. J. Therm. Spray Technol. **9**(4), 478–482 (2001)
53. Sevostianov, I., Kachanov, M.: Plasma-sprayed ceramic coatings: anisotropic elastic and conductive properties in relation to the microstructure; cross-property correlations. Mater. Sci. Eng. A **297**, 235–243 (2001)
54. Dwivedi, G., Wentz, T., Sampath, S., Nakamura, T.: Assessing process and coating reliability through monitoring of process and design relevant coating properties. J. Thermal Spray Tech. **19**, 695–712 (2010)
55. Liu, Y., Nakamura, T., Dwivedi, G., Valarezo, A., Sampath, S.: Anelastic behavior of plasma sprayed zirconia coatings. J. Am. Ceram. Soc. **91**, 4036–4043 (2008)
56. Liu, Y., Nakamura, T., Srinivasan, V., Vaidya, A., Gouldstone, A., Sampath, S.: Nonlinear elastic properties of plasma sprayed zirconia coatings and associated relationships to processing conditions. Acta Mater. **55**, 4667–4678 (2007)
57. Nakamura, T., Liu, Y.: Determination of nonlinear properties of thermal sprayed ceramic coatings via inverse analysis. Int. J. Solids Struct. **44**, 1990–2009 (2007)
58. Nakamura, T., Qian, G., Berndt, C.C.: Effects of pores on mechanical properties of plasma sprayed ceramic coatings. J. Am. Ceram. Soc. **83**, 578–584 (2000)
59. Qian, G., Nakamura, T., Berndt, C.C.: Effects of thermal gradient and residual stresses on thermal barrier coating fracture. Mech. Mater. **27**, 91–110 (1998)
60. Grekov, M.A., Kostyrko, S.A.: A multilayer film coating with slightly curved boundary. Int. J. Eng. Sci. **89**, 61–74 (2015)
61. Huang, Y., Leu, M.C., Mazumdar, J., Donmez, A.: Additive manufacturing: current state, future potential, gaps and needs, and recommendation. J. Manuf. Sci. Eng. **137**, 014001 (2015)

62. Kachanov, M., Abedian, B.: On the isotropic viscosity of suspensions containing particles of diverse shapes and orientations. **94**, 71–85 (2015)
63. Abedian, B., Kachanov, M.: On the effective viscosity of suspensions. Int. J. Eng. Sci. **48**(11), 962–965 (2010). https://doi.org/10.1016/j.ijengsci.2010.08.012
64. Sevostianov, I., Kachanov, M.: Effective properties of heterogeneous materials: proper application of the non-interaction and the "dilute limit" approximations. Int. J. Eng. Sci. **58**, 124–128 (2012)
65. Onate, E., Idelsohn, S.R., Celigueta, M.A., Rossi, R.: Advances in the particle finite element method for the analysis of fluid-multibody interaction and bed erosion in free surface flows. Comput. Method. Appl. Mech. Eng. **197**(19–20), 1777–1800 (2008)
66. Onate, E., Celigueta, M.A., Idelsohn, S.R., Salazar, F., Surez, B.: Possibilities of the particle finite element method for fluid-soil-structure interaction problems. Comput. Mech. **48**, 307–318 (2011)
67. Rojek, J., Labra, C., Su, O., Onate, E.: Comparative study of different discrete element models and evaluation of equivalent micromechanical parameters. Int. J. Solids Struct. **49**, 1497–1517 (2012). https://doi.org/10.1016/j.ijsolstr.2012.02.032
68. Carbonell, J.M., Onate, E., Suarez, B.: Modeling of ground excavation with the particle finite element method. J. Eng. Mech. ASCE **136**, 455–463 (2010)
69. Labra, C., Onate, E.: High-density sphere packing for discrete element method simulations. Commun. Numer. Methods Eng. **25**(7), 837–849 (2009)
70. Leonardi, A., Wittel, F.K., Mendoza, M., Herrmann, H.J.: Coupled DEM-LBM method for the free-surface simulation of heterogeneous suspensions. Comput. Part. Mech. **1**(1), 3–13 (2014)
71. Cante, J., Davalos, C., Hernandez, J.A., Oliver, J., Jonsen, P., Gustafsson, G., Haggblad, H.A.: PFEM-based modeling of industrial granular flows. Comput. Part. Mech. **1**(1), 47–70 (2014)
72. Rojek, J.: Discrete element thermomechanical modelling of rock cutting with valuation of tool wear. Comput. Part. Mech. **1**(1), 71–84 (2014)
73. Bolintineanu, D.S., Grest, G.S., Lechman, J.B., Pierce, F., Plimpton, S.J., Schunk, P.R.: Particle dynamics modeling methods for colloid suspensions. Comput. Part. Mech. **1**(3), 321–356 (2014)
74. Avci, B., Wriggers, P.: A DEM-FEM coupling approach for the direct numerical simulation of 3D particulate flows. J. Appl. Mech. **79**(010901), 1–7 (2012)
75. Onate, E., Celigueta, M.A., Latorre, S., Casas, G., Rossi, R., Rojek, J.: Lagrangian analysis of multiscale particulate flows with the particle finite element method. Comput. Part. Mech. **1**(1), 85–102 (2014)
76. Boussinesq, J.: Application des Potentials a l' etude de l' equilibre et du mouvement des solides elastiques. Paris: Gauthier-Villars, 45, 108 (1885)
77. Kachanov, M., Shafiro, B., Tsukrov, I.: Handbook of Elasticity Solutions. Kluwer (2003)
78. Widom, B.: Random sequential addition of hard spheres to a volume. J. Chem. Phy. **44**, 3888–3894 (1966)
79. Kansaal, A., Torquato, S., Stillinger, F.: Diversity of order & densities in jammed hard-particle packings. Phy. Rev. E **66**, 041109 (2002)
80. Donev, A., Cisse, I., Sachs, D., Variano, E.A., Stillinger, F., Connelly, R., Torquato, S., Chaikin, P.: Improving the density of jammed disordered packings using ellipsoids. Science **13, 303**, 990–993 (2004a)
81. Donev, A., Torquato, S., Stillinger, F.: Neighbor list collision-driven molecular dynamics simulation for nonspherical hard particles-I algorithmic details. J. Comput. Phy. **202**, 737 (2005a)
82. Donev, A., Torquato, S., Stillinger, F.: Neighbor list collision-driven molecular dynamics simulation for nonspherical hard particles-II. Application to ellipses and ellipsoids. J. Comput. Phys. **202**, 765 (2005b)
83. Torquato, S.: Random Heterogeneous Materials: Microstructure and Macroscopic Properties. Springer, New York (2002)
84. Wriggers, P.: Computational Contact Mechanics. Wiley (2002)

85. Afazov, S.M., Becker, A.A., Hyde, T.H.: Mathematical modeling and implementation of residual stress mapping from microscale to macroscale finite element models. J. Manuf. Sci. Eng. **134**(2), 021001 (2012)
86. Bagherifard, S., Giglio, M., Giudici, L. Guagliano, M.: Experimental and Numerical Analysis of Fatigue Properties Improvement in a Titanium Alloy by Shot Peening. In: Proceedings of the ASME. 49163; ASME 2010 10th Biennial Conference on Engineering Systems Design and Analysis, vol. 2, pp. 317–322 (2010)
87. Elbella, A., Fadul, F., Uddanda, S.H., Kasarla, N.R.: Influence of shot peening parameters on process effectiveness. In: Proceedings of the ASME. 45196; Volume 3: Design, Materials and Manufacturing, Parts A, B, and C, pp. 2015–2021 (2012)
88. Chen, Z., Yang, F. and Meguid, S.A.: Realistic Finite Element Simulations of Arc-Height Development in Shot-Peened Almen Strips. J. Eng. Mater. Technol. **136**(4), 041002–041002-7
89. Park, J.-U., Hardy, M., Kang, S.J., Barton, K., Adair, K., Mukhopadhyay, D.K., Lee, C.Y., Strano, M.S., Alleyne, A.G., Georgiadis, J.G., Ferreira, P.M., Rogers, J.A.: High-resolution electrohydrodynamic jet printing. Nat. Mater. **6**, 782–789 (2007)
90. Ko, S.H., Park, I., Pan, H., Grigoropoulos, C.P., Pisano, A.P., Luscombe, C.K., Frechet, J.M.J.: Direct nanoimprinting of metal nanoparticles for nanoscale electronics fabrication. Nan Letters. **7**, 1869–1877 (2007)
91. Ko, S.H., Park, I., Pan, H., Misra, N., Rogers, M.S., Grigoropoulos, C.P., Pisano, A.P.: ZnO nanowire network transistor fabrication by lowtemperature, allinorganic nanoparticle solution process. Appl. Phy. Lett. **92**, 154102 (2008)
92. Park, I., Ko, S.H., Pan, H., Grigoropoulos, C.P., Pisano, A.P., Frechet, J.M.J., Lee, E.S., Jeong, J.H.: Nanoscale patterning and electronics on flexible substrate by direct nanoimprinting of metallic nanoparticles. Adv. Mater. **20**, 489 (2008)

Chapter 10
DEM Extensions: Higher-Fidelity Laser Modeling

This chapter develops a computational model and corresponding solution algorithm for rapid simulation of the laser processing and targeted localized heating of materials composed of discrete particles that are packed together, which goes beyond a simple Beer–Lambert representation. Such materials possess a complex microstructure which contains gaps and interfaces. This type of process is extremely difficult to simulate using continuum-based methods, such as the Finite Difference Time Domain Method or the Finite Element Method. The model (components of which are shown in Fig. 10.1) that is developed captures the main physical effects. The features of the computational model are (1) a discretization of a concentrated laser beam into rays, (2) a discrete element representation of the particulate material microstructure, and (3) a discrete element transient heat transfer model that accounts for optical (laser) energy propagation (reflection and absorption), its conversion into heat, the subsequent conduction of heat and phase transformations, involving melting and vaporization. A discrete ray-tracking algorithm is developed, along with an embedded, staggered, iterative solution scheme, in order to calculate the optical-to-thermal conversion, particle-to-particle conduction, and phase transformations, implicitly. Numerical examples are provided to illustrate the model and algorithm.

Remarks: The characterization of the flow of concentrated high-frequency energy, the irradiance, through compacted particulate systems is a key objective of this work. It is assumed that the particles and surface features are at least an order of magnitude larger than the wavelength of the incident electromagnetic radiation; therefore, "geometrical" ray-tracing theory is applicable and is well-suited for the systems of interest. Our goal is to develop a computational tool by assembling relatively simple, physically meaningful, models directly at the particle scale, for many interacting particles, in order to allow for much more refined estimates of the resulting overall system temperature and, ultimately, its change of phase from a solid to a liquid to a gas. This will help guide the proper selection of the laser intensity, duration, etc.

T. I. Zohdi, *Modeling and Simulation of Functionalized Materials for Additive Manufacturing and 3D Printing: Continuous and Discrete Media*, Lecture Notes in Applied and Computational Mechanics 60, https://doi.org/10.1007/978-3-319-70079-3_10

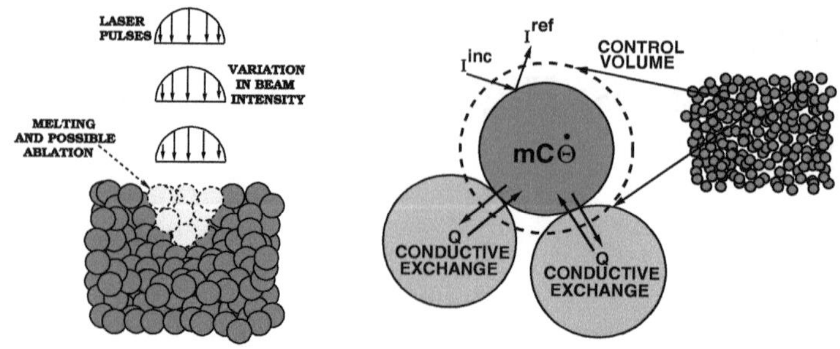

Fig. 10.1 Left: a laser pulse applied to a powdered material. Right: control volume for heat transfer, with an incident ray that encounters an idealized smooth particle scatterer

10.1 Propagation of Electromagnetic Energy

The interest here is on behavior of initially coherent beams (Fig. 10.1), composed of multiple collinear (collimated) rays (initially forming a planar wave front), where each ray is a vector in the direction of the flow of electromagnetic energy (the rays are parallel to the initial wave's propagation vector). Ray-tracing is a method that is employed to produce rapid approximate solutions to wave equations for high-frequency/small-wavelength applications where the primary interest is in the overall propagation of energy.[1] Essentially, ray-tracing methods proceed by initially representing wave fronts by an array of discrete rays. *Thereafter, the problem becomes one of primarily geometric character*, where one tracks the changing trajectories and magnitudes of individual rays which are dictated by the reflectivity and the Fresnel conditions (if a ray encounters a material interface). Ray-tracing methods are well-suited for computation of scattering in complex systems that are difficult to mesh/discretize, relative to procedures such as the Finite Difference Time Domain Method or the Finite Element Method, and therefore, they are frequently employed by analysts in such situations. For review of the state of the art in industrially oriented optics, see Gross [1]. The next section characterizes ray propagation through particulate media, building on approaches found in Zohdi [8–10].

10.1.1 Electromagnetic Wave Propagation

The propagation of electromagnetic waves in free space can be described by a simplified form of Maxwell's equations

$$\nabla \times \boldsymbol{E} = -\mu_o \frac{\partial \boldsymbol{H}}{\partial t}, \quad \text{and} \quad \nabla \times \boldsymbol{H} = \epsilon_o \frac{\partial \boldsymbol{E}}{\partial t}, \tag{10.1}$$

[1]Resolving diffraction (which ray theory is incapable of describing) is unimportant for the applications of interest.

where $\nabla \cdot \boldsymbol{H} = 0$, $\nabla \cdot \boldsymbol{E} = 0$, $\boldsymbol{E}$ is the electric field intensity, $\boldsymbol{H}$ is the magnetic flux intensity, ϵ_o is the free space permittivity, and μ_o is the free space permeability. Using standard vector identities, one can show that

$$\nabla \times (\nabla \times \boldsymbol{E}) = -\mu_o \epsilon_o \frac{\partial^2 \boldsymbol{E}}{\partial t^2}, \quad \text{and} \quad \nabla \times (\nabla \times \boldsymbol{H}) = -\mu_o \epsilon_o \frac{\partial^2 \boldsymbol{H}}{\partial t^2}, \tag{10.2}$$

and that

$$\nabla^2 \boldsymbol{E} = \frac{1}{c^2} \frac{\partial^2 \boldsymbol{E}}{\partial t^2}, \quad \text{and} \quad \nabla^2 \boldsymbol{H} = \frac{1}{c^2} \frac{\partial^2 \boldsymbol{H}}{\partial t^2}, \tag{10.3}$$

where the speed of electromagnetic waves is $c = \frac{1}{\sqrt{\epsilon_o \mu_o}}$. All electromagnetic radiation travels, in a vacuum, at the speed $c \approx 2.99792458 \times 10^8 \pm 1.1\,\text{m/s}$. In any another medium, for electromagnetic waves, the propagation speed is $v = \frac{1}{\sqrt{\epsilon \mu}}$, where ϵ and μ are the electric permittivity and magnetic permeability of that medium, respectively.[2]

10.1.2 Plane Harmonic Wave Fronts

Now consider the special case of plane harmonic waves, for example, of the form

$$\boldsymbol{E} = \boldsymbol{E}_o cos(\boldsymbol{k} \cdot \boldsymbol{x} - \omega t) \quad \text{and} \quad \boldsymbol{H} = \boldsymbol{H}_o cos(\boldsymbol{k} \cdot \boldsymbol{x} - \omega t), \tag{10.4}$$

where $\boldsymbol{x}$ is an initial position vector to the wave front and $\boldsymbol{k}$ is the direction of propagation. For plane waves, $\boldsymbol{k} \cdot \boldsymbol{x} = constant$. We refer to the phase as $\phi = \boldsymbol{k} \cdot \boldsymbol{x} - \omega t$, and $\omega = \frac{2\pi}{\tau}$ as the angular frequency, where τ is the period. For plane waves, the wave front is a plane on which ϕ is constant, which is orthogonal to the direction of propagation, characterized by $\boldsymbol{k}$. In the case of harmonic waves, we have

$$\boldsymbol{k} \times \boldsymbol{E} = \mu_o \omega \boldsymbol{H} \quad \text{and} \quad \boldsymbol{k} \times \boldsymbol{H} = -\epsilon_o \omega \boldsymbol{E}, \tag{10.5}$$

and $\boldsymbol{k} \cdot \boldsymbol{E} = 0$ and $\boldsymbol{k} \cdot \boldsymbol{H} = 0$. The three vectors $\boldsymbol{k}$, $\boldsymbol{E}$, and $\boldsymbol{H}$ constitute a mutually orthogonal triad.[3] The direction of wave propagation is given by $\frac{\boldsymbol{E} \times \boldsymbol{H}}{||\boldsymbol{E} \times \boldsymbol{H}||}$. Electromagnetic waves traveling through space carry electromagnetic energy which flows in the direction of wave propagation. The energy per unit area per unit time flowing perpendicularly into a surface in free space is given by the Poynting vector $\boldsymbol{S} = \boldsymbol{E} \times \boldsymbol{H}$.

---

[2]The free space electric permittivity is $\epsilon_o = \frac{1}{c^2 \mu_o} = 8.8542 \times 10^{-12}\ \text{C N}^{-1}\ \text{m}^{-1}$, and the free space magnetic permeability is $\mu_o = 4\pi \times 10^{-7}\ \text{Wb A}^{-1}\text{m}^{-1} = 1.2566 \times 10^{-6}\ \text{Wb A}^{-1}\text{m}^{-1}$.

[3]By combining the relations in Eq. 10.5, one obtains $||\boldsymbol{k}|| = \frac{\omega}{c}$.

10.1.3 Special Case: Natural (Random) Electromagnetic Energy Propagation

Since at high frequencies $\boldsymbol{E}$, $\boldsymbol{H}$, and $\boldsymbol{S}$ oscillate rapidly, it is impractical to measure instantaneous values of $\boldsymbol{S}$ directly. Consider the harmonic representations in Eq. 10.4 which leads to $\boldsymbol{S} = \boldsymbol{E}_o \times \boldsymbol{H}_o cos^2(\boldsymbol{k} \cdot \boldsymbol{x} - \omega t)$, and consequently the average value over a longer time interval ($\mathcal{T}$) than the timescale of rapid random oscillation,

$$\langle \boldsymbol{S} \rangle_{\mathcal{T}} = \boldsymbol{E}_o \times \boldsymbol{H}_o \langle cos^2(\boldsymbol{k} \cdot \boldsymbol{x} - \omega t) \rangle_{\mathcal{T}} = \tfrac{1}{2} \boldsymbol{E}_o \times \boldsymbol{H}_o, \tag{10.6}$$

leading to the definition of the *irradiance*

$$I \stackrel{\text{def}}{=} \langle ||\boldsymbol{S}|| \rangle_{\mathcal{T}} = \tfrac{1}{2} ||\boldsymbol{E}_o \times \boldsymbol{H}_o|| = \tfrac{1}{2} \sqrt{\tfrac{\epsilon_o}{\mu_o}} ||\boldsymbol{E}_o||^2. \tag{10.7}$$

Thus, the rate of flow of energy is proportional to the square of the amplitude of the electric field. Furthermore, in isotropic media, which we consider for the remainder of the work, the direction of energy is in the direction of $\boldsymbol{S}$ and in the same direction as $\boldsymbol{k}$.

10.1.4 Beam Decomposition into Rays

The appendix provides more details on the theory of ray representations of electromagnetic waves. Since I is the energy per unit area per unit time, if we multiply by the "cross-sectional" area of the ray (A^r), we obtain the energy associated with an entire beam by multiplying the irradiance by the cross-sectional area of a coherent beam, IA^b, where A^b is the cross-sectional area of the beam (comprising all of the rays). The energy for a ray in the beam is then given by $IA^r = IA^b/N_r$, where N_r is the number of rays in the beam (Fig. 10.2).

The angle between the point of contact of a ray (Fig. 10.3) and the outward normal to the surface at that point is the angle of incidence (θ_i). The classical reflection law ("Snell's law") states that the angle at which the ray is reflected is the same as the angle of incidence and that the incoming (incident, θ_i) and outgoing (reflected, θ_r) rays lay in the same plane, and that $\theta_i = \theta_r$. Furthermore, the refraction/absorption law states that if the ray passes from one medium into a second one (with a different index of refraction) and if the index of refraction of the second medium is less than that of the first, then the angle the ray makes with the normal to the interface is always less than the angle of incidence, where θ_a is the angle of the absorbed ray (Fig. 10.3), $n \stackrel{\text{def}}{=} \frac{c}{v} = \sqrt{\frac{\epsilon \mu}{\epsilon_o \mu_o}} = \frac{sin\theta_i}{sin\theta_a}$, c is the propagation speed in a vacuum, and v is the propagation speed in the incident medium. By using the classical Fresnel

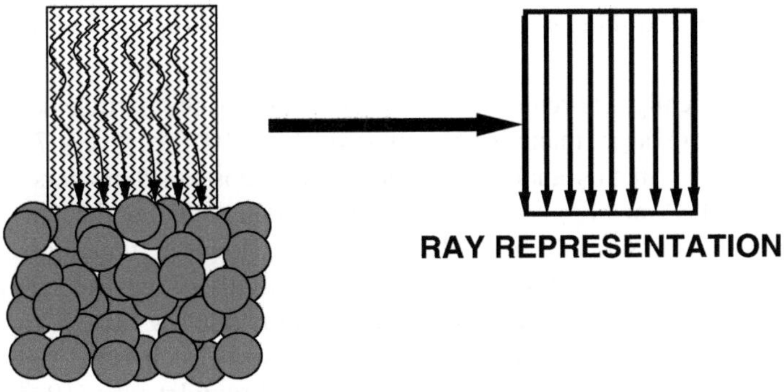

Fig. 10.2 Decomposition of an incoming beam into idealized "rays"

Fig. 10.3 Reflection and absorption of an incoming ray

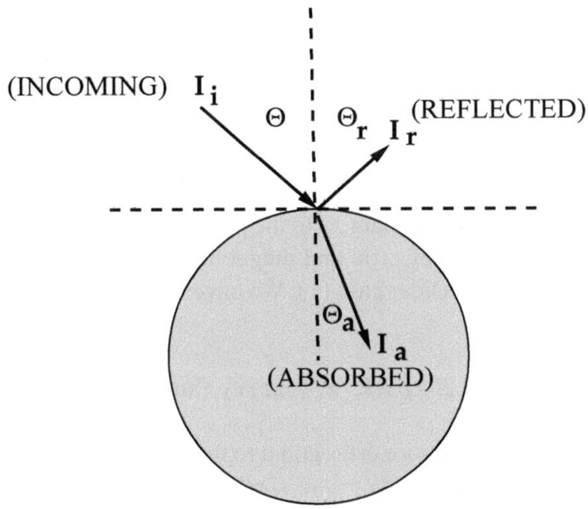

equations, one can also describe the changes in ray magnitude. For example, if we consider a ray incident upon a boundary separating two different materials, which produces a reflected ray and an absorbed (refracted) ray (Fig. 10.1), the amount of incident electromagnetic energy (I_i) that is reflected (I_r) is given by the total reflectance $I\!R \stackrel{\text{def}}{=} \frac{I_r}{I_i}$, where $0 \leq I\!R \leq 1$, $I\!R$ given by Eq. 10.25, for unpolarized electromagnetic radiation, where $\hat{n}$ is the ratio of the refractive indices of the ambient (incident) medium (n_i) and absorbing particle medium (n_a), $\hat{n} = n_a/n_i$, where $\hat{\mu}$ is the ratio of the magnetic permeabilities of the surrounding incident medium (μ_i) and absorbing particle medium (μ_a), $\hat{\mu} = \mu_a/\mu_i$. Although we will present a relatively general formulation, later we will consider applications where the magnetic permeability is, within experimental measurements, virtually the same for both the surroundings and particle. In other words, later in the work, we shall take $\hat{\mu} = 1$

($\mu_o = \mu_i = \mu_a$), and thus, $\hat{n} = \frac{n_a}{n_i} = \sqrt{\frac{\epsilon_a \mu_a}{\epsilon_i \mu_i}} \Rightarrow \epsilon_a \mu_a = (\hat{n})^2 \epsilon_i \mu_i \Rightarrow \epsilon_a = (\hat{n})^2 \epsilon_i$, where $\epsilon_i = \epsilon_o$.

Remark: Specifically, the regimes of interest are where the particle scatterers and surface features are larger than visible light rays: $3.8 \times 10^{-7}\,\mathrm{m} \leq \lambda \leq 7.2 \times 10^{-7}\mathrm{m}$. Thus, the particles in this analysis are assumed to possess diameters larger than approximately 10^{-5} m (10 μ). For particulate systems smaller than this, one can simply use the ensuing results as qualitative guides. However, the range of applicability of lasers is not limited to visible frequencies. Other high-frequency applications where the developed model can be employed include UV rays, X-rays, gamma rays, and correspondingly smaller particle scatterer sizes, such as (a) regimes where the scatterers and surface features are larger than ultraviolet rays ($10^{-9}\,\mathrm{m} \leq \lambda \leq 10^{-8}\,\mathrm{m}$), (b) regimes where the scatterers and surface features are larger than X-rays ($10^{-11}\,\mathrm{m} \leq \lambda \leq 10^{-9}\,\mathrm{m}$), and (c) regimes where the scatterers and surface features are larger than gamma rays ($10^{-12}\,\mathrm{m} \leq \lambda \leq 10^{-11}\,\mathrm{m}$).

Reflection and absorption of energy—Fresnel relations

We consider a plane harmonic wave incident upon a plane boundary separating two different materials, which produces a reflected wave and an absorbed (refracted) wave (Fig. 10.1). Two cases for the electric field vector are considered: (1) electric field vectors that are parallel ($||$) to the plane of incidence and (2) electric field vectors that are perpendicular ($\perp$) to the plane of incidence. In either case, the tangential components of the electric and magnetic fields are required to be continuous across the interface. Consider case (1). We have the following general vectorial representations

$$\boldsymbol{E}_{||} = E_{||} cos(\boldsymbol{k} \cdot \boldsymbol{x} - \omega t)\, \boldsymbol{e}_1 \quad \text{and} \quad \boldsymbol{H}_{||} = H_{||} cos(\boldsymbol{k} \cdot \boldsymbol{x} - \omega t)\, \boldsymbol{e}_2, \tag{10.8}$$

where $\boldsymbol{e}_1$ and $\boldsymbol{e}_2$ are orthogonal to the propagation direction $\boldsymbol{k}$. By employing the law of refraction ($n_i sin\theta_i = n_a sin\theta_a$), we obtain the following conditions relating the incident, reflected, and absorbed components of the electric field quantities

$$E_{||i} cos\theta_i - E_{||r} cos\theta_r = E_{||a} cos\theta_a \quad \text{and} \quad H_{\perp i} + H_{\perp r} = H_{\perp a}. \tag{10.9}$$

Since, for plane harmonic waves, the magnetic and electric field amplitudes are related by $H = \frac{E}{v\mu}$, we have

$$E_{||i} + E_{||r} = \frac{\mu_i}{\mu_a} \frac{v_i}{v_a} E_{||a} = \frac{\mu_i}{\mu_a} \frac{n_a}{n_i} E_{||a} \stackrel{\text{def}}{=} \frac{\hat{n}}{\hat{\mu}} E_{||a}, \tag{10.10}$$

where $\hat{\mu} \stackrel{\text{def}}{=} \frac{\mu_a}{\mu_i}$, $\hat{n} \stackrel{\text{def}}{=} \frac{n_a}{n_i}$, and v_i, v_r, and v_a are the values of the velocity in the incident, reflected, and absorbed directions.[4] By again employing the law of refraction, we

[4] Throughout the analysis, we assume that $\hat{n} \geq 1$.

obtain the Fresnel reflection and transmission coefficients, generalized for the case of unequal magnetic permeabilities

$$r_{||} = \frac{E_{||r}}{E_{||i}} = \frac{\frac{\hat{n}}{\mu}cos\theta_i - cos\theta_a}{\frac{\hat{n}}{\mu}cos\theta_i + cos\theta_a} \quad \text{and} \quad a_{||} = \frac{E_{||a}}{E_{||i}} = \frac{2cos\theta_i}{cos\theta_a + \frac{\hat{n}}{\mu}cos\theta_i}. \tag{10.11}$$

Following the same procedure for case (2), where the components of $\boldsymbol{E}$ are perpendicular to the plane of incidence, we have

$$r_{\perp} = \frac{E_{\perp r}}{E_{\perp i}} = \frac{cos\theta_i - \frac{\hat{n}}{\mu}cos\theta_a}{cos\theta_i + \frac{\hat{n}}{\mu}cos\theta_a} \quad \text{and} \quad a_{\perp} = \frac{E_{\perp a}}{E_{\perp i}} = \frac{2cos\theta_i}{cos\theta_i + \frac{\hat{n}}{\mu}cos\theta_a}. \tag{10.12}$$

Our primary interest is in the reflections. We define the reflectances as

$$I\!R_{||} \stackrel{\text{def}}{=} r_{||}^2 \quad \text{and} \quad I\!R_{\perp} \stackrel{\text{def}}{=} r_{\perp}^2. \tag{10.13}$$

Particularly convenient forms for the reflections are

$$r_{||} = \frac{\frac{\hat{n}^2}{\mu}cos\theta_i - (\hat{n}^2 - sin^2\theta_i)^{\frac{1}{2}}}{\frac{\hat{n}^2}{\mu}cos\theta_i + (\hat{n}^2 - sin^2\theta_i)^{\frac{1}{2}}} \quad \text{and} \quad r_{\perp} = \frac{cos\theta_i - \frac{1}{\mu}(\hat{n}^2 - sin^2\theta_i)^{\frac{1}{2}}}{cos\theta_i + \frac{1}{\mu}(\hat{n}^2 - sin^2\theta_i)^{\frac{1}{2}}}. \tag{10.14}$$

Thus, the total energy reflected can be characterized by

$$I\!R \stackrel{\text{def}}{=} \left(\frac{E_r}{E_i}\right)^2 = \frac{E_{\perp r}^2 + E_{||r}^2}{E_i^2} = \frac{I_{||r} + I_{\perp r}}{I_i}. \tag{10.15}$$

If the resultant plane of oscillation of the (polarized) wave makes an angle of γ_i with the plane of incidence, then

$$E_{||i} = E_i cos\gamma_i \quad \text{and} \quad E_{\perp i} = E_i sin\gamma_i, \tag{10.16}$$

and it follows from the previous definition of I that

$$I_{||i} = I_i cos^2\gamma_i \quad \text{and} \quad I_{\perp i} = I_i sin^2\gamma_i. \tag{10.17}$$

Substituting these expressions back into the expressions for the reflectances yields

$$I\!R = \frac{I_{||r}}{I_i} cos^2\gamma_i + \frac{I_{\perp r}}{I_i} sin^2\gamma_i = I\!R_{||} cos^2\gamma_i + I\!R_{\perp} sin^2\gamma_i. \tag{10.18}$$

For natural or unpolarized electromagnetic radiation, the angle γ_i varies rapidly in a random manner, as does the field amplitude. Thus, since

$$\langle cos^2\gamma_i(t)\rangle_T = \tfrac{1}{2} \text{ and } \langle sin^2\gamma_i(t)\rangle_T = \tfrac{1}{2}, \tag{10.19}$$

and therefore for natural electromagnetic radiation

$$I_{||i} = \tfrac{I_i}{2} \text{ and } I_{\perp i} = \tfrac{I_i}{2}. \tag{10.20}$$

and therefore

$$r_{||}^2 = \left(\tfrac{E_{||r}^2}{E_{||i}^2}\right)^2 = \tfrac{I_{||r}}{I_{||i}} \text{ and } r_{\perp}^2 = \left(\tfrac{E_{\perp r}^2}{E_{\perp i}^2}\right)^2 = \tfrac{I_{\perp r}}{I_{\perp i}}. \tag{10.21}$$

Thus, the total reflectance becomes

$$I\!R = \frac{1}{2}(I\!R_{||} + I\!R_{\perp}) = \frac{1}{2}(r_{||}^2 + r_{\perp}^2), \tag{10.22}$$

where $0 \leq I\!R \leq 1$. For the cases where $sin\theta_a = \frac{sin\theta_i}{\hat{n}} > 1$, one may rewrite reflection relations as

$$r_{||} = \frac{\frac{\hat{n}^2}{\hat{\mu}} cos\theta_i - j(sin^2\theta_i - \hat{n}^2)^{\frac{1}{2}}}{\frac{\hat{n}^2}{\hat{\mu}} cos\theta_i + j(sin^2\theta_i - \hat{n}^2)^{\frac{1}{2}}} \text{ and } r_{\perp} = \frac{cos\theta_i - \frac{1}{\hat{\mu}} j(sin^2\theta_i - \hat{n}^2)^{\frac{1}{2}}}{cos\theta_i + \frac{1}{\hat{\mu}} j(sin^2\theta_i - \hat{n}^2)^{\frac{1}{2}}}, \tag{10.23}$$

where $j = \sqrt{-1}$, and in this complex case[5]

$$I\!R_{||} \stackrel{\text{def}}{=} r_{||}\bar{r}_{||} = 1, \text{ and } I\!R_{\perp} \stackrel{\text{def}}{=} r_{\perp}\bar{r}_{\perp} = 1, \tag{10.24}$$

where $\bar{r}_{||}$ and $\bar{r}_{\perp}$ are complex conjugates. Thus, for angles above the critical angle θ_i^*, all of the energy is reflected. Notice that as $\hat{n} \to 1$ we have complete absorption, while as $\hat{n} \to \infty$ we have complete reflection. The amount of absorbed irradiance by the particles is $I_a = (1 - I\!R)I_i$.

[5]The limiting case $\frac{sin\theta_i^*}{\hat{n}} = 1$ is the critical angle (θ_i^*) case.

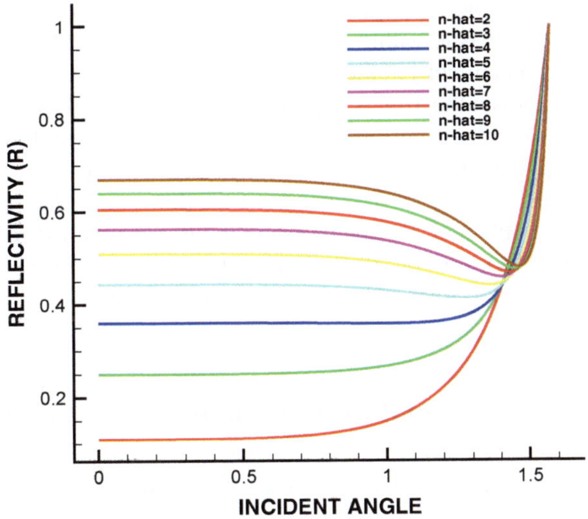

Fig. 10.4 Reflectance ($I\!R$) as a function of incident angle

Reflectivity

To observe the dependency of $I\!R$ on $\hat{n}$ and θ_i, we can explicitly write

$$I\!R = \frac{1}{2}\left(\left(\frac{\frac{\hat{n}^2}{\hat{\mu}}cos\theta_i - (\hat{n}^2 - sin^2\theta_i)^{\frac{1}{2}}}{\frac{\hat{n}^2}{\hat{\mu}}cos\theta_i + (\hat{n}^2 - sin^2\theta_i)^{\frac{1}{2}}}\right)^2 + \left(\frac{cos\theta_i - \frac{1}{\hat{\mu}}(\hat{n}^2 - sin^2\theta_i)^{\frac{1}{2}}}{cos\theta_i + \frac{1}{\hat{\mu}}(\hat{n}^2 - sin^2\theta_i)^{\frac{1}{2}}}\right)^2\right), \tag{10.25}$$

which is plotted in Fig. 10.4. Except for the case of $\hat{n} > 4$, there is discernible nonmonotone behavior. The nonmonotone behavior is slight for $\hat{n} = 4$, but nonetheless present. Clearly, as $\hat{n} \to \infty$, $I\!R \to 1$, no matter what the angle of incidence's value is. We note that as $\hat{n} \to 1$, provided that $\hat{\mu} = 1$, $I\!R \to 0$, i.e., all incident energy is absorbed (it is transparent). With increasing $\hat{n}$, the angle for minimum reflectance grows larger. For more details, we refer the reader to the relatively recent treatise of Gross [1] and the cited literature in the references.

Remark 1 From this point forth, we assume that the ambient and interstitial medium (surrounding the particles) behaves as a vacuum. Accordingly, there are no energetic losses as the rays move through the surrounding medium. Furthermore, we assume that absorbed rays that enter a particle are not re-emitted, but are converted into a heat source.

Remark 2 We note that the use of lasers for the related problem of dermal ablation is well-established and involves the use of special types of dyes to increase absorption of the tissue, referred to as sclerostomics. The dyes are applied by electrophoresis, i.e., electrical current is used to direct dye into tissue. There are five main types of interaction: (1) photochemical, (2) thermal, (3) photoablation, (4) plasma-induced ablation, and (5) photodisruption. Photochemical interaction stems from empirical observations that light can induce chemical effects and reactions within macromolecules or tissue within macromolecules or tissue. Biostimulation is also attributed to photochemical interaction. Photodynamics therapy is performed as follows: First, a photosynthesizer is injected into the tissue. It remains inactive until irradiated. This can be used for targeted interaction. Thermal interaction can be classified as (a) coagulation, (b) vaporization, (c) carbonization, and (d) melting. The probability of cells staying alive depends on the duration and temporal evolution of the temperature obtained. Photoablation etching was first applied to polymethyl methacrylate (PMMA), polyimide, Teflon, and other synthetic polymers. For more details, see Niemz [2], Steen [3], and Grigoropoulos [4].

10.2 Thermal Conversion of Beam (Optical) Losses

It is assumed that the particle scatterers are small enough to consider the temperature as being uniform within each particle. Furthermore, we assume that the space between the particle scatterers, i.e., the "ether," plays no role in the heat transfer process. Conduction is assumed to occur between particles in contact. An energy balance governing thermal storage in a particle (i), the absorption of optical energy from rays which come in contact (N_{rc}), and all conductive exchange with other particles in contact (N_{pc}) reads as

$$m_i C_i \dot{\theta}_i = \underbrace{\sum_{j=1}^{N_{pc}} \mathcal{Q}_{ij}}_{\text{conduction}} + \underbrace{\sum_{k=1}^{N_{rc}} \mathcal{H}_{ik}^{rays}}_{\text{ray sources}} \stackrel{\text{def}}{=} \mathcal{F}_i^{tot}. \tag{10.26}$$

We remark that the validity of using a lumped thermal model for each particle, i.e., ignoring temperature gradients and assuming a uniform temperature within a particle scatterer, is dictated by the magnitude of the Biot number. A small Biot number (significantly less than unity) indicates that such an approximation is reasonable. The Biot number for spheres scales with the ratio of particle scatterer volume (V) to particle scatterer surface area (A_s), $\frac{V}{A_s} = \frac{\mathcal{R}}{3}$ ($\mathcal{R}$ is the particle radius), which indicates that a uniform temperature distribution is appropriate, since the particle scatterers, by definition, are small. For the conductive contribution, we have[6]

$$\sum_{j=1}^{N_{pc}} \mathcal{Q}_{ij} \approx \sum_{j=1}^{N_{pc}} I\!K_{ij} A_{ij}^{c} \frac{\theta_j - \theta_i}{||\boldsymbol{r}_j - \boldsymbol{r}_i||} \stackrel{\text{def}}{=} \mathcal{Q}_i^{tot}. \tag{10.27}$$

The energy absorbed by a particle i from a ray in contact ($k = 1, \ldots, N_{pc}$) is defined as

[6] $I\!K_{ij}$ is approximated by the average interfacial value of the $i - j$ pair, $I\!K_{ij} \approx \frac{I\!K_i + I\!K_j}{2}$.

$$\Delta\mathcal{H}_i^{rays,tot} \stackrel{\text{def}}{=} \int_t^{t+\Delta t} \sum_{k=1}^{N_{rc}} \mathcal{H}_{ik}^{rays}\, dt \approx \sum_{k=1}^{N_{rc}} (I_{ik}^{inc} - I_{ik}^{ref}) A_r \Delta t = \sum_{k=1}^{N_{rc}} (1 - I\!R_{ik}) I_{ik}^{inc} A_r \Delta t, \tag{10.28}$$

where I^{inc} is the incoming ray's irradiance, I^{ref} is the reflected ray's irradiance, A_r is a ray-area parameter that is computed by taking the total initial cross-sectional area of the entire beam (collimated rays) and dividing it by the total number of rays. *As stated previously, explicitly, the ray-area parameter was calculated as* $A_r \stackrel{\text{def}}{=} A^b/N_r$, *where* N_r *is the number of rays in the beam and* A^b *is the initial cross-sectional area of the beam.*

Remark 1 Convective and infrared radiative effects are considered of secondary importance in the current analysis, but have been accounted for in Zohdi [7] for related applications.

Remark 2 From this point forth, we will denote I^{inc} instead of I_i and I^{ref} instead of I_r to avoid any possible confusion with subscripts to come later.

10.2.1 Algorithmic Details

After temporal integration with a finite difference time-step of Δt, we have, using a trapezoidal rule with variable ($0 \le \phi \le 1$) integration metric:

$$\begin{aligned}
\theta_i(t+\Delta t) &= \theta_i(t) + \frac{1}{m_i C_i} \int_t^{t+\Delta t} \mathcal{F}_i^{tot}\, dt \qquad (10.29)\\
&= \theta_i(t) + \frac{1}{m_i C_i} \left(\underbrace{\int_t^{t+\Delta t} \sum_{k=1}^{N_{pc}} \mathcal{Q}_{ik}\, dt}_{\mathcal{Q}_i^{tot}} + \underbrace{\int_t^{t+\Delta t} \sum_{j=1}^{N_{rc}} \mathcal{H}_{ij}^{rays}\, dt}_{\Delta\mathcal{H}_i^{rays,tot}} \right)\\
&\approx \theta_i(t) + \frac{1}{m_i C_i} \left(\underbrace{\left(\phi \mathcal{Q}_i^{tot}(t+\Delta t) + (1-\phi)\mathcal{Q}_i^{tot}(t)\right)\Delta t}_{smooth} + \underbrace{\Delta\mathcal{H}_i^{rays,tot}}_{unsmooth} \right),
\end{aligned}$$

where the term "smooth" indicates temporally smoothly varying and "unsmooth" denotes a much shorter discrete event (incident rays encountering and reflecting off of a particle). Thus, explicitly,

$$\theta_i^{K+1}(t+\Delta t) = \theta_i(t) + \frac{\Delta t \phi}{m_i C_i} \sum_{j=1}^{N_{pc}} \frac{I\!K_{ij} A_{ij}^c (\theta_j^K(t+\Delta t) - \theta_i^K(t+\Delta t))}{||\boldsymbol{r}_j - \boldsymbol{r}_i||}$$

$$+ \frac{\Delta t(1-\phi)}{m_i C_i} \sum_{j=1}^{N_{pc}} \frac{I\!K_{ij} A_{ij}^c (\theta_j(t) - \theta_i(t))}{||\boldsymbol{r}_j - \boldsymbol{r}_i||} + \frac{\Delta \mathcal{H}_i^{rays,tot}}{m_i C_i}. \quad (10.30)$$

This iterative procedure is embedded into the overall ray-tracing scheme. The overall algorithm is as follows, starting at $t = 0$ and ending at $t = T$:

1. **COMPUTE NEW RAY MAGNITUDES AND ORIENTATIONS AFTER REFLECTION:** I_j^{ref}, $j = 1, 2, \ldots, Rays$.
2. **COMPUTE ABSORPTION CONTRIBUTIONS TO THE PARTICLE SCATTERERS:** $\Delta \mathcal{H}_i^{rays,tot}$, $i = 1, 2, \ldots, Particles$.
3. **COMPUTE CONDUCTION OF THE PARTICLE SCATTERERS:** $\mathcal{Q}_i^{tot}$, $i = 1, 2, \ldots, Particles$.
4. **COMPUTE PARTICLE TEMPERATURES RECURSIVELY K=1,2,..., UNTIL CONVERGENCE, USING EQUATION** 10.30 **FOR** $\theta_i, i = 1, 2, \ldots, Particles$, **REPEATING STEPS 1–4.**
5. **INCREMENT ALL RAY POSITIONS:** $\boldsymbol{x}_j(t+\Delta t) = \boldsymbol{x}_j(t) + \Delta t \boldsymbol{v}_j(t)$, $j = 1, 2, \ldots, Rays$.
6. **INCREMENT TIME** ($t = t + \Delta t$), **GO TO STEP 1 AND REPEAT STEPS 1–5.**

Remark: In order to capture all of the ray reflections that occur, the time-step size Δt is dictated by the size of the particle scatterers. A somewhat ad hoc approach is to scale the time-step size by the speed of the ray propagation according to $\Delta t = \xi \frac{\mathcal{R}}{||\boldsymbol{v}||}$, where $\mathcal{R}$ is the median radius of the particle scatterers (if the particle sizes are not uniform) and $0.05 \le \xi \le 0.1$. Typically, the results are insensitive to ξ that are smaller than this range. The subsequent convergence of the thermal calculation is rather quick, since the time-steps are extremely small. For more details on convergence on iterative time-stepping schemes, see Zohdi [5–27].

10.3 Phase Transformations: *Solid* ⇒ *Liquid* ⇒ *Vapor*

As before, to include phase transformations, we consider five cases, which are implemented in a predictor-corrector manner by first solving the governing equations to obtain predicted temperature, and then checking the following:

- *Solid → solid-no melting with* $C_i = C_S$: If $\theta(t) < \theta_m$ and $\theta(t+\Delta t) < \theta_m$ then retain $C(\theta) = C_S$,
- *Solid → liquid-melting with* $C_i = C_S$: If $\theta(t) < \theta_m$ and $\theta(t+\Delta t) \ge \theta_m$ then re-solve the governing equations with $C(\theta) = C_S + \frac{\delta \mathcal{P}^{S \to L}}{\delta \theta}$,
- *Liquid → liquid-melted with* $C_i = C_L$: If $\theta(t) \ge \theta_m$ and $\theta(t+\Delta t) \ge \theta_m$ then retain $C(\theta) = C_L$,

- *Liquid → solid-solidification with* $C_i = C_L$: If $\theta(t) \geq \theta_m$ and $\theta(t + \Delta t) < \theta_m$ then re-solve the governing equations with $C(\theta) = C_L + \frac{\delta \mathcal{P}^{L \to S}}{\delta \theta}$,
- *Liquid → vapor-vaporization with* $C_i = C_L$: If $\theta(t) < \theta_v$ and $\theta(t + \Delta t) \geq \theta_v$ then re-solve the governing equations with $C(\theta) = C_L + \frac{\delta \mathcal{P}^{L \to V}}{\delta \theta}$,

where C_S is the heat capacity of the solid and C_L is the heat capacity of the liquid and

- $0 < \delta \mathcal{P}^{S \to L}$ is the latent heat of melting,
- $0 < \delta \mathcal{P}^{L \to S}$ is the latent heat of solidification,
- $0 < \delta \mathcal{P}^{L \to V}$ is the latent heat of vaporization,
- $0 < \delta \theta$ is small and can be thought of as a "bandwidth" for a phase transformation. For more details on melting processes, see Davis [28].

Remark 1 Latent heats have a tendency to resist the phase transformations, achieved by adding the positive terms in the denominator, thus enforcing reduced temperature (during the phase transformation).[7] This approach is relatively straightforward to include within the staggering framework.

Remark 2 We note that, because of changes in the heat capacity (due to phase transformations), the thermal storage term is no longer simply $m_i C_i \dot{\theta}_i$ but is

$$m_i \frac{d}{dt}(C_i \theta_i - \theta_o) = m_i \dot{C}_i (\theta_i - \theta_o) + m_i C_i \dot{\theta}_i. \tag{10.31}$$

Remark 3 As a consequence, the number of particles in the system and their heat capacities will also change in the algorithm, and Eq. 10.30 becomes

$$\begin{aligned}
\theta_i^{K+1}(t + \Delta t) = \theta_i(t) &+ \frac{\Delta t \phi}{m_i C_i^K (t + \phi \Delta t)} \sum_{j=1}^{N_{pc}^K} \frac{I\!K_{ij} A_{ij}^c (\theta_j^K (t + \Delta t) - \theta_i^K (t + \Delta t))}{||\boldsymbol{r}_j - \boldsymbol{r}_i||} \\
&+ \frac{\Delta t (1 - \phi)}{m_i C_i^K (t + \phi \Delta t)} \sum_{j=1}^{N_{pc}} \frac{I\!K_{ij} A_{ij}^c (\theta_j(t) - \theta_i(t))}{||\boldsymbol{r}_j - \boldsymbol{r}_i||} + \frac{\Delta \mathcal{H}_i^{rays,tot,K}}{m_i C_i^K (t + \phi \Delta t)} \\
&- \Delta t \frac{m_i \dot{C}_i (t + \phi \Delta t)(\theta_i (t + \phi \Delta t) - \theta_o)}{m_i C_i^K (t + \phi \Delta t)}.
\end{aligned} \tag{10.32}$$

Remark 4 Vaporized particles are removed from the simulation for subsequent time-steps.

10.3.1 Optional Time Scaling and Simulation Acceleration

Some scaling arguments can be made to accelerate computations. Consider a system that is subjected to a source $\mathcal{F}(t)$

[7] In the idealized limit, the temperature would be constant.

$$mC\dot{\theta} = \mathcal{F}(t). \tag{10.33}$$

Integrating overall time interval T and breaking it up into M subintervals yield $(\tau = \frac{T}{M})$

$$\theta(T) = \theta(t=0) + \underbrace{\frac{1}{mC}\int_{o}^{T} \mathcal{F}\,dt}_{\text{heat input}} \tag{10.34}$$

$$= \theta(t=0) + \underbrace{\frac{1}{mC}\left(\int_{0}^{\tau} \mathcal{F}\,dt + \int_{\tau}^{2\tau} \mathcal{F}\,dt + \cdots + \int_{(M-1)\tau}^{M\tau} \mathcal{F}\,dt\right)}_{\text{heat input}}.$$

If $\mathcal{F}(t)$ is periodic over each subinterval $((i-1)\tau, i\tau)$, $i = 1, 2, \ldots, M$, then

$$\theta(T) = \theta(t=0) + \underbrace{\frac{1}{mC}\left(\int_{0}^{\tau} \mathcal{F}\,dt + \int_{\tau}^{2\tau} \mathcal{F}\,dt + \cdots + \int_{(M-1)\tau}^{M\tau} \mathcal{F}\,dt\right)}_{\text{heat input}}$$

$$= \theta(t=0) + \frac{M}{mC}\int_{0}^{\tau} \mathcal{F}\,dt$$

$$= \theta(t=0) + \frac{1}{\underbrace{\left(\frac{mC}{M}\right)}_{pseudo\,thermal\,mass}}\int_{0}^{\tau} \mathcal{F}\,dt, \tag{10.35}$$

where the last line is an approximation of the heat input "dumped" into the system. One can interpret $\frac{mC}{M}$ as a pseudo (reduced) thermal mass. Specifically, for the applications of interest here, if absorbed irradiance loading $\Delta\mathcal{H}^{rays,tot}$ was perfectly periodic (repeated over shorter time intervals), with no phase changes (which would change C and the $\Delta\mathcal{H}^{rays,tot}$ absorbed), and if the conductive flux terms were constant (it is not in the general case), then running the simulation over $(0, \tau)$, using $\frac{mC}{M}$ as a pseudo (reduced) thermal mass, would be equivalent to a simulation over the interval $(0, T)$ using mC. *Thus, as an optional approximation*, either by magnifying I or reducing mC, one can deliver the approximate "heat dump," provided one fully resolves the movement of a pulse through the entire system to accurately capture the share of the energy reflected and absorbed by the particles.

Remark: In the simple case

$$m C\dot{\theta} = \mathcal{F} = I A^b \Rightarrow \theta(T) = \theta(0) + \frac{I A^b T}{mC} = \theta(0) + M\frac{I A^b \tau}{mC} = \theta(0) + \frac{I A^b \tau}{\left(\frac{mC}{M}\right)}, \tag{10.36}$$

the periodic approximation is exact.

10.4 Numerical Examples

As a model problem, we considered a group of N_p overlapping randomly packed spherical particles, of equal size, in a cubical domain of dimensions, $D \times D \times D$. The particle size was determined by a particle/sample size ratio, which was defined via a subvolume size $V \stackrel{\text{def}}{=} \frac{D\times D\times D}{N_p}$, where N_p was the number of particles in the entire cubical domain. The nondimensional ratio between the radius ($\mathcal{R}$) and the subvolume was denoted by $\mathcal{L} \stackrel{\text{def}}{=} \frac{\mathcal{R}}{V^{\frac{1}{3}}}$. If the particles were perfectly arranged in a cubical array, a value of $\mathcal{L} = 0.25$ would mean that they just touch, and values of $\mathcal{L} > 0.25$ would indicate overlapping. Later, in the simulations was used $\mathcal{L} = 0.325$ as an example. We steadily increased the number of rays in the beam from $N_r = 100, 200, etc$ until the results were insensitive to further refinements. This approach indicated that approximately $N_r = 1000$ parallel, but randomly placed, rays in the circular cross-sectional plane of the beam (Fig. 10.5), corresponding to unpolarized incoming

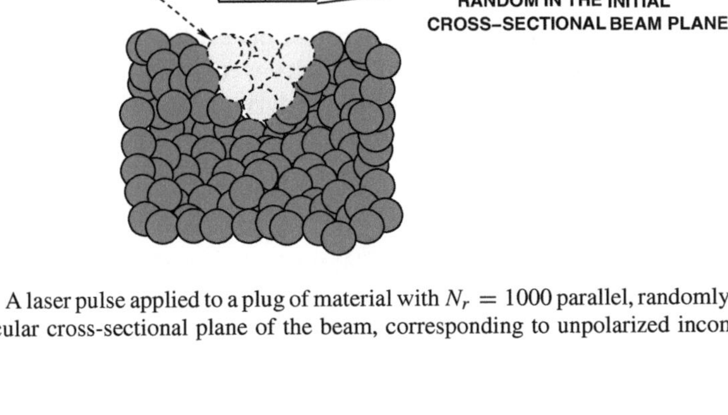

Fig. 10.5 A laser pulse applied to a plug of material with $N_r = 1000$ parallel, randomly placed rays in the circular cross-sectional plane of the beam, corresponding to unpolarized incoming optical energy

optical energy, yielded stable results.[8] Therefore, we consider the responses to be, for all practical purposes, independent of the ray-grid density. This particle/ray system provided convergent results, i.e., increasing the number of rays and/or the number of particles surrounding the beam resulted in negligibly different overall system responses. Of course, there can be cases where much higher resolution may be absolutely necessary.[9] In such a direct numerical approach, one can easily introduce nonuniform beam profiles such as shown in Fig. 10.1

$$I(d) = I(d = 0)e^{-ad}, \tag{10.37}$$

where d is the distance from the center of the initial beam line. In the case of $a = 0$, we recapture a flat beam, $I(d) = I(d = 0)$. We specifically used $a = 2$. We set total initial irradiance via $\sum_{i=1}^{N_r} I_i^{inc}(t = 0)A_r = 6000$ W.[10] Other simulation parameters of importance were:

- The dimensions of the sample were 10^{-3} m $\times$ 10^{-3} m $\times$ 10^{-3} m.
- The initial velocity vector for all initially collinear rays comprising the beam was $\boldsymbol{v} = (c, 0, 0)$, where $c = 3 \times 10^8$ m/s is the speed of light in a vacuum.
- Twenty pulses were applied, allowing each pulse to completely enter and exit the system.
- We used a refractive index ratio of $\hat{n} = 1.4$ for the set of particle scatterers and surrounding environment and enforced a condition that if the material melted, we reduced the refractive index ratio to approximately $\hat{n} = 1$.
- We used a melting temperature of $\theta_m = 600°$ K and a vaporization temperature of $\theta_v = 700°$ K.
- The time-scaling factor, $M = 10^6$
- The material density, $\rho = 1000\,\mathrm{kg/m^3}$.
- The solid heat capacity, $C_S = 1000\,\mathrm{J/kg-K}$.
- The liquid heat capacity, $C_L = 2000\,\mathrm{J/kg-K}$.
- The latent heat of melting, $\frac{\delta \mathcal{P}^{S\to L}}{\delta\theta} = 1000\,\mathrm{J/kg-K}$.
- The latent heat of solidification, $\frac{\delta \mathcal{P}^{L\to S}}{\delta\theta} = 1000\,\mathrm{J/kg-K}$.
- The latent heat of vaporization, $\frac{\delta \mathcal{P}^{L\to V}}{\delta\theta} = 1000\,\mathrm{J/kg-K}$.

We considered a series of numerical experiments starting at $I\!K = 0$ (no conductivity), where a maximum number of particles were vaporized, until a $I\!K$ was met where virtually no particles were vaporized during the process. While it is not the objective of this work to carry out a series of parameter studies, we show some

[8] We repeatedly refined the "ray density" up to 10000 rays and found no significant difference compared to the 1000-ray result.

[9] In order to adequately compare between different tests in Table 10.1, the same random 1000-particle ensemble was used each time.

[10] To achieve this distribution, we first placed rays randomly in the plane, and then scaled the individual I^{inc} by e^{-ad} and normalized the average so that the total was 6000 W.

Table 10.1 Response sensitivity to 1000 rays for 1000 randomly dispersed particle scatterers for increasing $I\!K$ after a sufficiently long time $t = T$, allowing all of the rays to have exited the scattering system ($\mathcal{L} = 0.325$). Twenty pulses were applied. In order to adequately compare between different tests, the same random 1000-particle ensemble was used each time

$I\!K$	Vaporized (v_v)
0	0.162
1	0.159
10	0.149
20	0.135
40	0.095
60	0.071
80	0.040
100	0.037
120	0.022
140	0.020
200	0.017
300	0.008
400	0.006

trends in Fig. 10.7 and Table 10.1. Specifically, Fig. 10.7 illustrates the volume fraction of particles that are vaporized for increasing $I\!K$. The final results are given in Table 10.1. As theoretically predicted, with low $I\!K$ values, the heat is retained, and virtually all of the particles underneath the beam were vaporized. As $I\!K$ increases, the heat conducts away from the target zone, and less particles meet the criteria for melting and vaporization. In Fig. 10.6, starting from left to right and top to bottom, the progressive movement of rays comprising a beam whose initial diameter is 30 percent of the length of the sample, as it is being absorbed by the material, is shown. We have the following general observations, which are consistent with the analytical results:

- No conductivity retains heat for vaporization—it concentrates the heat in the target zone and more particles are vaporized in a cascading effect because of the reduction of the refractive index, which absorbs more optical energy.
- As the conductivity increases, heat is conducted away from the target zone, and less particles are vaporized.

The utility of the numerical simulation tool is, of course, that one can probe deeper into the spatial evolution of the temperature, phase changes, and vaporization.
Remark: *The simulations take on the order of two minutes on a laptop.* For extremely large-scale systems, it is important to note that for the ray-tracing method, there are two natural ways to proceed to exploit parallelism: (1) By assigning each processor its share of the rays, and checking which particle scatterers interact with those rays or (2) By assigning each processor its share of particle scatterers, and checking which rays

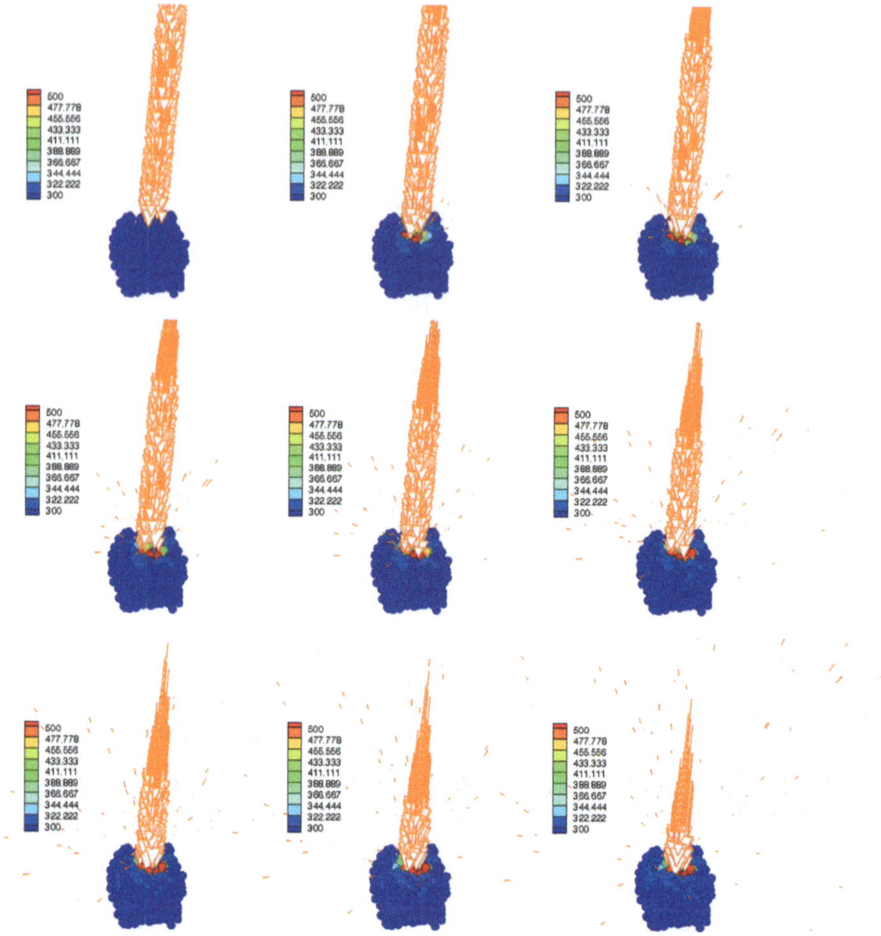

Fig. 10.6 Frames for the progressive absorption of a laser beam and subsequent heating for conductivity $I\!K = 100$

interact with those particle scatterers. High-performance computational methods for the determination of ray/particle intersection can be developed by slightly modifying fast ray-particle contact detection algorithms found in, for example, Donev et al. [29–33] or Pöschel and Schwager [34], for general particle shapes.

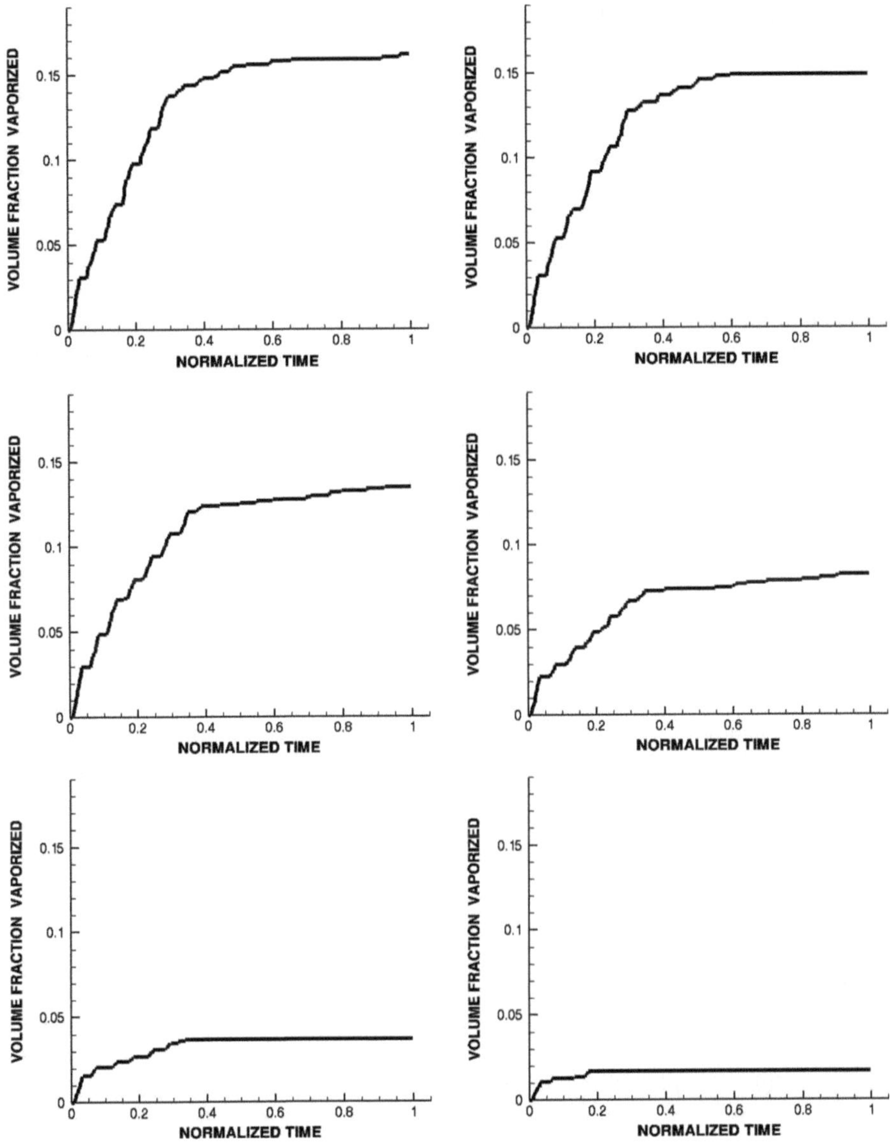

Fig. 10.7 Volume fraction of vaporized material for $I\!K = 0$, $I\!K = 10$, $I\!K = 20$, $I\!K = 50$, $I\!K = 100$, and $I\!K = 200$

10.5 Summary and Extensions

This work developed a computational model and a corresponding solution algorithm for the rapid simulation of the laser processing of materials. The objective was to simulate targeted zonal heating of discrete particles that are packed together, in particular focusing on phase transformations and vaporization of particles in the target zone. Because of the complex microstructure, containing gaps and interfaces, this type of system is extremely difficult to simulate using continuum-based methods, such as the Finite Difference Time Domain Method or the Finite Element Method. The model that was developed captured the primary effects, namely reflection and absorption of optical energy, its conversion into heat via (1) a discretization of a concentrated laser beam into rays, (2) a discrete element representation of the particulate material microstructure, and (3) a discrete element transient heat transfer model that accounts for optical (laser) energy propagation (reflection and absorption), its conversion into heat, the subsequent conduction of heat and phase transformations involving melting and vaporization. A discrete ray-tracking algorithm was developed, along with an embedded, staggered, iterative solution scheme, in order to calculate the optical-to-thermal conversion, particle-to-particle conduction, and phase transformations, implicitly. The current work is focusing on laser treatment of different types of microstructures that arise in other processing methods of particulate materials, such as self-assembly, which leads to very regular patterns that are induced by inserting the particles in solution, depositing them onto surfaces, and evaporating the interstitial liquid, thus allowing hydrodynamic forces to induce the mentioned patterns (Choi et al. [35–37], Demko et al. [59, 60]). Another example where structured media can occur is spray processing of small-scale particles, which can induce chainlike microstructures, due to surface tension and alignment with electrical field lines during deposition (Zohdi [17]). The application of laser processing to these materials is under investigation by the author.[11]

An overall technological goal is to develop computational tools to accelerate the manufacturing of printed electronics. *Lasers can play a central role in precisely processing these systems.* As mentioned earlier, print-based methods are ideal for large-scale applications and provide a framework for assembling electronic circuits by mounting printed electronic devices on flexible plastic substrates, such as polyimide and "PEEK" (polyether ether ketone, a flexible thermoplastic polymer) film. There are many types of this technology, often referred to as flexible electronics or flex circuits. Flex circuits can be, for example, screen-printed silver circuits on polyester.[12] In order to develop flexible microelectronics for large area deployment, traditional methods of fabrication using silicon-based approaches have become limited, due to high cost of materials and equipment (which frequently needs a vacuum environment). For flexibility and lower cost, the ability to develop these electronics on plastics is necessary. Accordingly, print-based technologies are starting to

[11] For review of the wide range of particle deposition processes, in particular sprays, see Martin [38, 39].

[12] For an early history of the printed electronics field, see Gamota et al. [40].

become popular for these applications. In many cases, this requires the development of high volume fraction particle-functionalized "inks." These particles include germanium (which has higher mobility and a more tailorable absorption spectrum for ambient light than silicon) and silver (which is being studied due to the possibility to sinter the particles even without the need high-intensity heating). Other semiconductor particle-based materials, including zinc- and cadmium-based compounds and metals, such as gold and copper, can be considered. Precise patterning of (particle-functionalized) prints is critical for a number of different applications. For recent applications that include low-temperature electrode deposition, see Huang et al. [41], for optical coatings and photonics, see Nakanishi et al. [42] and Maier and Atwater [43], for biosensors, see Alivisatos [44], for catalysts, see Haruta [45], and for MEMS applications, see Full et al. [46] and Ho et al. [47]. Several methods are available for patterning of particles, such as inkjet printing (see Ridley et al. [48] and Sirringhaus et al. [49]), which is attractive due to its simplicity, high throughput, and low material loss. However, patterning with inkjet printing is limited to a resolution of around 20–50 microns with current printers (Ridley et al. [48]), with higher resolution possible by adding complexity to the substrate prior to printing (Wang et al. [50]). Electrohydrodynamic printing has also been proposed to increase the resolution beyond the limits of inkjet printing, achieving a line resolution as small as 700 nm (Park et al. [51]).[13] Unfortunately, inkjet and electrohydrodynamic printing do not allow precise control over the structure of the printed lines. This often results in lines with scalloped edges or nonuniform width and offer only limited control over the height of the printed features (Huang et al. [41], Sirringhaus et al. [49], Ahmad et al. [52], and Samarasinghe et al. [53]). Also, recently, imprint lithography has been proposed as a means of decreasing the feature size of patterned particles while allowing more precise control over the structure of the printed lines (Ko et al. [54, 55], and Park et al. [56]). In this fabrication method, the particle-laden inks are patterned by pressing with an elastomer mold and the particles are dried into their final configuration. While the resolution of imprint lithography is improved over inkjet printing, there exists a residual layer on the substrate that must be etched away after patterning. Control over the height of features can be corrupted by capillary action between the mold and the drying ink, in particular along the length of longer features. Thus, as a possible alternative to imprint lithography, particle self-assembly methods, based on capillary filling of photoresist templates, have been proposed (Demko et al. [59]) and appear to be promising. *Clearly, laser post-processing can play a role in all of the previously mentioned approaches, in order to add a degree of precision that is otherwise missing.*

[13] In such applications, lasers can be used to heat-treat and dry particle-based inks.

10.6 Chapter Appendix: Geometrical Ray Theory

Following a somewhat classical analysis found in, for example, Elmore and Heald [57], Cerveny et al. [58], and others, we consider the propagation of a general disturbance, ψ, governed by a generic wave equation:

$$\nabla^2\psi = \frac{1}{c^2(\boldsymbol{x})}\frac{\partial^2\psi}{\partial t^2}. \tag{10.38}$$

Here, $c(\boldsymbol{x})$ is a spatially varying wave speed corresponding to a general inhomogeneous medium, where $c(\boldsymbol{x}) = c_o$ in a homogeneous reference medium and the refractive index is defined as $n = c_o/c(\boldsymbol{x})$. Consider a trial solution of the form

$$\psi(\boldsymbol{x}, t) = A(\boldsymbol{x})e^{j(k_o S(\boldsymbol{x})-\omega t)}, \tag{10.39}$$

where $A(\boldsymbol{x})$ is the amplitude of the disturbance and $k_o = \omega/c_o = 2\pi/\lambda$ is the wave number in the reference medium. The function $S(\boldsymbol{x})$ (dimensions of length) is known as the "Eikonal," which in Greek means "image." One can interpret a set of waves as simply a family of surfaces for which the values of $k_o S(\boldsymbol{x})$ differ in increments of 2π. Substituting the trial solution into the wave equation, one obtains

$$k_o^2 A(n^2 - \nabla S \cdot \nabla S) + jk_o(2\nabla A \cdot \nabla S + A\nabla^2 S) + \nabla^2 A = 0. \tag{10.40}$$

There are a variety of arguments to motivate so-called ray theory. Probably, the simplest is simply to require that, as $k_o \to \infty$, each of the k_o-terms, the zeroth-order k_o-term, the first-order k_o-term, and the second-order k_o-term, must vanish. Applying this requirement to the second-order k_o-term yields

$$n^2 = \nabla S \cdot \nabla S = ||\nabla S||^2. \tag{10.41}$$

For a uniform medium, $n = const$, provided $\nabla^2 A = 0$ and an initial plane wave surface $S = const$, then Eq. 10.40 implies

$$S(\boldsymbol{x}) = n(\alpha x + \beta y + \phi z), \tag{10.42}$$

where α, β, and ϕ are direction cosines. More generally, when $n \neq 0$, then Eq. 10.41 implies

$$\nabla S(\boldsymbol{x}) = n(\boldsymbol{x})\hat{\boldsymbol{s}}(\boldsymbol{x}), \tag{10.43}$$

where $\hat{\boldsymbol{s}}(\boldsymbol{x})$ is a unit (direction) vector. From elementary calculus, recall that ∇S is perpendicular to $S = const$. This allows for the natural definition of continuous curves, called rays, that are everywhere parallel to the local direction $\hat{\boldsymbol{s}}(\boldsymbol{x})$. Rearranging first-order k_o-term of Eq. 10.40

$$\frac{1}{A}\nabla A \cdot \nabla S = -\frac{1}{2}\nabla^2 S = -\frac{1}{2}\nabla \cdot (n\hat{s}). \tag{10.44}$$

Recalling the directional derivative, $\frac{d(\cdot)}{ds} \stackrel{\text{def}}{=} \hat{s} \cdot \nabla(\cdot)$, we have

$$(\frac{\nabla S}{||\nabla S||}) \cdot \nabla A = (\frac{\nabla S}{n}) \cdot \nabla A = \frac{dA}{ds}, \tag{10.45}$$

where s is the arc-length coordinate along the ray. With this definition, once $S(\boldsymbol{x})$ is known, the component of ∇A in the $\hat{s}(\boldsymbol{x})$ can be found from Eqs. 10.44 and 10.45:

$$\frac{1}{A}\frac{dA}{ds} = -\frac{1}{2n}\nabla \cdot (n\hat{s}). \tag{10.46}$$

Thus, we are able to determine how the amplitude of the trial solution changes along a ray, but not perpendicular to the trajectory.

Geometrical "ray-tracing" deals directly with the ray trajectories, rather than finding them as a by-product of the solution of the wave equation for the Eikonal function S and the resulting wave front. To eliminate S, we need to look at the rate of change of the quantity $n\hat{s}$ along the ray. Making repeated use of Eq. 10.43, we have

$$\frac{d(n\hat{s})}{ds} = \hat{s} \cdot \nabla(\nabla S) = \frac{\nabla S}{n} \cdot \nabla(\nabla S) = \frac{1}{2n}\nabla(\nabla S \cdot \nabla S) = \frac{1}{2n}\nabla n^2 = \nabla n, \tag{10.47}$$

where $\frac{d(\cdot)}{ds} \stackrel{\text{def}}{=} \hat{s} \cdot \nabla(\cdot)$. The previous equation allows us to find the trajectories of a ray $(\hat{s})$, given only the refractive index $n(\boldsymbol{x})$ and the initial direction $\hat{s}_i$ of the desired ray.
Remark: A more general derivation of the Eikonal equation can be found in a variety of textbooks, for example, Cerveny et al. [58], and starts by assuming a trial solution of the form

$$\psi(\boldsymbol{x}, t) = A(\boldsymbol{x})\Phi(t - \Lambda(\boldsymbol{x})) \tag{10.48}$$

where Λ is an Eikonal function, and the waveform function α is assumed to be of high frequency.[14] This function is then substituted into the wave equation to yield

$$\nabla^2 A\Phi + 2\nabla A \cdot \nabla\Phi + A\nabla^2\Phi = \frac{1}{c^2}A\frac{\partial^2\Phi}{\partial t^2}. \tag{10.49}$$

After using the chain rule of differentiation, this can be written as

[14] This is a more general case than the one considered in Eq. 10.39, where $\Phi(t - \Lambda(\boldsymbol{x})) = e^{j(k_o S(\boldsymbol{x}) - \omega t)}$.

$$\frac{\partial^2 \Phi}{\partial \Lambda^2} A \left(\nabla \Lambda \cdot \nabla \Lambda - \frac{1}{c^2}\right) + \frac{\partial \Phi}{\partial \Lambda} \left(2\nabla A \cdot \nabla \Lambda + A\nabla^2 \Lambda\right) + \Phi \nabla^2 A = 0. \quad (10.50)$$

Analogous to the special case considered before, to motivate so-called ray theory, one requires that the coefficients of $\frac{\partial^2 \Phi}{\partial \Lambda^2}$, $\frac{\partial \Phi}{\partial \Lambda}$, and Φ are satisfied separately, in other words, the following holds

$$\nabla \Lambda \cdot \nabla \Lambda - \frac{1}{c^2} = 0 \quad (10.51)$$

and

$$2\nabla A \cdot \nabla \Lambda + A\nabla^2 \Lambda = 0 \quad (10.52)$$

and

$$\nabla^2 A = 0. \quad (10.53)$$

References

1. Gross, H.: Handbook of optical systems. In: Gross, H. (ed.) Fundamental of Technical Optics. Wiley-VCH (2005)
2. Niemz, M.H.: Laser Tissue Interactions-Fundamentals and Applications. Springer (2004)
3. Steen, W.M.: Laser Material Processing. Springer (1998)
4. Grigoropoulos, C.P.: Transport in Laser Microfabrication. Press, Cambridge Uni (2009)
5. Zohdi, T.I.: Modeling and simulation of a class of coupled thermo-chemo-mechanical processes in multiphase solids. Comput. Methods Appl. Mech. Eng. **193**(6–8), 679–699 (2004)
6. Zohdi, T.I.: Modeling and direct simulation of near-field granular flows. Int. J. Solids Struct. **42**(2), 539–564 (2004)
7. Zohdi, T.I.: Charge-induced clustering in multifield particulate flow. Int. J. Num. Methods Eng. **62**(7), 870–898 (2005)
8. Zohdi, T.I.: On the optical thickness of disordered particulate media. Mech. Mater. **38**, 969–981 (2006)
9. Zohdi, T.I., Kuypers, F.A.: Modeling and rapid simulation of multiple red blood cell light scattering. Proc. R. Soc. Interface **3**(11), 823–831 (2006)
10. Zohdi, T.I.: Computation of the coupled thermo-optical scattering properties of random particulate systems. Comput. Methods Appl. Mech. Eng. **195**, 5813–5830 (2006)
11. Zohdi, T.I.: Computation of strongly coupled multifield interaction in particle-fluid systems. Comput. Methods Appl. Mech. Eng. **196**, 3927–3950 (2007)
12. Zohdi, T.I.: On the computation of the coupled thermo-electromagnetic response of continua with particulate microstructure. Int. J. Num. Methods Eng. **76**, 1250–1279 (2008)
13. Zohdi, T.I.: On the dynamics of charged electromagnetic particulate jets. Arch. Comput. Methods Eng. **17**(2), 109–135 (2010)
14. Zohdi, T.I., Kuypers, F.A., Lee, W.C.: Estimation of red blood cell volume fraction from overall permittivity measurement. Int. J. Eng. Sci. **48**, 1681–1691 (2010)
15. Zohdi, T.I.: Dynamics of clusters of charged particulates in electromagnetic fields. Int. J. Num. Methods Eng. **85**, 1140–1159 (2011)
16. Zohdi, T.I.: Modeling and simulation of the optical response rod-functionalized reflective surfaces. Comput. Mech. **50**(2), 257–268 (2012)

17. Zohdi, T.I.: Numerical simulation of charged particulate cluster-droplet impact on electrified surfaces. J. Comput. Phys. **233**, 509–526 (2013)
18. Zohdi, T.I.: Rapid simulation of laser processing of discrete particulate materials. Arch. Comput. Methods Eng. **20**, 309–325 (2013)
19. Zohdi, T.I.: A direct particle-based computational framework for electrically-enhanced thermo-mechanical sintering of powdered materials. Math. Mech. Solids. **19**(1), 93–113
20. Zohdi, T.I.: Additive particle deposition and selective laser processing-a computational manufacturing framework. Comput. Mech. **54**, 171–191 (2014)
21. Zohdi, T.I.: Embedded electromagnetically sensitive particle motion in functionalized fluids. Comput. Part. Mech. **1**, 27–45 (2014)
22. Zohdi, T.I.: A computational modelling framework for high-frequency particulate obscurant cloud performance. Int. J. Eng. Sci. **89**, 75–85 (2015)
23. Zohdi, T.I.: Modeling and simulation of cooling-induced residual stresses in heated particulate mixture depositions. Comput. Mech. **56**, 613–630 (2015)
24. Zohdi, T.I.: Modeling and efficient simulation of the deposition of particulate flows onto compliant substrates. Int. J. Eng. Sci. **99**, 74–91 (2015). https://doi.org/10.1016/j.ijengsci.2015.10.012
25. Zohdi, T.I.: Modeling and simulation of laser processing of particulate-functionalized materials. Arch. Comput. Methods Eng. (2015). https://doi.org/10.1007/s11831-015-9160-1, 1–25
26. Zohdi, T.I., Wriggers, P.: Introduction to Computational Micromechanics. Springer, Second Reprinting (Peer Reviewed, 2005) (2008)
27. Zohdi, T.I.: Simulation of coupled microscale multiphysical-fields in particulate-doped dielectrics with staggered adaptive FDTD. Comput. Methods Appl. Mech. Eng. **199**, 79–101 (2010)
28. Davis, S.H.: Theory of Solidification. Cambridge University Press (2001)
29. Donev, A., Cisse, I., Sachs, D., Variano, E.A., Stillinger, F., Connelly, R., Torquato, S., Chaikin, P.: Improving the density of jammed disordered packings using ellipsoids. Science 13 **303**, 990–993 (2004a)
30. Donev, A., Stillinger, F.H., Chaikin, P.M., Torquato, S.: Unusually dense crystal ellipsoid packings. Phys. Rev. Lett. **92**, 255506 (2004b)
31. Donev, A., Torquato, S., Stillinger, F.: Neighbor list collision-driven molecular dynamics simulation for nonspherical hard particles-I algorithmic details. J. Comput. Phys. **202**, 737 (2005a)
32. Donev, A., Torquato, S., Stillinger, F.: Neighbor list collision-driven molecular dynamics simulation for nonspherical hard particles-II. Application to ellipses and ellipsoids. J. Comput. Phys. **202**, 765 (2005b)
33. Donev, A., Torquato, S., Stillinger, F.H.: Pair correlation function characteristics of nearly jammed disordered and ordered hard-sphere packings. Phys. Rev. E **71**, 011105 (2005c)
34. Pöschel, T., Schwager, T.: Computational Granular Dynamics. Springer (2004)
35. Choi, S., Park, I., Hao, Z., Holman, H.Y., Pisano, A.P., Zohdi, T.I.: Ultra-fast self-assembly of micro-scale particles by open channel flow. Langmuir **26**(7), 4661–4667 (2010)
36. Choi, S., Stassi, S., Pisano, A.P., Zohdi, T.I.: Coffee-ring effect-based three dimensional patterning of micro, nanoparticle assembly with a single droplet. Langmuir **26**(14), 11690–11698 (2010)
37. Choi, S., Jamshidi, A., Seok, T.J., Zohdi, T.I., Wu., M. C., Pisano, A.P.: Fast, High-throughput creation of size-tunable micro, nanoparticle clusters via evaporative self-assembly in picoliter-scale droplets of particle suspension. Langmuir **28**(6), 3102–11 (2012)
38. Martin, P.: Handbook of Deposition Technologies for Films and Coatings, 3rd ed. Elsevier (2009)
39. Martin, P.: Introduction to Surface Engineering and Functionally Engineered Materials. Scrivener and Elsevier (2011)
40. Gamota, D., Brazis, P., Kalyanasundaram, K., Zhang, J.: Printed Organic and Molecular Electronics. Kluwer Academic Publishers, New York (2004)
41. Huang, D., Liao, F., Molesa, S., Redinger, D., Subramanian, V.: Plastic-compatible low-resistance printable gold nanoparticle conductors for flexible electronics. J. Electrochem. Soc. **150**(7), G412–417 (2003)

42. Nakanishi, H., Bishop, K. J. M., Kowalczyk, B., Nitzan, A., Weiss, E.A., Tretiakov, K.V., Apodaca, M.M., Klajn, R., Stoddart, J.F., Grzybowski, B.A.: Photoconductance and inverse photoconductance in thin films of functionalized metal nanoparticles. Nature **460**, 371–375 (2009)
43. Maier, S.A., Atwater, H.A.: Plasmonics: localization and guiding of electromagnetic energy in metal/dielectric structures. J. Appl. Phys. **2005**(98), 011101 (2005)
44. Alivisatos, P.: The use of nanocrystals in biological detection. Nat. Biotechnol. **22**(1), 47–52 (2004)
45. Haruta, M.: Catalysis of gold nanoparticles deposited on metal oxides. Cattech **6**(3), 102–115 (2002)
46. Fuller, S.B., Wilhelm, E.J., Jacobson, J.M.: Ink-jet printed nanoparticle microelectromechanical systems. J. Microelectromech. Syst. **11**, 54–60 (2002)
47. Ho, C., Steingart, D., Salminent, J., Sin, W., Rantala, T., Evans, J., Wright, P.: Dispenser printed electrochemical capacitors for power management of millimeter scale lithium ion polymer microbatteries for wireless sensors. In: 6th International Workshop on Micro and Nanotechnology for Power Generation and Energy Conversion Applications (PowerMEMS 2006), Berkeley, CA (2006)
48. Ridley, B.A., Nivi, B., Jacobson, J.M.: All-inorganic field effect transistors fabricated by printing. Science **286**, 746–749 (1999)
49. Sirringhaus, H., Kawase, T., Friend, R.H., Shimoda, T., Inbasekaran, M., Wu, W., Woo, E.P.: High-resolution inkjet printing of all-polymer transistor circuits. Science **290**, 2123–2126 (2000)
50. Wang, J.Z., Zheng, Z.H., Li, H.W., Huck, W.T.S., Sirringhaus, H.: Dewetting of conducting polymer inkjet droplets on patterned surfaces. Nat. Mater. **3**, 171–176 (2004)
51. Park, J.-U., Hardy, M., Kang, S.J., Barton, K., Adair, K., Mukhopadhyay, D.K., Lee, C.Y., Strano, M.S., Alleyne, A.G., Georgiadis, J.G., Ferreira, P.M., Rogers, J.A.: High-resolution electrohydrodynamic jet printing. Nat. Mater. **6**, 782–789 (2007)
52. Ahmad, Z., Rasekh, M., Edirisinghe, M.: Electrohydrodynamic direct writing of biomedical polymers and composites. Macromol. Mater. Eng. **295**, 315–319 (2010)
53. Samarasinghe, S.R., Pastoriza-Santos, I., Edirisinghe, M.J., Reece, M.J., Liz-Marzan, L.M.: Printing gold nanoparticles with an electrohydrodynamic direct write device. Gold Bull. **39**, 48–53 (2006)
54. Ko, S.H., Park, I., Pan, H., Grigoropoulos, C.P., Pisano, A.P., Luscombe, C.K., Frechet, J.M.J.: Direct nanoimprinting of metal nanoparticles for nanoscale electronics fabrication. Nan Lett. **7**, 1869–1877 (2007)
55. Ko, S.H., Park, I., Pan, H., Misra, N., Rogers, M.S., Grigoropoulos, C.P., Pisano, A.P.: ZnO nanowire network transistor fabrication by lowtemperature, allinorganic nanoparticle solution process. Appl. Phys. Lett. **92**, 154102 (2008)
56. Park, I., Ko, S.H., Pan, H., Grigoropoulos, C.P., Pisano, A.P., Frechet, J.M.J., Lee, E.S., Jeong, J.H.: Nanoscale patterning and electronics on flexible substrate by direct nanoimprinting of metallic nanoparticles. Adv. Mater. **20**, 489 (2008)
57. Elmore, W.C., Heald, M.A.: Physics of Waves. Dover Publishers (1985)
58. Cerveny, V., Molotkov, I.A., Psencik., I.: Ray Methods in Seismology. Univerzita Karlova, Praha (1977)
59. Demko, M., Choi, S., Zohdi, T.I., Pisano, A.P.: High resolution patterning of nanoparticles by evaporative self-assembly enabled by in-situ creation and mechanical lift-off of a polymer template. Appl. Phys. Lett. **99**, 253102-1–253102-3 (2012)
60. Demko, M.T., Cheng, J.C., Pisano, A.P.: High-resolution direct patterning of gold nanoparticles by the microfluidic molding process. Langmuir, 412–417 (2010)

Chapter 11
DEM Extensions: Acoustical Pre-Processing

11.1 Introduction

In numerous industries, particle-laden fluids are a key part of the fabrication of products such as (1) casted machine parts, (2) additively manufactured and 3D printed electronics and medical devices, and even (3) slurry processed food to name a few. Common to many of these areas are new types of solidified heterogeneous materials, comprised of particulates in a binding matrix. The fluid precursor to the final solid product is a particle-laden fluid, which is delivered through piping, nozzles, and channels (Choi et al. [1–3], Demko et al. [4, 5] and Martin [6, 7]).[1] Typically, such materials start in particulate form and are then mixed with a binder and delivered as a flowing particle-laden fluid which is to be cast into their final shape.[2]

Often, in the mixing of these materials, agglomerations of particles occur, which need to be broken up and dispersed in order to ensure production of high-quality products (Figs. 11.1 and 11.2). A unique feature of small-scale particulate systems is that they exhibit a strong sensitivity to inter-particle near-field forces, stemming from stray electrostatic charges, process ionization, and mechanical adhesion, leading to agglomeration and cluster formation, which can strongly affect manufactured product quality. These agglomerations can also remain coherent as suspended clusters in the fluid material during processing. Inadequate elimination of these clusters can lead to manufacturing inconsistencies/variability which can strongly affect the overall solidified product quality, in particular if the manufactured devices have small

[1] Also, printed electronics, using processes such as high-resolution electrohydrodynamic jet printing, are also emerging as viable methods. For overviews, see Wei and Dong [8], who also develop specialized processes employing phase-change inks. Such processes are capable of producing micron-level footprints for high-resolution additive manufacturing.

[2] Over 50% of man-made materials start in granulated form (Duran [9] and Torquato [10]).

T. I. Zohdi, *Modeling and Simulation of Functionalized Materials for Additive Manufacturing and 3D Printing: Continuous and Discrete Media*, Lecture Notes in Applied and Computational Mechanics 60, https://doi.org/10.1007/978-3-319-70079-3_11

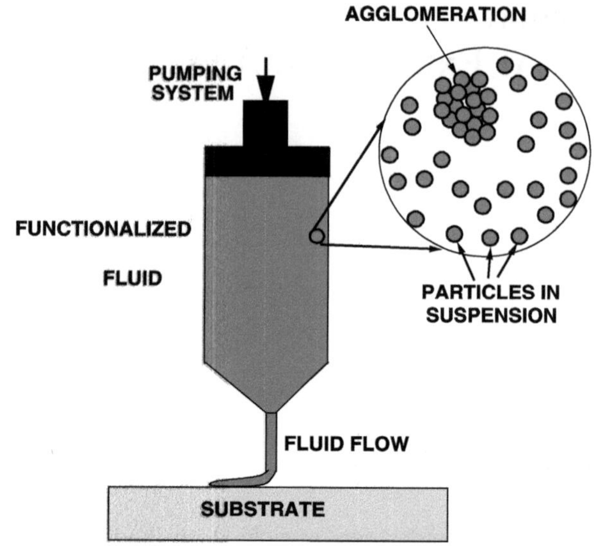

Fig. 11.1 A particle-laden fluid in a channel with an agglomeration

dimensions. Furthermore, in other cases, the agglomerations can accumulate on the surfaces of manufacturing devices. For particle-laden fluids delivered through channels, surface particulate agglomeration can lead to system malfunction, primarily due to clogging.

One approach to remove or break up the agglomerations is to acoustically pressure pulse them (Fig. 11.2). Acoustical pulsing consists of an acoustical driver, which is a sound source that is powered by compressed air and an acoustical resonator which amplifies and guides the energy. Essentially, an acoustical pulser will produce a series of rapid pressure-peak pulses in a very targeted manner. Acoustical cleaning has been utilized in macroscale bulk processes, but its application on the microscale for precision manufacturing, as well as modeling and simulation, has been lacking. The key parameters are (a) the strength of the pulse and (b) the repetition of the pulse.

A primary simulation issue is that the interaction of pressure pulses with the agglomeration leads to a set of coupled differential equations for the dynamics of the particles and the propagation and breakup of the pressure pulse. The objective of this chapter is to develop a computational model and corresponding solution algorithm to enable rapid simulation of a pressure pulse's effect on an agglomeration, composed of aggregated discrete particles. Because of the complex agglomeration microstructure, containing gaps and interfaces, this type of system is extremely difficult to mesh and simulate the pressure pulse dynamics using continuum-based methods, such as the Finite Difference Time Domain Method or the Finite Element Method. Accordingly, a computationally amenable discrete element/discrete ray model is developed which captures the primary physical events, such as the reflection and absorption of acoustical energy, and the forces induced onto the particulate microstructure, resulting in dynamics of large numbers of particles involving particle-to-particle contact. In the

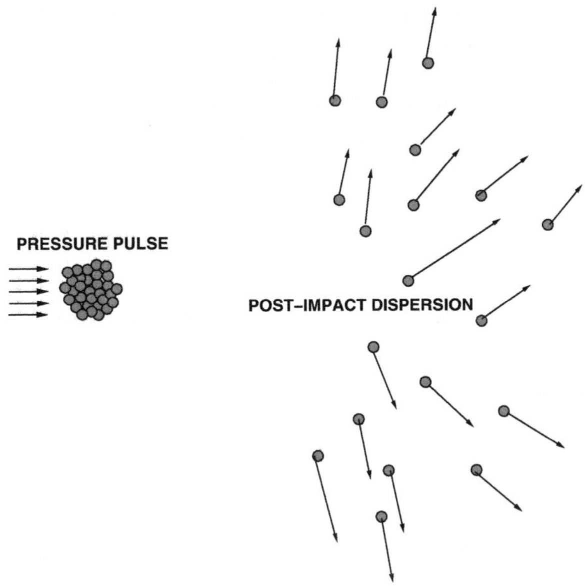

Fig. 11.2 Pulsing to break up and disperse an agglomeration

model, the agglomeration is represented by a Discrete Element Method representation and the pressure pulse/shock wave is described by a collection of propagating rays with prescribed acoustical power content, direction, and velocity. The approach also develops a staggered, iterative solution scheme, which is needed to calculate the power transfer from the acoustical pulse to the particles and the subsequent changes (breakup) of the pulse due to the particles. Three-dimensional examples are provided to illustrate the approach.

Remark 1 There is a variety of discrete element-like methods, for example, see Onate et al. [11, 12], Rojek et al. [13], Carbonell et al. [14], Labra and Onate [15], Leonardi et al. [16], Cante et al. [17], Rojek [18], Onate et al. [19], Bolintineanu et al. [20], Avci and Wriggers [21], and Zohdi [22–48]. In particular, we note the Discrete Element Method has been used for particle agglomerations in acoustic fields, for instance in Markov [49]. The ray representation of waves can be found in a variety of works, for example, in Virovlyanskii [50], Borejko et al. [51] and Zohdi [26–43, 47] among others.

Remark 2 This work has direct applicability to the subject of "Design For Cleanability" which concerns itself with the advantageous manufacturing of devices, in particular those with interior keyways and channels, by assessing the ability to perform maintenance during the design phase. In many industrial systems, buildup of unwanted particulate material is inaccessible for removal and ultimately leads to system malfunction. For more details, we refer the reader to Aurich and Dornfeld [52], Garg et al. [53], Avila et al. [54, 55]. Furthermore, residual particulate matter can also be present after a manufacturing process involving particles (see, for example,

Luo and Dornfeld [56–59], Arbelaez et al. [60, 61], Campello and Zohdi [62, 63], Gomes-Ferreira et al. [64], Ghobeity et al. [65, 66]).[3]

Remark 3 We remark that in biological systems, oftentimes it is a goal to destroy certain types of agglomerations, such as kidney stones in shock wave lithotripsy (Zohdi [29]). The approach developed in this chapter has direct applicability in this domain as well.

11.2 Dynamic Response of an Agglomeration

In order to represent the group of interacting particles, we follow a relatively flexible formulation found in the previous chapters and in Zohdi [22–48]. We consider a group of nonintersecting particles ($i = 1, 2, \ldots, N_p$). The equation of motion for the ith particle in system is

$$m_i \ddot{\boldsymbol{r}}_i = \boldsymbol{\Psi}_i^{tot}(\boldsymbol{r}_1, \boldsymbol{r}_2, \ldots, \boldsymbol{r}_{N_p}) = \boldsymbol{\Psi}_{ti}^{ray} + \boldsymbol{\Psi}_i^{con} + \boldsymbol{\Psi}_i^{bond} + \boldsymbol{\Psi}_i^{damp} + \boldsymbol{\Psi}_i^{e+m}, \quad (11.1)$$

where $\boldsymbol{r}_i$ is the position vector of the ith particle and where $\boldsymbol{\Psi}_i^{tot}$ represents all forces acting on particle i, which is decomposed into the sum of forces due to:

- Shock (transmitted ray) forces ($\boldsymbol{\Psi}_{ti}^{ray}$),
- Inter-particle forces ($\boldsymbol{\Psi}_i^{con}$) generated by contact with other particles,
- Adhesive bonding forces ($\boldsymbol{\Psi}_i^{bond}$) with other particles,
- Damping forces arising from the surrounding interstitial environment ($\boldsymbol{\Psi}_i^{damp}$) occurring from potentially viscous, surrounding, interstitial fluids, surfactants, and
- External electromagnetic forces ($\boldsymbol{\Psi}_i^{e+m}$) which can play a key role in small charged or magnetized particles.

In the next section, we examine particle-shock wave contact and draw upon the previous chapters for all other forces.

11.3 Particle-Shock Wave Contact

We consider cases where the wavelengths of incident high-frequency acoustical pressure waves (p-waves) are at least one order of magnitude smaller than the diameter (d) of the particle scatterers ($10^{-6}\,\text{m} \leq d \leq 10^{-4}\,\text{m}$). This length scale ordering indicates that diffraction is minimal and makes (geometric) ray-tracing techniques

[3]Even techniques associated with shot peening can leave residual particulate matter. We also refer the reader to Afazov et al. [67], Bagherifard et al. [68], Elbella et al. [69], and Chen et al. [70]).

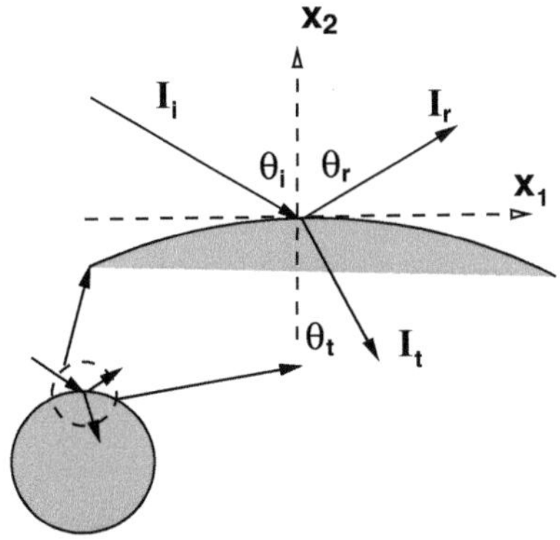

Fig. 11.3 Zoom on a ray-particle contact

accurate.[4] In such cases, geometric ray-tracing can be used to determine the amount of propagating incident energy that is reflected and the amount that is absorbed by multiple particles. As alluded to in the previous chapters, for the benefit of readers unfamiliar with ray-tracing, we remark that it is essentially an approximate solution to the wave equation, based on the Eikonal equation, which is the limiting case of wave phenomena as the wave length tends toward zero.

11.3.1 Ray-Tracing: Incidence, Reflection, and Transmission

The reflection of a ray at an interface is given by enforcing continuity of the (acoustical) pressure and disturbance velocity at that location; this yields the ratio between the incident and reflected pressures. We use a local coordinate system (Fig. 11.3) and enforce (1) that the number of waves per unit length in the x_1 − direction must be the same for the incident, reflected, and refracted (transmitted) waves and (2) a pressure balance at the interface. After some algebra, this yields the reflectance for the (acoustical) power (energy per unit time)

$$\mathcal{R} = \frac{I_r}{I_i} = \left(\frac{\hat{A} cos\theta_i - cos\theta_t}{\hat{A} cos\theta_i + cos\theta_t} \right)^2, \tag{11.2}$$

where I_i is the incident (power) ray, I_r is the reflected (power) ray, I_t is the transmitted (power) ray, $\hat{A} \stackrel{\text{def}}{=} \frac{A_t}{A_i} \stackrel{\text{def}}{=} \frac{\rho_t c_t}{\rho_i c_i}$, ρ_t is the density of the medium which the ray encounters (transmitted), c_t is corresponding sound speed in that medium, A_t is the corresponding acoustical impedance, ρ_i is the density of the medium in which the ray was traveling (incident), c_i is corresponding sound speed in that medium A_i is

[4]Even if this wavelength to particle size ratio is not present, ray representation of p-waves is still often used and can be considered as a way to approximately track the propagation of energy, however, without the ability to capture diffraction properly.

the corresponding acoustical impedance. The relationship (the law of refraction) between the incident and transmitted angles is $c_t sin\theta_t = c_i sin\theta_i$. Defining $\tilde{c} \stackrel{\text{def}}{=} \frac{c_i}{c_t}$, and studying asymptotic cases where $sin\theta_t \rightarrow 1$, we have $sin\theta_i \rightarrow \tilde{c}$, which identifies a so-called critical angle, where no energy is transmitted. A simple way of observing complete reflection is to set $\theta_t = \pi/2$ in

$$\mathcal{R} = \frac{I_r}{I_i} = \left(\frac{\hat{A}cos\theta_i - cos\theta_t}{\hat{A}cos\theta_i + cos\theta_t}\right)^2 = \left(\frac{\hat{A}cos\theta_i - 0}{\hat{A}cos\theta_i + 0}\right)^2 = 1. \tag{11.3}$$

Remark 1 A more rigorous way of analyzing the critical angle phenomena is to rewrite the reflection relation as $\mathcal{R} \stackrel{\text{def}}{=} r * \bar{r} = 1$, where $\bar{r}$ is the complex conjugate, where

$$r = \frac{\tilde{c}\hat{A}cos\theta_i - j(sin^2\theta_i - \tilde{c}^2)^{\frac{1}{2}}}{\tilde{c}\hat{A}cos\theta_i + j(sin^2\theta_i - \tilde{c}^2)^{\frac{1}{2}}}, \tag{11.4}$$

where $j = \sqrt{-1}$. For angles above the critical angle $\theta_i \geq \theta_i^*$, all of the energy is reflected. We note that when $A_t = A_i$ and $c_i = c_t$, then there is no reflection. Also, when $A_t >> A_i$, then $r \rightarrow 1$, and when $A_t << A_i$, then $r \rightarrow -1$.

Remark 2 The power input (transmitted) from incidence and reflection is

$$I_t = I_i - I_r = I_i(1 - \mathcal{R}). \tag{11.5}$$

The overall force magnitude imparted by the ray can be determined by

$$\text{power transmitted} = ||\boldsymbol{\Psi}_t^{ray}||||\boldsymbol{v}_t|| = \frac{||\boldsymbol{\Psi}_t^{ray}||||\boldsymbol{\Psi}_t^{ray}||}{\rho_t c_t a_r} = I_i(1 - \mathcal{R}), \tag{11.6}$$

thus

$$||\boldsymbol{\Psi}_t^{ray}|| = \sqrt{I_i(1 - \mathcal{R})\rho_t c_t a_r}, \tag{11.7}$$

where the velocity in the transmitted medium is $||\boldsymbol{v}_t|| = \frac{P_t}{\rho_t c_t}$, where P_t is the pressure content of the ray, and $||\boldsymbol{\Psi}_t^{ray}|| \propto \frac{P_t}{a_r}$, where a_r is a (pseudo) contact area parameter. The contact area parameter can be approximated from the total area of the original plane wave divided by the total number of rays. This is discussed further in the next section. The force's magnitude is projected in the direction of the transmitted ray, i.e., at the angle θ_t in the plane of incidence. This force is assumed to be completely absorbed by the particle and is not re-transmitted. The reflected ray is simply projected in the direction at the reflected angle, θ_r.

Remark 3 The initial plane wave of rays is generated by (Fig. 11.4):

- Generating a set of vectors in a "master template" domain,
- Rotating the master template to the desired direction of the pulse,

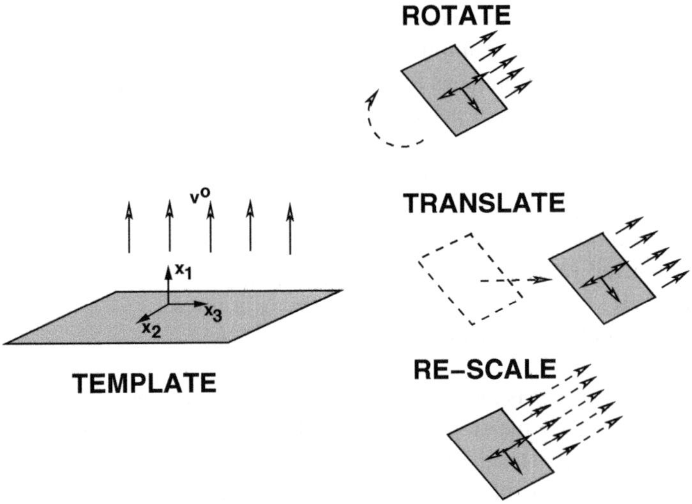

Fig. 11.4 Construction of the initial plane wave of rays

- Translating the rotated master template to the desired starting location, and
- The vectors are then scaled according to the initial starting power content.

11.3.2 Acoustical-Pulse Computational Algorithm

We now consider an initially coherent plane wave composed of multiple (initially) parallel rays (Fig. 11.3). Each ray is a vector in the direction of the flow of energy, which, in isotropic media, corresponds to the normal to the wave front. For isotropic media, the rays are parallel to the wave propagation vector. It is of particular interest to describe the breakup of initially highly directional coherent beams which do not spread out into multidirectional rays unless they encounter scatterers. It is convenient to define vectors for a ray by projecting the power onto the unit vector associated with the velocity

$$\boldsymbol{I} \stackrel{\text{def}}{=} I\frac{\boldsymbol{v}^{ray}}{||\boldsymbol{v}^{ray}||}, \tag{11.8}$$

where $\boldsymbol{v}^{ray}$ is the velocity of the ray, $I = ||\boldsymbol{I}||$ is the magnitude of the power, and initially, $I_o \stackrel{\text{def}}{=} ||\boldsymbol{I}(t = 0)||$. To obtain the ray contributions from a beam, we simply partition the initial beam cross section into equal rays, $Ia_r = Ia_b/N_r$, where N_r is the number of rays in the beam and where a_b is cross-sectional area of the beam. Later in the chapter, we will elaborate more on the selection of the number of rays, which is determined by successive refinement of the number of rays in a pulse, until no appreciable differences occur. Clearly, $I_o a_b$ can be considered the total power associated with the beam. The parameter a_r provides us with a way of appropriately distributing or "lumping" the amount of energy in the overall wave (beam) into rays. The computational algorithm is as follows, starting at $t = 0$ and ending at $t = T$, for an instantaneously fixed set of particles:

$$\begin{array}{l}(1)\ \text{Compute ray reflections (Fresnel relation).}\\ (2)\ \text{Compute energy absorbed by each particle} : \Delta I = (I_i - I_r) = (1 - \mathcal{R})I_i.\\ (3)\ \text{Compute induced forces for each particle} : \boldsymbol{\Psi}_t^{\mathrm{ray}}.\\ (4)\ \text{Increment all ray positions} :\\ \quad \boldsymbol{r}_q^{ray}(t + \Delta t) = \boldsymbol{r}_q^{ray}(t) + \Delta t \boldsymbol{v}_q^{ray}(t), \quad q = 1, ..., rays.\\ (5)\ \text{Go to (1) and repeat with } (t = t + \Delta t).\end{array} \tag{11.9}$$

The time-step size Δt is dictated by the size of the particles. A somewhat ad hoc approach is to scale the time-step size to be no larger than $\Delta t \propto \frac{\xi R}{||\boldsymbol{v}^{ray}||}$, where R is the nominal radius of the particles, $||\boldsymbol{v}^{ray}||$ is the magnitude of the velocity of the rays, and ξ is a scaling factor, typically $0.05 \leq \xi \leq 0.1$. The required time-step limitation is then also compared against other time-step needed to integrate the particle dynamics properly, which we discuss next. This ensures that meaningful interactions are not skipped. *We now elaborate on the dynamics of the particles in STEP (3) above.*

11.3.3 Iterative (Implicit) Solution Method Algorithm

Following the basic framework in Zohdi [22–46] for particle i

$$\boldsymbol{r}_i^{L+1} = \boldsymbol{r}_i^L + \boldsymbol{v}_i^L \Delta t + \frac{\phi(\Delta t)^2}{m_i}\left(\phi(\boldsymbol{\Psi}_i^{tot,L+1}) + (1 - \phi)(\boldsymbol{\Psi}^{tot,L})\right), \tag{11.10}$$

which leads to a coupled set equations for $i = 1, 2, \ldots, N_p$ particles, where the superscript L is a time interval counter. The solution STEPS are, within a time-step:

(0) Determine the ray force contribution for each particle according to the previously explained algorithm (BOX 11.9).
(1) Start a global fixed iteration (set $i = 1$ (particle counter) and $K = 0$ (iteration counter)).
(2) If $i > N_p$, then go to (4).
(3) If $i \leq N_p$, then:

   (a) Compute the position $\boldsymbol{r}_i^{L+1,K}$,
   (b) Go to (2) for the next particle ($i = i + 1$).

(4) Measure error (normalized) quantities

   (a) $\varpi_K \stackrel{\text{def}}{=} \dfrac{\sum_{i=1}^{N_p} ||\boldsymbol{r}_i^{L+1,K} - \boldsymbol{r}_i^{L+1,K-1}||}{\sum_{i=1}^{N_p} ||\boldsymbol{r}_i^{L+1,K} - \boldsymbol{r}_i^{L}||}$

(b) $Z_K \stackrel{\text{def}}{=} \frac{\varpi_K}{TOL_r}$

(c) $\Lambda_K \stackrel{\text{def}}{=} \left(\frac{(\frac{TOL}{\varpi_0})^{\frac{1}{pK_d}}}{(\frac{\varpi_K}{\varpi_0})^{\frac{1}{pK}}} \right).$

(5) If the tolerance is met: $Z_K \leq 1$ and $K < K_d$, then

(a) Increment time: $t = t + \Delta t$,
(b) Construct the next time-step: $\Delta t^{new} = \Lambda_K \Delta t^{old}$,
(c) Select the minimum size: $\Delta t = \mathbf{MIN}(\Delta t^{lim}, \Delta t^{new})$, and
(d) Update the ray positions and go to (0).

(6) If the tolerance is not met: $Z_K > 1$ and $K < K_d$, then

(a) Update the iteration counter: $K = K + 1$,
(b) Reset the particle counter: $i = 1$, and
(c) Go to (2).

(7) If the tolerance is not met ($Z_K > 1$) and $K = K_d$, then

(a) Construct a new time-step: $\Delta t^{new} = \Lambda_K \Delta t^{old}$,
(b) Restart at time t and go to (1).

Time-step size adaptivity is critical, since the system's dynamics and configuration can dramatically change over the course of time, possibly requiring quite different time-step sizes to control the iterative error. However, to maintain the accuracy of the time-stepping scheme, one must respect an upper bound dictated by the discretization error, i.e., $\Delta t \leq \Delta t^{lim}$. Note that in STEP (5), Λ_K may enlarge the time-step if the error is lower than the preset tolerance.

11.4 Numerical Example

As an example, we consider a group of N_p randomly dispersed spherical particles, of equal size, generated within in a spherical (aggregate) domain of diameter D. For illustration purposes only (Fig. 11.5), the radius of the pulse cross section and the agglomeration radius were both set to unity (one meter). The ratio of particle diameter, d, to total domain diameter, D, was $d/D = 0.05$. The total initial energy (again for illustration purposes) of each pulse was set to $I(t = 0) = I_o = 7.06 \times 10^{13}$ J, which roughly corresponded to a pressure pulse of 150 MPa traveling at a velocity of 1500 m/s over the area of the pulse. The initial energy for each ray was calculated as I/N_r, where $N_r = 10000$ was the number of rays in the beam. We remark that in the example calculation, we used $N_p = 2000$ particles and $N_r = 10000$ rays, initially parallel and placed in a circular domain (pulse cross section), but randomly distributed (Fig. 11.6). This system provided stable results, i.e., increasing the number of rays and/or the number of particles beyond these levels resulted in negligibly

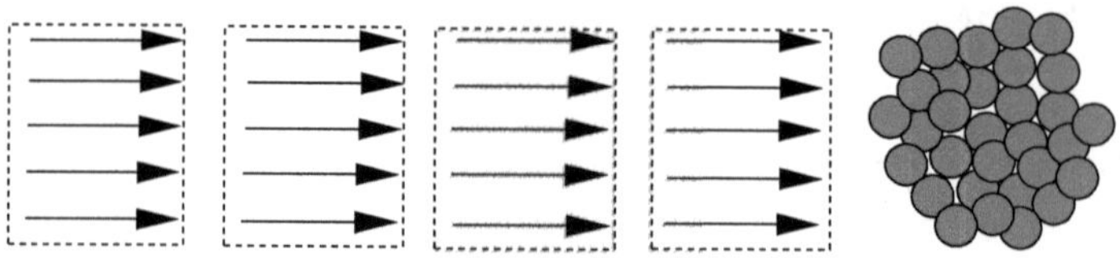

Fig. 11.5 Pulsing a free-standing agglomeration

different overall system responses. We note successive refinement of the ray density from 1000, 2000, 3000, ..., 10000 total rays was found to produce no noticeable differences beyond 10000 rays. Thus, we can consider, for all practical purposes that the results are independent of ray-grid density. One can consider the representation of a beam by multiple rays as simply taking a large "sampling" of the diffraction by the beam (wave front) over the portion of the scatterer where the beam is incident, as opposed to a discretization technique. It is relatively simple to scale the geometry down to agglomerations of practical interest, and to use corresponding pulse energy contents. The ratio of refractive indices was set to $\tilde{c} \stackrel{\text{def}}{=} \frac{c_i}{c_t} = 1500/4500$, where $c_i = 1500\,\text{m/s}$ (water) and $c_t = 4500\,\text{m/s}$ (ceramic). The densities were set to $\rho_i = 1000\,\text{kg/m}^3$ (water) and $\rho_t = 6000\,\text{kg/m}^3$ (ceramic). This leads to a ratio of acoustical impedances of $\hat{A} = 18$ and, consequently, a reflectivity, at a zero angle of incidence, from Eq. 11.2, of

$$\mathcal{R}(\theta = 0) = \left(\frac{18 - 1}{18 + 1}\right)^2 = 0.944. \tag{11.11}$$

Thus, at a zero angle of incidence, the amount of energy transmitted is approximately $1 - 0.944 = 0.056$. In order to generate the random particle positions, the classical Random Sequential Addition (RSA) algorithm was used to place nonoverlapping particles into the domain of interest (Widom [71]), initially a sphere, which was then allowed to dynamically converge to an equilibrium state.[5] The following relevant other simulation parameters chosen were (in SI units if not explicitly stated):

- The normal contact parameter was $K_{po} = 10^7\,\text{N/m}^2$, at a fixed temperature, $K_p = \mathbf{MAX}(K_{po}\left(e^{-(a\frac{\Theta}{\Theta^*}-1)}\right), K_p^{lim})$, where $\Theta^* = 500\,°\text{K}$, $K_p^{lim} = 10^6\,\text{N/m}^2$, the exponent in the contact law was set to $p_p = 2$, the temperature was fixed to be $\Theta = 300\,°\text{K}$, and the thermal sensitivity parameter was set to $a = 1$,
- The contact damping parameter, $c^{cd} = 10^5$,
- The friction contact parameter, $K^f = 10^7$,
- The coefficient of static friction, $\mu_s = 0.4$,
- The coefficient of dynamic friction, $\mu_d = 0.3$,

[5]See Torquato [10] and Torquato and coworkers (see, e.g., Kansaal et al. [72] and Donev et al. [73–77]) for a detailed review of particle packing algorithms.

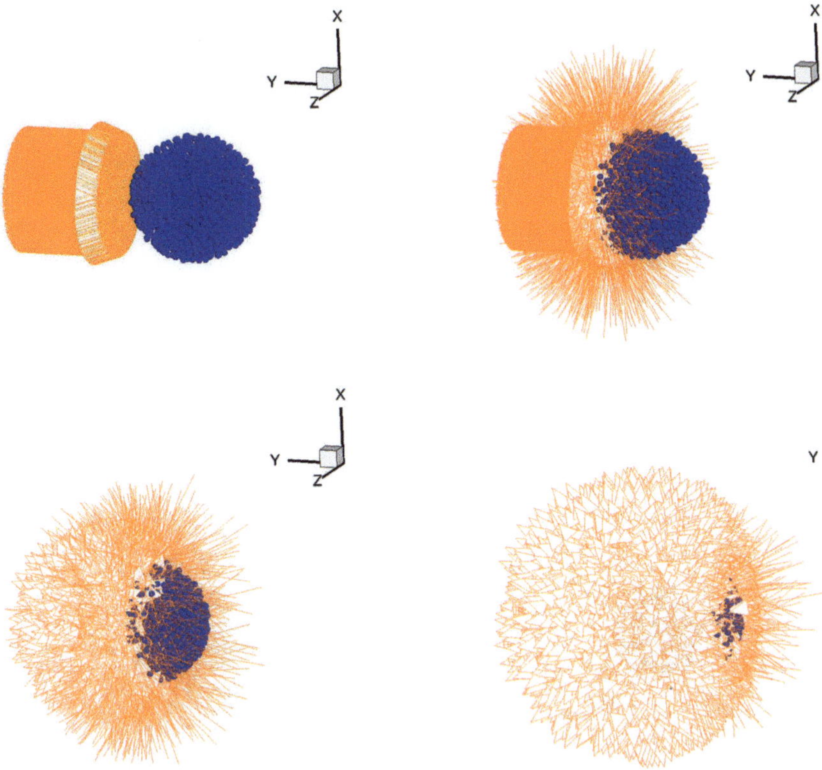

Fig. 11.6 From left to right and top to bottom, the progressive movement of rays comprising a beam. The vector lengths indicate the energy associated with the ray

- The normal bond parameter, $K^{nb} = 10^6$ N/m^2, and the exponent in the binding law was set to $p_b = 2$,
- The rotational/tangential bond parameter, $K^{rb} = 10^3$,
- The interstitial damping coefficient, $c^e = 1$ (assumed Stokesian-like),
- The target number of fixed-point iterations, $K_d = 10$,
- The trapezoidal time-stepping parameter, $\phi = 0.5$,
- The simulation duration, 2 s,
- The initial time-step size, 0.00001 s,
- The time-step upper bound, 0.00025 s, and
- The tolerance for the fixed-point iteration, 5×10^{-4}.

Figures 11.6 and 11.7 illustrate the results after a single pulse. After approximately 4 pulses, for the parameters chosen in this chapter, this agglomeration was completely broken up and dispersed. The number of reflections was 59238 for this sample problem. Several other statistical realizations were generated with on the order of 58,000–62,000 reflections incurred. Figure 11.8 shows the accumulated number of reflections after four pulses.

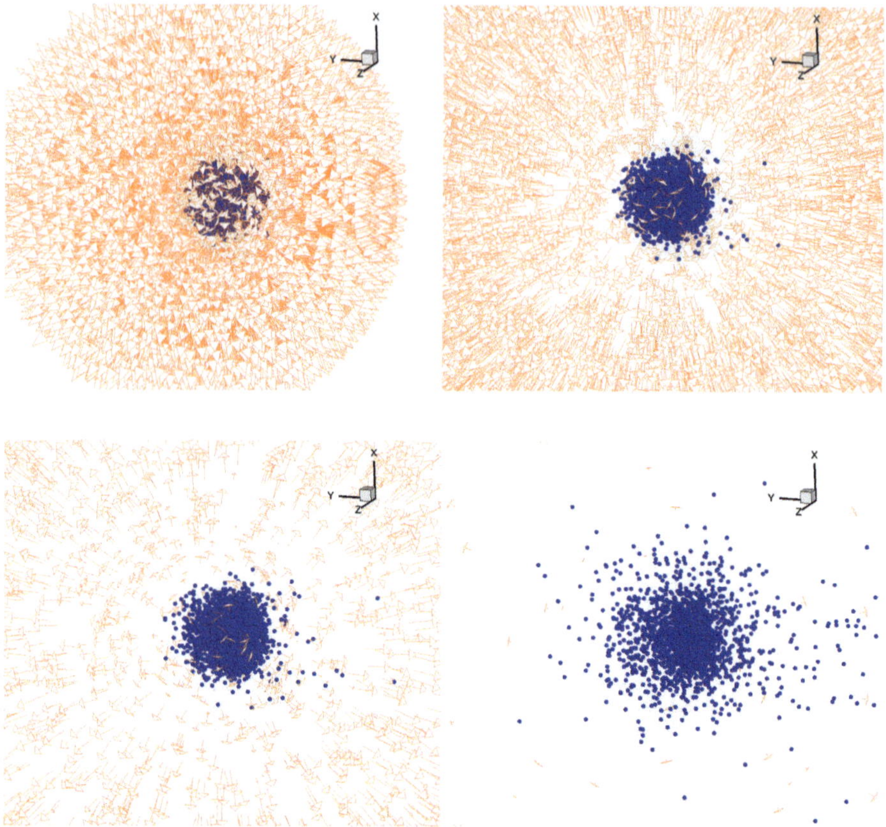

Fig. 11.7 From left to right and top to bottom, the progressive movement of rays comprising a beam. The vector lengths indicate the energy associated with the ray

11.5 Closing Statements

In summary, this work developed a discrete ray/discrete particle model to characterize the acoustical energy associated with pulses with agglomerations of discrete particles. The approach provides a simpler alternative to a direct computationally intensive discretization of a continuum description if it is even possible. Because of the complex microstructure, containing gaps and interfaces, this type of system is extremely difficult to simulate using continuum-based methods, such as the Finite Difference Time Domain Method or the Finite Element Method. The simplified model captures the primary effects, namely, reflection and absorption of acoustical energy via: (1) a discrete element representation of the particle system, (2) a discretization of a concentrated pulse into rays, (3) a discrete ray-tracking algorithm is developed to track the propagation of rays, and (4) a Discrete Element Method to track the break up and dynamics of the agglomeration. *The simulations take on the*

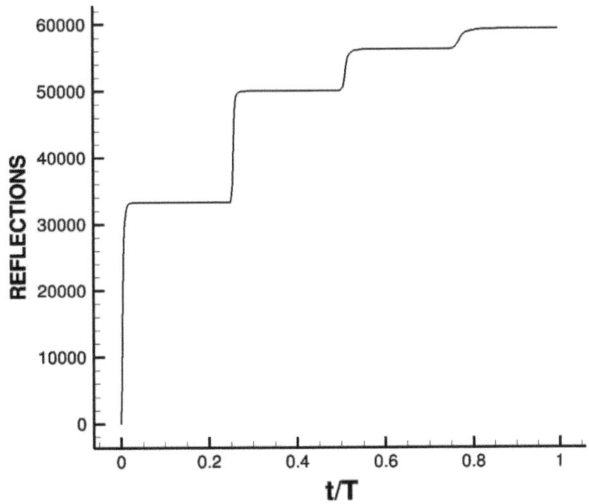

Fig. 11.8 Number of ray reflections as a function of time for 4 successive pulses

order of two minutes on a laptop. This technique was used to determine the amount of propagating incident energy that was reflected and the amount that was absorbed by each particle. Three-dimensional examples were given to illustrate the method. In closing, there are limitations in the use of ray theory. In particular, for small particles with size on the order of the wavelength, ray theory is inappropriate.

The chapter focused only on the propagation of acoustical energy in the system. The energy that was absorbed was assumed "available" to move the material. Clearly, not all of the absorbed energy would be "converted" into motion; however, this serves as a simple conservative estimate of the effects. The purpose of the present work was only to isolate one aspect of the complex series of events associated with shock-type loading of agglomerated microstructures mainly the determination of the energy absorbed and the system dynamics that would result. The approach developed provides a fast computational tool to analyze particulate agglomerations. It can be used on virtually any type of agglomeration domain. Since the results are derived from a direct numerical simulation, one can also post-process detailed statistical information for the breakup metrics. It is relatively straightforward to track quantities of interest related to the rays that comprise the acoustical disturbance, for example, the (normalized) acoustical energy retained in the beam, defined similarly:

$$\mathcal{I}_x \stackrel{\text{def}}{=} \frac{\sum_{q=1}^{N_r} \boldsymbol{I}_q^{ray} \cdot \boldsymbol{e}_x}{\sum_{q=1}^{N_r} ||\boldsymbol{I}_q^{ray}(t=0)||} \quad \text{and} \quad \mathcal{I}_y \stackrel{\text{def}}{=} \frac{\sum_{q=1}^{N_r} \boldsymbol{I}_q^{ray} \cdot \boldsymbol{e}_y}{\sum_{q=1}^{N_r} ||\boldsymbol{I}_q^{ray}(t=0)||} \quad \text{and}$$

$$\mathcal{I}_z \stackrel{\text{def}}{=} \frac{\sum_{q=1}^{N_r} \boldsymbol{I}_q^{ray} \cdot \boldsymbol{e}_z}{\sum_{q=1}^{N_r} ||\boldsymbol{I}_q^{ray}(t=0)||}. \tag{11.12}$$

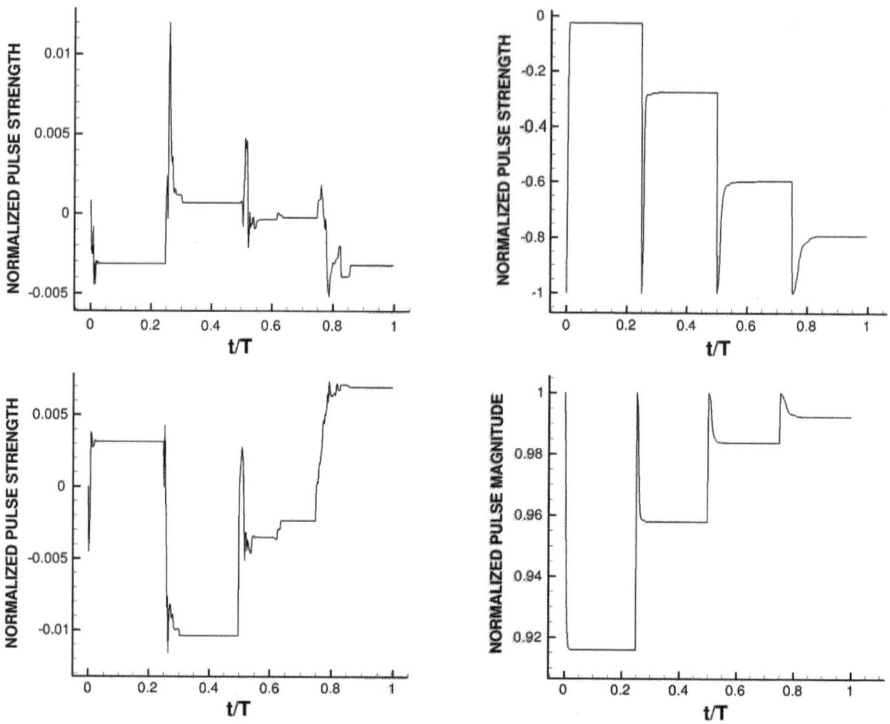

Fig. 11.9 Top to bottom and left to right: **a** normalized pulse strength in the x-direction: $\mathcal{I}_x \stackrel{\text{def}}{=} \frac{\sum_{q=1}^{N_r} \boldsymbol{I}_q^{ray} \cdot \boldsymbol{e}_x}{\sum_{q=1}^{N_r} ||\boldsymbol{I}_q^{ray}(t=0)||}$. **b** Normalized pulse strength in the y-direction: $\mathcal{I}_y \stackrel{\text{def}}{=} \frac{\sum_{q=1}^{N_r} \boldsymbol{I}_q^{ray} \cdot \boldsymbol{e}_y}{\sum_{q=1}^{N_r} ||\boldsymbol{I}_q^{ray}(t=0)||}$. **c** Normalized pulse strength in the z-direction: $\mathcal{I}_z \stackrel{\text{def}}{=} \frac{\sum_{q=1}^{N_r} \boldsymbol{I}_q^{ray} \cdot \boldsymbol{e}_z}{\sum_{q=1}^{N_r} ||\boldsymbol{I}_q^{ray}(t=0)||}$. **d** Normalized pulse magnitude: $\frac{\sqrt{\sum_{q=1}^{N_r}((\boldsymbol{I}_q^{ray} \cdot \boldsymbol{e}_x)^2 + (\boldsymbol{I}_q^{ray} \cdot \boldsymbol{e}_y)^2 + (\boldsymbol{I}_q^{ray} \cdot \boldsymbol{e}_z)^2)}}{\sum_{q=1}^{N_r} ||\boldsymbol{I}_q^{ray}(t=0)||}$

where $\boldsymbol{I}_q^{ray}$ are the individual ray contributions (Fig. 11.9). This can be extended, for example, for any quantity of interest, $\boldsymbol{Q}$ (e.g., the positions of the particles and their velocities), with a distribution of values ($\boldsymbol{Q}_i$, $i = 1, 2, \ldots, N_p =$ particles) about an arbitrary reference point, denoted $\boldsymbol{Q}^\star$, as follows:

$$\boldsymbol{M}_r^{\boldsymbol{Q}_i - \boldsymbol{Q}^\star} \stackrel{\text{def}}{=} \frac{\sum_{i=1}^{N_p} a_i (\boldsymbol{Q}_i - \boldsymbol{Q}^\star)^r}{\sum_{i=1}^{N_p} a_i} \stackrel{\text{def}}{=} \overline{(\boldsymbol{Q}_i - \boldsymbol{Q}^\star)^r}. \tag{11.13}$$

The various moments characterize the distribution, for example:

1. $\boldsymbol{M}_1^{\boldsymbol{Q}_i - \boldsymbol{A}}$ measures the first deviation from the average, which equals zero.
2. $\boldsymbol{M}_1^{\boldsymbol{Q}_i - \boldsymbol{0}} \stackrel{\text{def}}{=} \frac{\sum_{i=1}^{N_p} a_i (\boldsymbol{Q}_i - \boldsymbol{0})}{\sum_{i=1}^{N_p} a_i} \stackrel{\text{def}}{=} \overline{(\boldsymbol{Q}_i - \boldsymbol{0})} = \boldsymbol{A}$.

3. $M_2^{Q_i-A}$ is the standard deviation.
4. $M_3^{Q_i-A}$ is the skewness, which measures the bias, or asymmetry of the distribution of data.
5. $M_4^{Q_i-A}$ is the kurtosis (fourth moment), which measures the "tightness" of the distribution.

For the purposes of particle dynamics, $a_i = m_i$. This is straightforward to implement and can provide much more detailed information on post-impact system characteristics.

11.6 Chapter Appendix: Basics of Acoustics

In our approach, we model the individual particles as being rigid, and the material surrounding the particles as being isotropic and having a relatively low shear modulus, in the zero limit becoming an acoustical medium. Generally, for an isotropic material, one has the classical relationship between the components of infinitesimal strain (ϵ) to the Cauchy stress ($\boldsymbol{\sigma}$)

$$\boldsymbol{\sigma} = \mathbb{E} : \boldsymbol{\epsilon} = 3\kappa \frac{\text{tr}\boldsymbol{\epsilon}}{3}\mathbf{1} + 2\mu\boldsymbol{\epsilon}', \tag{11.14}$$

where $\mathbb{E}$ is the elasticity tensor and where $\boldsymbol{\epsilon}'$ is the strain deviator. The corresponding strain energy density is

$$W = \frac{1}{2}\boldsymbol{\epsilon} : \mathbb{E} : \boldsymbol{\epsilon} = \frac{1}{2}\left(9\kappa(\frac{\text{tr}\boldsymbol{\epsilon}}{3})^2 + 2\mu\boldsymbol{\epsilon}' : \boldsymbol{\epsilon}'\right). \tag{11.15}$$

We focus on the dilatational deformation in the low shear modulus matrix surrounding the particles. This naturally leads to an idealized "acoustical" material approximation, $\mu \approx 0$. Hence, Eq. 11.14 collapses to $\boldsymbol{\sigma} = -p\mathbf{1}$, where the pressure is $p = -3\,\kappa\,\frac{\text{tr}\boldsymbol{\epsilon}}{3}\mathbf{1}$ and with a corresponding strain energy of $W = \frac{1}{2}\frac{p^2}{\kappa}$. By inserting the simplified expression of the stress $\boldsymbol{\sigma} = -p\mathbf{1}$ into the equation of equilibrium, we obtain

$$\nabla \cdot \boldsymbol{\sigma} = -\nabla p = \rho\ddot{\boldsymbol{u}}, \tag{11.16}$$

where $\boldsymbol{u}$ is the displacement. By taking the divergence of both sides, and recognizing that $\nabla \cdot \boldsymbol{u} = -\frac{p}{\kappa}$, we obtain

$$\nabla^2 p = \frac{\rho}{\kappa}\ddot{p} = \frac{1}{c^2}\ddot{p}. \tag{11.17}$$

If we assume a harmonic solution, we obtain

$$p = Pe^{j(\boldsymbol{k}\cdot\boldsymbol{r}-\omega t)} \Rightarrow \dot{p} = Pj\omega e^{j(\boldsymbol{k}\cdot\boldsymbol{r}-\omega t)} \Rightarrow \ddot{p} = -P\omega^2 e^{j(\boldsymbol{k}\cdot\boldsymbol{r}-\omega t)}, \tag{11.18}$$

and

$$\begin{aligned}\nabla p &= Pj(k_x\boldsymbol{e}_x + k_y\boldsymbol{e}_y + k_z\boldsymbol{e}_z)e^{j(\boldsymbol{k}\cdot\boldsymbol{r}-\omega t)} \Rightarrow \nabla\cdot\nabla p = \nabla^2 p \\ &= -P\underbrace{(k_x^2 + k_y^2 + k_z^2)}_{||\boldsymbol{k}||^2} e^{j(\boldsymbol{k}\cdot\boldsymbol{r}-\omega t)}.\end{aligned} \tag{11.19}$$

We insert these relations into Eq. 11.17 and obtain an expression for the magnitude of the wave number vector

$$-P||\boldsymbol{k}||^2 e^{j(\boldsymbol{k}\cdot\boldsymbol{r}-\omega t)} = -\frac{\rho}{\kappa}P\omega^2 e^{j(\boldsymbol{k}\cdot\boldsymbol{r}-\omega t)} \Rightarrow ||\boldsymbol{k}|| = \frac{\omega}{c}. \tag{11.20}$$

Equation 11.16 (balance of linear momentum) implies

$$\rho\ddot{\boldsymbol{u}} = -\nabla p = -Pj(k_x\boldsymbol{e}_x + k_y\boldsymbol{e}_y + k_z\boldsymbol{e}_z)e^{j(\boldsymbol{k}\cdot\boldsymbol{r}-\omega t)}. \tag{11.21}$$

Now we integrate once, which is equivalent to dividing by $-j\omega$, and obtain the velocity

$$\dot{\boldsymbol{u}} = \frac{Pj}{\rho\omega}(k_x\boldsymbol{e}_x + k_y\boldsymbol{e}_y + k_z\boldsymbol{e}_z)e^{j(\boldsymbol{k}\cdot\boldsymbol{r}-\omega t)}, \tag{11.22}$$

and do so again for the displacement

$$\boldsymbol{u} = \frac{Pj}{\rho\omega^2}(k_x\boldsymbol{e}_x + k_y\boldsymbol{e}_y + k_z\boldsymbol{e}_z)e^{j(\boldsymbol{k}\cdot\boldsymbol{r}-\omega t)}. \tag{11.23}$$

Thus, we have

$$||\dot{\boldsymbol{u}}|| = \frac{P}{c\rho}. \tag{11.24}$$

The reflection of a plane harmonic pressure wave at an interface is given by enforcing continuity of the (acoustical) pressure and disturbance velocity at that location; this yields the ratio between the incident and reflected pressures. We use a local coordinate system (Fig. 11.3) and require that the number of waves per unit length in the $x-$direction must be the same for the incident, reflected, and refracted (transmitted) waves,

$$\boldsymbol{k}_i\cdot\boldsymbol{e}_x = \boldsymbol{k}_r\cdot\boldsymbol{e}_x = \boldsymbol{k}_t\cdot\boldsymbol{e}_x. \tag{11.25}$$

From the pressure balance at the interface, we have

$$P_i e^{j(\boldsymbol{k}_i\cdot\boldsymbol{r}-\omega t)} + P_r e^{j(\boldsymbol{k}_r\cdot\boldsymbol{r}-\omega t)} = P_t e^{j(\boldsymbol{k}_t\cdot\boldsymbol{r}-\omega t)}, \tag{11.26}$$

where P_i is the incident pressure ray, P_r is the reflected pressure ray, and P_t is the transmitted pressure ray. This forces a time-invariant relation to hold at all parts on the boundary, because the arguments of the exponential must be the same. This leads to ($k_i = k_r$)

$$k_i sin\theta_i = k_r sin\theta_r \Rightarrow \theta_i = \theta_r, \tag{11.27}$$

and

$$k_i sin\theta_i = k_t sin\theta_t \Rightarrow \frac{k_i}{k_t} = \frac{sin\theta_t}{sin\theta_i} = \frac{\omega/c_t}{\omega/c_i} = \frac{c_i}{c_t} = \frac{v_i}{v_t} = \frac{n_t}{n_i}. \tag{11.28}$$

Equations 11.25 and 11.26 imply

$$P_i e^{j(\boldsymbol{k}_i \cdot \boldsymbol{r})} + P_r e^{j(\boldsymbol{k}_r \cdot \boldsymbol{r})} = P_t e^{j(\boldsymbol{k}_t \cdot \boldsymbol{r})}. \tag{11.29}$$

The continuity of the displacement, and hence the velocity

$$\boldsymbol{v}_i + \boldsymbol{v}_r = \boldsymbol{v}_t, \tag{11.30}$$

leads to, after use of Eq. 11.24,

$$-\frac{P_i}{\rho_i c_i} cos\theta_i + \frac{P_r}{\rho_r c_r} cos\theta_r = -\frac{P_t}{\rho_t c_t} cos\theta_t. \tag{11.31}$$

We solve for the ratio of the reflected and incident pressures to obtain

$$r = \frac{P_r}{P_i} = \frac{\hat{A} cos\theta_i - cos\theta_t}{\hat{A} cos\theta_i + cos\theta_t}, \tag{11.32}$$

where $\hat{A} \stackrel{\text{def}}{=} \frac{A_t}{A_i} = \frac{\rho_t c_t}{\rho_i c_i}$, where ρ_t is the medium which the ray encounters (transmitted), c_t is corresponding sound speed in that medium, A_t is the corresponding acoustical impedance, ρ_i is the medium in which the ray was traveling (incident), c_i is corresponding sound speed in that medium A_i is the corresponding acoustical impedance. The relationship (the law of refraction) between the incident and transmitted angles is $c_t sin\theta_t = c_i sin\theta_i$. Thus, we may write the Fresnel relation

$$r = \frac{\tilde{c}\hat{A} cos\theta_i - (\tilde{c}^2 - sin^2\theta_i)^{\frac{1}{2}}}{\tilde{c}\hat{A} cos\theta_i + (\tilde{c}^2 - sin^2\theta_i)^{\frac{1}{2}}}, \tag{11.33}$$

where $\tilde{c} \stackrel{\text{def}}{=} \frac{c_i}{c_t}$. The reflectance for the (acoustical) energy $\mathcal{R} = r^2$ is

$$\mathcal{R} = \left(\frac{P_r}{P_i}\right)^2 = \left(\frac{\hat{A} cos\theta_i - cos\theta_t}{\hat{A} cos\theta_i + cos\theta_t}\right)^2 = \left(\frac{I_r}{I_i}\right)^2. \tag{11.34}$$

For the cases where $sin\theta_t = \frac{sin\theta_i}{\tilde{c}} > 1$, one may rewrite the reflection relation as

$$r = \frac{\tilde{c}\hat{A}cos\theta_i - j(sin^2\theta_i - \tilde{c}^2)^{\frac{1}{2}}}{\tilde{c}\hat{A}cos\theta_i + j(sin^2\theta_i - \tilde{c}^2)^{\frac{1}{2}}}. \tag{11.35}$$

where $j = \sqrt{-1}$. The reflectance is $\mathcal{R} \stackrel{\text{def}}{=} r\bar{r} = 1$, where $\bar{r}$ is the complex conjugate. Thus, for angles above the critical angle $\theta_i \geq \theta_i^*$, all of the energy is reflected. We note that when $A_t = A_i$ and $c_i = c_t$, then there is no reflection. Also, when $A_t >> A_i$ or when $A_t << A_i$, then $r \rightarrow 1$.

Remark If one considers for a moment an incoming pressure wave (ray), which is incident on an interface between two general elastic media ($\mu \neq 0$), reflected shear waves must be generated in order to satisfy continuity of the traction, $[\![\boldsymbol{\sigma} \cdot \boldsymbol{n}]\!] = \mathbf{0}$. This is due to the fact that

$$[\![\left(3\kappa \text{tr}\frac{\boldsymbol{\epsilon}}{3}\mathbf{1} + 2\mu\boldsymbol{\epsilon}'\right) \cdot \boldsymbol{n}]\!] = \mathbf{0}. \tag{11.36}$$

For an idealized acoustical medium, $\mu = 0$, no shear waves need to be generated to satisfy Eq. 11.36.

References

1. Choi, S., Park, I., Hao, Z., Holman, H.Y., Pisano, A.P., Zohdi, T.I.: Ultra-fast self-assembly of micro-scale particles by open channel flow. Langmuir **26**(7), 4661–4667 (2010)
2. Choi, S., Stassi, S., Pisano, A.P., Zohdi, T.I.: Coffee-ring effect-based three dimensional patterning of micro, nanoparticle assembly with a single droplet. Langmuir **26**(14), 11690–11698 (2010)
3. Choi, S., Jamshidi, A., Seok, T.J., Zohdi, T.I., Wu., M.C., Pisano, A.P.: Fast, High-throughput creation of size-tunable micro, nanoparticle clusters via evaporative self-assembly in picoliter-scale droplets of particle suspension. Langmuir **28**(6), 3102–3111 (2012)
4. Demko, M., Choi, S., Zohdi, T.I., Pisano, A.P.: High resolution patterning of nanoparticles by evaporative self-assembly enabled by in-situ creation and mechanical lift-off of a polymer template. Appl. Phys. Lett. **99**, 253102-1–253102-3 (2012)
5. Demko, M.T., Cheng, J.C., Pisano, A.P.: High-resolution direct patterning of gold nanoparticles by the microfluidic molding process. Langmuir 412–417 (2010)
6. Martin, P.: Handbook of Deposition Technologies for Films and Coatings, 3rd edn. Elsevier (2009)
7. Martin, P.: Introduction to Surface Engineering and Functionally Engineered Materials. Scrivener and Elsevier (2011)
8. Wei, C., Dong, J.: Development and modeling of melt electrohydrodynamic-jet printing of phase-change inks for high-resolution additive manufacturing. J. Manuf. Sci. Eng. **136**, 061010 (2014). Paper no: MANU-14-1179. https://doi.org/10.1115/1.4028483
9. Duran, J.: Sands, Powders and Grains: An Introduction to the Physics of Granular Matter. Springer (1997)
10. Torquato, S.: Random Heterogeneous Materials: Microstructure and Macroscopic Properties. Springer, New York (2002)
11. Onate, E., Idelsohn, S.R., Celigueta, M.A., Rossi, R.: Advances in the particle finite element method for the analysis of fluid-multibody interaction and bed erosion in free surface flows. Comput. Methods Appl. Mech. Eng. **197**(19–20), 1777–1800 (2008)

12. Onate, E., Celigueta, M.A., Idelsohn, S.R., Salazar, F., Surez, B.: Possibilities of the particle finite element method for fluid-soil-structure interaction problems. Comput. Mech. **48**, 307–318 (2011)
13. Rojek, J., Labra, C., Su, O., Onate, E.: Comparative study of different discrete element models and evaluation of equivalent micromechanical parameters. Int. J. Solids Struct. **49**, 1497–1517 (2012). https://doi.org/10.1016/j.ijsolstr.2012.02.032
14. Carbonell, J.M., Onate, E., Suarez, B.: Modeling of ground excavation with the particle finite element method. J. Eng. Mech. ASCE **136**, 455–463 (2010)
15. Labra, C., Onate, E.: High-density sphere packing for discrete element method simulations. Commun. Numer. Methods Eng. **25**(7), 837–849 (2009)
16. Leonardi, A., Wittel, F.K., Mendoza, M., Herrmann, H.J.: Coupled DEM-LBM method for the free-surface simulation of heterogeneous suspensions. Comput. Part. Mech. **1**(1), 3–13 (2014)
17. Cante, J., Davalos, C., Hernandez, J.A., Oliver, J., Jonsen, P., Gustafsson, G., Haggblad, H.A.: PFEM-based modeling of industrial granular flows. Comput. Part. Mech. **1**(1), 47–70 (2014)
18. Rojek, J.: Discrete element thermomechanical modelling of rock cutting with valuation of tool wear. Comput. Part. Mech. **1**(1), 71–84 (2014)
19. Onate, E., Celigueta, M.A., Latorre, S., Casas, G., Rossi, R., Rojek, J.: Lagrangian analysis of multiscale particulate flows with the particle finite element method. Comput. Part. Mech. **1**(1), 85–102 (2014)
20. Bolintineanu, D.S., Grest, G.S., Lechman, J.B., Pierce, F., Plimpton, S.J., Schunk, P.R.: Particle dynamics modeling methods for colloid suspensions. Comput. Part. Mech. **1**(3), 321–356 (2014)
21. Avci, B., Wriggers, P.: A DEM-FEM coupling approach for the direct numerical simulation of 3D particulate flows. J. Appl. Mech. **79**(1–7), 010901 (2012)
22. Zohdi, T.I.: Staggering error control for a class of inelastic processes in random microheterogeneous solids. Int. J. Nonlinear Mech. **39**, 281–297 (2004)
23. Zohdi, T.I.: Modeling and simulation of a class of coupled thermo-chemo-mechanical processes in multiphase solids. Comput. Methods Appl. Mech. Eng. **193**(6–8), 679–699 (2004)
24. Zohdi, T.I.: Modeling and direct simulation of near-field granular flows. Int. J. Solids Struct. **42**(2), 539–564 (2004)
25. Zohdi, T.I.: Charge-induced clustering in multifield particulate flow Int. J. Numer. Methods Eng. **62**(7), 870–898 (2005)
26. Zohdi, T.I.: On the optical thickness of disordered particulate media. Mech. Mater. **38**, 969–981 (2006)
27. Zohdi, T.I., Kuypers, F.A.: Modeling and rapid simulation of multiple red blood cell light scattering. Proc. R. Soc. Interface **3**(11), 823–831 (2006)
28. Zohdi, T.I.: Computation of the coupled thermo-optical scattering properties of random particulate systems. Comput. Methods Appl. Mech. Eng. **195**, 5813–5830 (2006)
29. Zohdi, T.I.: Computation of strongly coupled multifield interaction in particle-fluid systems. Comput. Methods Appl. Mech. Eng. **196**, 3927–3950 (2007)
30. Zohdi, T.I.: On the computation of the coupled thermo-electromagnetic response of continua with particulate microstructure. Int. J. Numer. Methods Eng. **76**, 1250–1279 (2008)
31. Zohdi, T.I.: Mechanistic modeling of swarms. Comput. Methods Appl. Mech. Eng. **198**(21–26), 2039–2051 (2009)
32. Zohdi, T.I.: On the dynamics of charged electromagnetic particulate jets. Arch. Comput. Methods Eng. **17**(2), 109–135 (2010)
33. Zohdi, T.I., Kuypers, F.A., Lee, W.C.: Estimation of red blood cell volume fraction from overall permittivity measurement. Int. J. Eng. Sci. **48**, 1681–1691 (2010)
34. Zohdi, T.I.: Simulation of coupled microscale multiphysical-fields in particulate-doped dielectrics with staggered adaptive FDTD. Comput. Methods Appl. Mech. Eng. **199**, 79–101 (2010)
35. Zohdi, T.I.: Dynamics of clusters of charged particulates in electromagnetic fields. Int. J. Numer. Methods Eng. **85**, 1140–1159 (2011)

36. Zohdi, T.I.: Modeling and simulation of the optical response rod-functionalized reflective surfaces. Comput. Mech. **50**(2), 257–268 (2012)
37. Zohdi, T.I.: Numerical simulation of charged particulate cluster-droplet impact on electrified surfaces. J. Comput. Phys. **233**, 509–526 (2013)
38. Zohdi, T.I.: On inducing compressive residual stress in microscale print-lines for flexible electronics. Int. J. Eng. Sci. **62**, 157–164 (2013)
39. Zohdi, T.I.: Rapid simulation of laser processing of discrete particulate materials. Arch. Comput. Methods Eng. **20**, 309–325 (2013)
40. Zohdi, T.I.: A direct particle-based computational framework for electrically-enhanced thermo-mechanical sintering of powdered materials. Math. Mech. Solids **19**(1), 93–113 (2014)
41. Zohdi, T.I.: Additive particle deposition and selective laser processing-a computational manufacturing framework. Comput. Mech. **54**, 171–191 (2014)
42. Zohdi, T.I.: Embedded electromagnetically sensitive particle motion in functionalized fluids. Comput. Part. Mech. **1**, 27–45 (2014)
43. Zohdi, T.I.: A computational modelling framework for high-frequency particulate obscurant cloud performance. Int. J. Eng. Sci. **89**, 75–85 (2015)
44. Zohdi, T.I.: Modeling and simulation of cooling-induced residual stresses in heated particulate mixture depositions. Comput. Mech. **56**, 613–630 (2015)
45. Zohdi, T.I.: Modeling and efficient simulation of the deposition of particulate flows onto compliant substrates. Int. J. Eng. Sci. **99**, 74–91 (2015). https://doi.org/10.1016/j.ijengsci.2015.10.012
46. Zohdi, T.I.: Modeling and simulation of laser processing of particulate-functionalized materials. Arch. Comput. Methods Eng. 1–25 (2015). https://doi.org/10.1007/s11831-015-9160-1
47. Zohdi, T.I.: A discrete element and ray framework for rapid simulation of acoustical dispersion of microscale particulate agglomerations. Comp. Mech. https://doi.org/10.1007/s0046-015-1256-2
48. Zohdi, T.I., Wriggers, P.: Introduction to Computational Micromechanics. Springer, Second Reprinting (Peer Reviewed) (2005)
49. Markov, K.Z.: Elementary micromechanics of heterogeneous media. In: Markov, K.Z., Preziozi, L. (eds.) Heterogeneous Media: Micromechanics Modeling Methods and Simulations, pp. 1–162. Birkhauser, Boston (2000)
50. Virovlyanskii, A.L.: Interrelation between ray and mode field representations in an acoustic waveguide. Radiophys. Quant. Electron. **38**, 76 (1995)
51. Borejko, P., Chen, C.F., Pao, Y.H.: Application of the method of generalized rays to acoustic waves in a liquid wedge over elastic bottom. J. Comput. Acoust. **9**, 41–68 (2011)
52. Aurich, J.C., Dornfeld, D.A.: Burrs—analysis, control and removal. In: Jan Aurich, C., David, A. (eds.) Proceedings of the CIRP International. Springer, Dornfeld (2010)
53. Garg, S., Dornfeld, D., Berger, K.: Formulation of the chip cleanability mechanics from fluid transport. In: Aurich, J.C., Dornfeld, D.A. (eds.) Burrs Analysis, Control and Removal, pp. 229–236. Springer, Germany (2010)
54. Avila, M., Gardner, J., Reich-Weiser, C., Tripathi, S., Vijayaraghavan, A., Dornfeld, D.A.: Strategies for Burr minimization and cleanability in aerospace and automotive manufacturing. SAE Tech. (2005). Paper 2005-01-3327. https://doi.org/10.4271/2005-01-3327
55. Avila, M. Reich-Weiser, C., Dornfeld, D.A., McMains, S.: Design and manufacturing for cleanability in high performance cutting. In: CIRP 2nd International Conference on High Performance Cutting, Vancouver, BC, June 2006, Paper number 63 (2006)
56. Luo, L., Dornfeld, D.A.: Material removal mechanism in chemical mechanical polishing: theory and modeling. IEEE Trans. Semicond. Manuf. **14**(2), 112–133 (2001)
57. Luo, L., Dornfeld, D.A.: Effects of abrasive size distribution in chemical-mechanical planarization: modeling and verification. IEEE Trans. Semicond. Manuf. **16**, 469–476 (2003)
58. Luo, L., Dornfeld, D.A.: Material removal regions in chemical mechanical planarization for sub-micron integration for sub-micron integrated circuit fabrication: coupling effects of slurry chemicals, abrasive size distribution, and wafer-pad contact area. IEEE Trans. Semicond. Manuf. **16**, 45–56 (2003)

59. Luo, L., Dornfeld, D.A.: Integrated Modeling of Chemical Mechanical Planarization of Submicron IC Fabrication. Springer (2004)
60. Arbelaez, D., Zohdi, T.I., Dornfeld, D.: Modeling and simulation of material removal with particulate flow. Comput. Mech. **42**, 749–759 (2008)
61. Arbelaez, D., Zohdi, T.I., Dornfeld, D.: On impinging near-field granular jets. Int. J. Numer. Methods Eng. **80**(6), 815–845 (2009)
62. Campello, E.M.B., Zohdi, T.I.: A computational framework for simulation of the delivery of substances into cells. Int. J. Numer. Methods Biomed. Eng. **30**(11), 1132–1152 (2014)
63. Campello, E.M.B., Zohdi, T.I.: Design evaluation of a particle bombardment system to deliver substances into cells. Comput. Mech. Eng. Sci. **98**(2), 221–245 (2014)
64. Gomes-Ferreira, C., Ciampini, D., Papini, M.: The effect of inter-particle collisions in erosive streams on the distribution of energy flux incident to a flat surface. Tribol. Int. **37**, 791–807 (2004)
65. Ghobeity, A., Spelt, J.K., Papini, M.: Abrasive jet micro-machining of planar areas and transitional slopes. J. Micromech. Microeng. **18**, 055014 (2008)
66. Ghobeity, A., Krajac, T., Burzynski, T., Papini, M., Spelt, J.K.: Surface evolution models in abrasive jet micromachining. Wear **264**, 185–198 (2008)
67. Afazov, S.M., Becker, A.A., Hyde, T.H.: Mathematical modeling and implementation of residual stress mapping from microscale to macroscale finite element models. J. Manuf. Sci. Eng. **134**(2), 021001-021001-11 (2012)
68. Bagherifard, S., Giglio, M., Giudici, L., Guagliano, M.: Experimental and numerical analysis of fatigue properties improvement in a titanium alloy by shot peening. In: Proceedings of the ASME. 49163; ASME 2010 10th Biennial Conference on Engineering Systems Design and Analysis, vol. 2, pp. 317–322 (2010)
69. Elbella, A., Fadul, F., Uddanda, S.H., Kasarla, N.R.: Influence of shot peening parameters on process effectiveness. In: Proceedings of the ASME. 45196; Volume 3: Design, Materials and Manufacturing, Parts A, B, and C: 2015–2021 (2012)
70. Chen, Z., Yang, F., Meguid, S.A.: Realistic finite element simulations of arc-height development in shot-peened Almen strips. J. Eng. Mater. Technol. **136**(4), 041002-041002-7 (2014)
71. Widom, B.: Random sequential addition of hard spheres to a volume. J. Chem. Phys. **44**, 3888–3894 (1966)
72. Kansaal, A., Torquato, S., Stillinger, F.: Diversity of order and densities in jammed hard-particle packings. Phys. Rev. E. **66**, 041109 (2002)
73. Donev, A., Cisse, I., Sachs, D., Variano, E.A., Stillinger, F., Connelly, R., Torquato, S., Chaikin, P.: Improving the density of jammed disordered packings using ellipsoids. Science **13**(303), 990–993 (2004)
74. Donev, A., Stillinger, F.H., Chaikin, P.M., Torquato, S.: Unusually dense crystal ellipsoid packings. Phys. Rev. Lett. **92**, 255506 (2004)
75. Donev, A., Torquato, S., Stillinger, F.: Neighbor list collision-driven molecular dynamics simulation for nonspherical hard particles-I: Algorithmic details. J. Comput. Phys. **202**, 737 (2005)
76. Donev, A., Torquato, S., Stillinger, F.: Neighbor list collision-driven molecular dynamics simulation for nonspherical hard particles-II: Application to ellipses and ellipsoids. J. Comput. Phys. **202**, 765 (2005)
77. Donev, A., Torquato, S., Stillinger, F.H.: Pair correlation function characteristics of nearly jammed disordered and ordered hard-sphere packings. Phys. Rev. E **71**, 011105 (2005)

Chapter 12
Summary and Closing Remarks

The adoption of detailed material models and computational methods in additive manufacturing and 3D printing has the potential to bring a level of systematic analysis that can make it a reliable large-scale manufacturing process. However, there are a number of challenges, from the point of view of the end user of a simulation tool to be addressed in the development of computational approaches for additive manufacturing including (see [1–6]):

- validated quality and integrity of the products designed using computational methods for critical applications which address (a) thermal effects and distortion and (b) mechanical and other physical properties, defects, inclusions/lack of proper bonding,
- validated precision surface features including form and dimensions (as the slice thickness and deposition angles are critical parameters): (a) production rates and (b) energy and material consumption issues and the environmental impact of the technology.

From the point of view of computational challenges, the types of numerical methods needed to simulate such processes are still evolving rapidly. One objective of future research is the development of Discrete Element Models and codes for high-fidelity additive manufacturing processes, guided by careful experimentation. The salient computational concerns are the development of modeling and simulation methods which can be seamlessly integrated into design, coupling all the various subprocesses together. Any realistic simulation of an advanced additive manufacturing process will involve coupling many of the subprocesses mentioned in the monograph [7–49]. If systematic approaches can be developed meeting industry standards, this is a game-changer. This needs to draw upon the expertise of researchers in a number of disciplines in engineering: (1) computational science, (2) precision manufacturing, (3) materials science, (4) fluid mechanics and rheology, (5) multidegree of freedom

T. I. Zohdi, *Modeling and Simulation of Functionalized Materials for Additive Manufacturing and 3D Printing: Continuous and Discrete Media*, Lecture Notes in Applied and Computational Mechanics 60, https://doi.org/10.1007/978-3-319-70079-3_12

precision machines, robotics, and control theory, (6) heat transfer, (7) tribology, and (8) computer-aided design (CAD), geometric part representation, computer-aided manufacturing (CAM), and tool path generation. As of 2015, there are a variety of companies in this area, in particular in the plastic-based domain:

- Stratasys: http://www.stratasys.com/
- 3D Systems: http://www.3dsystems.com/
- Type A Machines: https://www.typeamachines.com
- Materialise: http://www.materialise.com/
- ExOne: http://www.exone.com/
- Arcam: http://www.arcam.com/
- SLM Solutions: http://www.stage.slm-solutions.com/index.php?index_de
- Alphaform: http://www.alphaform.de/
- Voxeljet: http://www.voxeljet.de/en/

For a more complete listing, which is certainly not comprehensive, see:

- http://3dprintingindustry.com/
- http://3dprintingindustry.com/2015/03/19/fiscal-2014-revenue-results-3d-printings-top-10-guns/
- https://en.wikipedia.org/wiki/Category:3D-printer-companies

We remark that the development of new robotic systems is critical to the advancement of this field. Recently, most industrialized countries have stressed the importance of robotics for the development of next-generation additive manufacturing machines and processes. However, because of the entirely new manufacturing processes that have developed over the last forty years, there is a lag between what robots, who have not changed much fundamentally in this domain, were intended to do and what industry now needs them to do. For example, because of the new types of materials being handled, with completely new processes, the challenges range from

- Handling and deposition of powdered materials which have the appropriate properties for printing, in order to yield high-performance nanostructured heterogeneous interfaces,
- Simultaneous laser-system-control parameters and environmental-chamber conditions that precisely create heterogeneous materials,
- The in situ assembly of materials with sensors,
- Detailed monitoring of the thermo-mechanical behavior of the complete system during processing,
- Statistical analysis of the performance of printed heterogeneous systems, and
- Assessing environmental and resource issues associated with these materials.

Thus, the demands are at a level of robotic precision at unprecedented accuracy for large volume of products.

In closing, many of the challenges facing additive manufacturing processes, such as heated filament-based methods, stem from the fact that one wishes to extend the present approaches, which use simple geometries and polymers (PLA and ABS binders, ink-based processes) to more sophisticated complex geometries comprised of metals, ceramics, and composite materials. Currently, rapid printing does not allow precise control over the structure of the printed lines. This often results in lines with scalloped edges or nonuniform width and offers only limited control over the height of the printed features. See [50–58] for details. Recently, nanoimprint lithography has been proposed as a means of decreasing the feature size of patterned nanoparticles while allowing more precise control over the structure of the printed lines [59–62]. In this fabrication method, the nanoparticle inks are patterned by pressing with an elastomer mold and the particles are dried into their final configuration. While the resolution of nanoimprint lithography is improved over inkjet printing, there exists a residual layer on the substrate that must be etched away after patterning. Control over the height of features can be corrupted by capillary action between the mold and the drying ink, in particular along the length of longer features. Thus, as a possible alternative to nanoimprint lithography, nanoparticle self-assembly methods, based on capillary filling of photoresist templates, have been proposed by Demko et al. [55, 56] and appear to be promising. This leads to an obvious fact, namely additive manufacturing alone is inadequate and needs to be combined with classical manufacturing processes. *Explicitly stated, despite the attractive features of additive manufacturing, it alone rarely produces the surface quality needed for structural integrity. Often classical subtractive, intermediate, high-precision surface milling is needed to create components with acceptable toughness and fatigue life.* It is imperative that additive and subtractive processes be combined, guided by simulation software for deposition and removal of material with sensitivity to part quality, dimension, tolerances, and surface finish. This must draw on a combination of existing and new numerical methods for additive manufacturing, multiaxis machines, and computational modeling of process performance. Organizations such as the American Society for Precision Engineering (ASPE) are exploring a number of issues that will need to be addressed if additive manufacturing is to realize its full potential for real components used in critical systems. These include issues related to: (a) dimensional control needed for AM to be used in precision applications, (b) design for manufacturing including design rules for additive manufacturing and the impact of dimensional errors on structures designed using optimization methodologies, (c) standards including certifying additive manufacturing equipment capabilities and artifacts for assessing machine performance, (d) using AM-fabricated components in precision assemblies and component-to-component relationships, stack-up tolerances, friction, robotic grip-ability, and (e) metrology and quality of external surfaces and internal features including material validation. Some of these issues can be addressed by hybridizing AM with appropriate other processes ranging from precision machining to electrical discharge machining (EDM) to industrial-scale polishing methods such as chemical–mechanical planarization (CMP). Industry has been quick to react to the advances in technology. For example, one of the largest machine manufacturers in the world, DMG Mori, has just introduced a five-axis CNC machine tool with an array of conventional milling and drilling tools (subtractive processes), in addition to a Sauer laser powder fusion head (additive processes) that can be mounted in the machine tool spindle, in place of cutting tools, in order to offer a build and machine option. Industries stand to make great gains by understanding and adopting such tools and processes in manufacturing. From an educational point of view, a central goal is to provide researchers with the skills to innovate and improve margins through the combined use of cutting edge, advanced manufacturing, modeling and simulation techniques. Furthermore, the development of a modern advanced manufacturing curriculum, with a strong materials and mechanics foundation that can meet the needs of students across multiple departments, is crucial. If successful, students will gain a critical understanding of the issues currently at the forefront of advanced manufacturing technologies and, in particular, the ability to model and simulate such processes with high fidelity, thus enabling additive manufacturing to live up to its potential as a viable advanced manufacturing technology.

References

1. Avila, M., Gardner, J., Reich-Weiser, C., Tripathi, Vijayaraghavan, A., Dornfeld, D.A.: Strategies for burr minimization and cleanability in aerospace and automotive manufacturing. SAE Technical Paper 2005-01-3327 (2005). https://doi.org/10.4271/2005-01-3327
2. Dornfeld, D., Wright, P., Vijayaraghavan, A., Helu, M.: Enabling manufacturing research through interoperability. Trans. North Am. Manuf. Res. Inst. SME **37**, 443–450 (2009)
3. Vijayaraghavan, A., Huet, L., Dornfeld, D., Sobel, W., Blomquist, B., Conley, M.: Addressing process planning and verification issues with MTConnect. Trans. North Am. Manuf. Res. Inst. SME **38**, 557–564 (2010)
4. Vijayaraghavan, Dornfeld, D.: Automated energy monitoring of machine tools. CIRP Ann. **59**(1), 21–24 (2010)
5. Dornfeld, D.A., Lee, D.E.: Precision Manufacturing. Springer, New York (2008)
6. Dornfeld, D. (ed.): Green Manufacturing Fundamentals and Applications. Springer, New York (2013)
7. Zohdi, T.I.: Genetic design of solids possessing a random-particulate microstructure. Philos. Trans. R. Soc. Math. Phys. Eng. Sci. **361**(1806), 1021–1043 (2003)
8. Zohdi, T.I.: Simulation of coupled microscale multiphysical-fields in particulate-doped dielectrics with staggered adaptive FDTD. Comput. Method. Appl. Mech. Eng. **199**, 79–101 (2010)
9. Zohdi, T.I.: On the compaction of cohesive hyperelastic granules at finite strains. Proc. R. Soc. **454**(2034), 1395–1401 (2003)
10. Zohdi, T.I.: Computational design of swarms. Int. J. Numer. Method. Eng. **57**, 2205–2219 (2003)
11. Zohdi, T.I.: Constrained inverse formulations in random material design. Comput. Method. Appl. Mech. Eng. 1–20. **192**(28–30), 18, 3179–3194
12. Zohdi, T.I.: Staggering error control for a class of inelastic processes in random microheterogeneous solids. Int. J. Nonlinear Mech. **39**, 281–297 (2004)
13. Zohdi, T.I.: Modeling and simulation of a class of coupled thermo-chemo-mechanical processes in multiphase solids. Comput. Method. Appl. Mech. Eng. **193**(6), 679–699 (2004)
14. Zohdi, T.I.: Modeling and direct simulation of near-field granular flows. Int. J. Solids Struct. **42**(2), 539–564 (2004)
15. Zohdi, T.I.: A computational framework for agglomeration in thermo-chemically reacting granular flows. Proc. R. Soc. **460**(2052), 3421–3445 (2004)
16. Zohdi, T.I.: Statistical ensemble error bounds for homogenized microheterogeneous solids. J. Appl. Math. Phys. (Zeitschrift für Angewandte Mathematik und Physik) **56**(3), 497–515 (2005)
17. Zohdi, T.I.: Charge-induced clustering in multifield particulate flow. Int. J. Numer. Method. Eng. **62**(7D), 870–898 (2005)

18. Zohdi, T.I.: On the optical thickness of disordered particulate media. Mech. Mater. **38**, 969–981 (2006)
19. Zohdi, T.I., Kuypers, F.A.: Modeling and rapid simulation of multiple red blood cell light scattering. Proc. R. Soc. Interface **3**(11), 823–831 (2006)
20. Zohdi, T.I.: Computation of the coupled thermo-optical scattering properties of random particulate systems. Comput. Method. Appl. Mech. Eng. **195**, 5813–5830 (2006)
21. Zohdi, T.I.: Computation of strongly coupled multifield interaction in particle-fluid systems. Comput. Method. Appl. Mech. Eng. **196**, 3927–3950 (2007)
22. Zohdi, T.I.: On the computation of the coupled thermo-electromagnetic response of continua with particulate microstructure. Int. J. Numer. Method. Eng. **76**, 1250–1279 (2008)
23. Zohdi, T.I.: Mechanistic modeling of swarms. Comput. Method. Appl. Mech. Eng. **198**(21–26), 2039–2051 (2009)
24. Zohdi, T.I.: Charged wall-growth in channel-flow. Int. J. Eng. Sci. **48**, 1520 (2010)
25. Zohdi, T.I.: On the dynamics of charged electromagnetic particulate jets. Arch. Comput. Method. Eng. **17**(2), 109–135 (2010)
26. Zohdi, T.I., Kuypers, F.A., Lee, W.C.: Estimation of Red Blood Cell volume fraction from overall permittivity measurement. Int. J. Eng. Sci. **48**, 1681–1691 (2010)
27. Zohdi, T.I.: Simulation of coupled microscale multiphysical-fields in particulate-doped dielectrics with staggered adaptive FDTD. Comput. Method. Appl. Mech. Eng. **199**, 79–101 (2010)
28. Zohdi, T.I.: Dynamics of clusters of charged particulates in electromagnetic fields. Int. J. Numer. Method. Eng. **85**, 1140–1159 (2011)
29. Zohdi, T.I.: Joule-heating field phase-amplification in particulate-doped dielectrics. Int. J. Eng. Sci. **49**, 30–40 (2011)
30. Zohdi, T.I.: Estimation of electrical-heating load-shares for sintering of powder mixtures. Proc. R. Soc. **468**, 2174–2190 (2012)
31. Zohdi, T.I.: Modeling and simulation of the optical response rod-functionalized reflective surfaces. Comput. Mech. **50**(2), 257–268 (2012)
32. Zohdi, T.I.: On the reduction of heat generation in lubricants using microscale additives. Int. J. Eng. Sci. **62**, 84–89 (2013)
33. Zohdi, T.I.: Numerical simulation of charged particulate cluster-droplet impact on electrified surfaces. J. Comput. Phys. **233**, 509–526 (2013)
34. Zohdi, T.I.: On inducing compressive residual stress in microscale print-lines for flexible electronics. Int. J. Eng. Sci. **62**, 157–164 (2013)
35. Zohdi, T.I.: Rapid simulation of laser processing of discrete particulate materials. Arch. Comput. Method. Eng. **20**, 309–325 (2013)
36. Zohdi, T.I.: A direct particle-based computational framework for electrically-enhanced thermo-mechanical sintering of powdered materials. Math. Mech. Solids **19**(1), 93–113 (2014)
37. Zohdi, T.I.: On cross-correlation between thermal gradients and electric fields. Int. J. Eng. Sci. **74**, 143–150 (2014)
38. Zohdi, T.I.: A computational modeling framework for heat transfer processes in laser-induced dermal tissue removal. Comput. Mech. Eng. Sci. **98**(3), 261–277 (2014)
39. Zohdi, T.I.: Additive particle deposition and selective laser processing-a computational manufacturing framework. Comput. Mech. **54**, 171–191 (2014)
40. Zohdi, T.I.: Embedded electromagnetically sensitive particle motion in functionalized fluids. Comput. Part. Mech. **1**, 27–45 (2014)
41. Zohdi, T.I.: Rapid computation of statistically-stable particle/feature ratios for consistent substrate stresses in printed flexible electronics. J. Manuf. Sci. Eng. ASME MANU-14-1476 (2015). https://doi.org/10.1115/1.4029327
42. Zohdi, T.I.: A computational modelling framework for high-frequency particulate obscurant cloud performance. Int. J. Eng. Sci. **89**, 75–85 (2015)
43. Zohdi, T.I.: On necessary pumping pressures for industrial process-driven particle-laden fluid flows. J. Manuf. Sci. Eng. ASME (2015). https://doi.org/10.1115/1.4030620

44. Zohdi, T.I.: On the thermal response of a laser-irradiated powder particle in additive manufacturing. CIRP J. Manuf. Sci. Technol. **10**, 7783 (2015)
45. Zohdi, T.I.: Modeling and simulation of the post-impact trajectories of particles in oblique precision shot-peening. Comput. Part. Mech. (2015). https://doi.org/10.1007/s40571-015-0048-5
46. Zohdi, T.I.: Modeling and simulation of cooling-induced residual stresses in heated particulate mixture depositions. Comput. Mech. **56**, 613–630 (2015)
47. Zohdi, T.I.: Modeling and efficient simulation of the deposition of particulate flows onto compliant substrates. Int. J. Eng. Sci. **99**, 74–91 (2015). https://doi.org/10.1016/j.ijengsci.2015.10.012
48. Zohdi, T.I.: Modeling and simulation of laser processing of particulate-functionalized materials. Arch. Comput. Method. Eng. 1–25 (2015). https://doi.org/10.1007/s11831-015-9160-1
49. Zohdi, T.I., Wriggers, P.: Introduction to computational micromechanics. Springer, Second Reprinting (Peer Reviewed) (2005, 2008)
50. Sirringhaus, H., Kawase, T., Friend, R.H., Shimoda, T., Inbasekaran, M., Wu, W., Woo, E.P.: High-resolution inkjet printing of all-polymer transistor circuits. Science **290**, 2123–2126 (2000)
51. Wang, J.Z., Zheng, Z.H., Li, H.W., Huck, W.T.S., Sirringhaus, H.: Dewetting of conducting polymer inkjet droplets on patterned surfaces. Nat. Mater. **3**, 171–176 (2004)
52. Huang, D., Liao, F., Molesa, S., Redinger, D., Subramanian, V.: Plastic-compatible low-resistance printable gold nanoparticle conductors for flexible electronics. J. Electrochem. Soc. **150**(7), G412–417 (2003)
53. Choi, S., Park, I., Hao, Z., Holman, H.Y., Pisano, A.P., Zohdi, T.I.: Ultra-fast self-assembly of micro-scale particles by open channel flow. Langmuir **26**(7), 4661–4667 (2010)
54. Choi, S., Pisano, A.P., Zohdi, T.I.: An analysis of evaporative self-assembly of micro particles in printed picoliter suspension droplets. J. Thin Solid Films **537**(30), 180–189 (2013)
55. Demko, M., Choi, S., Zohdi, T.I., Pisano, A.P.: High resolution patterning of nanoparticles by evaporative self-assembly enabled by in-situ creation and mechanical lift-off of a polymer template. Appl. Phys. Lett. **99**, 253102-1–253102-3 (2012)
56. Demko, M.T., Cheng, J.C., Pisano, A.P.: High-resolution direct patterning of gold nanoparticles by the microfluidic molding process. Langmuir, 412–417 (2010)
57. Choi, S., Stassi, S., Pisano, A.P., Zohdi, T.I.: Coffee-ring effect-based three dimensional patterning of micro, nanoparticle assembly with a single droplet. Langmuir **26**(14), 11690–11698 (2010)
58. Choi, S., Jamshidi, A., Seok, T.J., Zohdi, T.I., Wu., M.C., Pisano, A.P.: Fast, high-throughput creation of size-tunable micro, nanoparticle clusters via evaporative self-assembly in picoliter-scale droplets of particle suspension. Langmuir **28**(6), 3102–3111 (2012)
59. Park, J.-U., Hardy, M., Kang, S.J., Barton, K., Adair, K., Mukhopadhyay, D.K., Lee, C.Y., Strano, M.S., Alleyne, A.G., Georgiadis, J.G., Ferreira, P.M., Rogers, J.A.: High-resolution electrohydrodynamic jet printing. Nat. Mater. **6**, 782–789 (2007)
60. Ko, S.H., Park, I., Pan, H., Grigoropoulos, C.P., Pisano, A.P., Luscombe, C.K., Frechet, J.M.J.: Direct nanoimprinting of metal nanoparticles for nanoscale electronics fabrication. Nan Lett. **7**, 1869–1877 (2007)
61. Ko, S.H., Park, I., Pan, H., Misra, N., Rogers, M.S., Grigoropoulos, C.P., Pisano, A.P.: ZnO nanowire network transistor fabrication by lowtemperature, allinorganic nanoparticle solution process. Appl. Phys. Lett. **92**, 154102 (2008)
62. Park, I., Ko, S.H., Pan, H., Grigoropoulos, C.P., Pisano, A.P., Frechet, J.M.J., Lee, E.S., Jeong, J.H.: Nanoscale Patterning and Electronics on Flexible Substrate by Direct Nanoimprinting of Metallic Nanoparticles. Adv. Mater. **20**, 489 (2008)

Appendices

Monograph Appendix A: Elementary Notation and Mathematical Operations

A.1 Vectors, Products, and Norms

In this work, boldface symbols imply vectors or tensors. *A fixed Cartesian coordinate system will be used throughout this monograph.* The unit vectors for such a system are given by the (fixed) mutually orthogonal triad $(\boldsymbol{e}_1, \boldsymbol{e}_2, \boldsymbol{e}_3)$. For the inner product of two vectors $\boldsymbol{u}$ and $\boldsymbol{v}$, we have in three dimensions

$$\boldsymbol{u} \cdot \boldsymbol{v} = \sum_{i=1}^{3} v_i u_i = u_1 v_1 + u_2 v_2 + u_3 v_3 = ||\boldsymbol{u}|| ||\boldsymbol{v}|| cos\theta, \tag{A.1}$$

where

$$||\boldsymbol{u}|| = \sqrt{u_1^2 + u_2^2 + u_3^2} \tag{A.2}$$

represents the Euclidean norm in $I\!R^3$ and θ is the angle between them. We recall that a norm has three main characteristics for any two bounded vectors $\boldsymbol{u}$ and $\boldsymbol{v}$ ($||\boldsymbol{u}|| < \infty$ and $||\boldsymbol{v}|| < \infty$):

- $||\boldsymbol{u}|| \geq 0$, $||\boldsymbol{u}|| = 0$ if and only if $\boldsymbol{u} = \boldsymbol{0}$,
- $||\boldsymbol{u} + \boldsymbol{v}|| \leq ||\boldsymbol{u}|| + ||\boldsymbol{v}||$, and
- $||\gamma \boldsymbol{u}|| = |\gamma| ||\boldsymbol{u}||$, where γ is a scalar.

Two vectors are said to be orthogonal if $\boldsymbol{u} \cdot \boldsymbol{v} = 0$. The cross (vector) product of two vectors is

$$\boldsymbol{u} \times \boldsymbol{v} = -\boldsymbol{v} \times \boldsymbol{u} = \begin{vmatrix} \boldsymbol{e}_1 & \boldsymbol{e}_2 & \boldsymbol{e}_3 \\ u_1 & u_2 & u_3 \\ v_1 & v_2 & v_3 \end{vmatrix} = ||\boldsymbol{u}|| || \boldsymbol{v}|| sin\theta \, \boldsymbol{n}, \tag{A.3}$$

where $\boldsymbol{n}$ is the unit normal to the plane formed by the vectors $\boldsymbol{u}$ and $\boldsymbol{v}$.

T. I. Zohdi, *Modeling and Simulation of Functionalized Materials for Additive Manufacturing and 3D Printing: Continuous and Discrete Media*, Lecture Notes in Applied and Computational Mechanics 60, https://doi.org/10.1007/978-3-319-70079-3

The temporal differentiation of a vector-valued function is given by

$$\frac{d}{dt}\boldsymbol{u}(t) = \frac{du_1(t)}{dt}\boldsymbol{e}_1 + \frac{du_2(t)}{dt}\boldsymbol{e}_2 + \frac{du_3(t)}{dt}\boldsymbol{e}_3 = \dot{u}_1\boldsymbol{e}_1 + \dot{u}_2\boldsymbol{e}_2 + \dot{u}_3\boldsymbol{e}_3. \tag{A.4}$$

The spatial gradient of a scalar-valued function (a dilation to a vector) is given by

$$\nabla_x \phi = \left(\frac{\partial \phi}{\partial x_1}\boldsymbol{e}_1 + \frac{\partial \phi}{\partial x_2}\boldsymbol{e}_2 + \frac{\partial \phi}{\partial x_3}\boldsymbol{e}_3 \right). \tag{A.5}$$

The gradient of a vector-valued function is a direct extension of the preceding definition. For example, $\nabla_x \boldsymbol{u}$ has components of $\frac{\partial u_i}{\partial x_j}$. The divergence of a vector (a contraction to a scalar) is defined by

$$\begin{aligned} \nabla_x \cdot \boldsymbol{u} &= \left(\boldsymbol{e}_1 \frac{\partial}{\partial x_1} + \boldsymbol{e}_2 \frac{\partial}{\partial x_2} + \boldsymbol{e}_3 \frac{\partial}{\partial x_3} \right) \cdot (u_1\boldsymbol{e}_1 + u_2\boldsymbol{e}_2 + u_3\boldsymbol{e}_3) \\ &= \left(\frac{\partial u_1}{\partial x_1} + \frac{\partial u_2}{\partial x_2} + \frac{\partial u_3}{\partial x_3} \right). \end{aligned} \tag{A.6}$$

The curl of a vector is defined as

$$\nabla_x \times \boldsymbol{u} = \begin{vmatrix} \boldsymbol{e}_1 & \boldsymbol{e}_2 & \boldsymbol{e}_3 \\ \frac{\partial}{\partial x_1} & \frac{\partial}{\partial x_2} & \frac{\partial}{\partial x_3} \\ u_1 & u_2 & u_3 \end{vmatrix}. \tag{A.7}$$

The triple product of three vectors is

$$\boldsymbol{w} \cdot (\boldsymbol{u} \times \boldsymbol{v}) = \left(\begin{vmatrix} w_1 & w_2 & w_3 \\ u_1 & u_2 & u_3 \\ v_1 & v_2 & v_3 \end{vmatrix} \right) = (\boldsymbol{w} \times \boldsymbol{u}) \cdot \boldsymbol{v}. \tag{A.8}$$

This represents the volume of a parallelepiped formed by the three vectors.

A.2 Basic Linear Algebra

If we consider the second-order tensor $\boldsymbol{A}$ with its matrix representation,

$$[\boldsymbol{A}] \stackrel{\text{def}}{=} \begin{bmatrix} A_{11} & A_{12} & A_{13} \\ A_{21} & A_{22} & A_{23} \\ A_{31} & A_{32} & A_{33} \end{bmatrix}. \tag{A.9}$$

The matrix $[\boldsymbol{A}]$ is said to be symmetric if $[\boldsymbol{A}] = [\boldsymbol{A}]^T$ and skew-symmetric if $[\boldsymbol{A}] = -[\boldsymbol{A}]^T$. A first-order contraction (inner product) of two matrices is defined by

$$\boldsymbol{A}\cdot\boldsymbol{B}=[\boldsymbol{A}][\boldsymbol{B}] \text{ which has components of } \sum_{j=1}^{N} A_{ij}B_{jk}=C_{ik}, \tag{A.10}$$

where it is clear that the range of the inner index j must be the same for $[\boldsymbol{A}]$ and $[\boldsymbol{B}]$. The second-order inner product of two matrices is

$$\boldsymbol{A}:\boldsymbol{B}=A_{ij}B_{ij}=tr([\boldsymbol{A}]^T[\boldsymbol{B}]). \tag{A.11}$$

Some properties of the determinant are (where, for example, $[\boldsymbol{A}]$ is a 3×3 matrix):

- $det[\boldsymbol{A}]=A_{11}(A_{22}A_{33}-A_{32}A_{23})-A_{12}(A_{21}A_{33}-A_{31}A_{23})+A_{13}(A_{21}A_{32}-A_{31}A_{22})$,
- $det[\boldsymbol{I}]=1$, $det\,\alpha[\boldsymbol{A}]=\alpha^3 det\,[\boldsymbol{A}]$, where α is a scalar,
- $det[\boldsymbol{A}][\boldsymbol{B}]=det[\boldsymbol{A}]det[\boldsymbol{B}]$, $det[\boldsymbol{A}]^T=det[\boldsymbol{A}]$, and $det[\boldsymbol{A}]^{-1}=\frac{1}{det[\boldsymbol{A}]}$.

An important use of the determinant is in forming the inverse by

$$[\boldsymbol{A}]^{-1}=\frac{adj[\boldsymbol{A}]}{det[\boldsymbol{A}]},\quad adj[\boldsymbol{A}]\stackrel{\text{def}}{=}\begin{bmatrix} C_{11} & C_{12} & C_{13}\\ C_{21} & C_{22} & C_{23}\\ C_{31} & C_{32} & C_{33}\end{bmatrix}^T, \tag{A.12}$$

where the so-called cofactors are

$$\boxed{\begin{array}{ll} C_{11}=A_{22}A_{33}-A_{32}A_{23} & C_{12}=-(A_{21}A_{33}-A_{31}A_{23})\\ C_{13}=A_{21}A_{32}-A_{31}A_{22} & C_{21}=-(A_{12}A_{33}-A_{32}A_{13})\\ C_{22}=A_{11}A_{33}-A_{31}A_{13} & C_{23}=-(A_{11}A_{32}-A_{31}A_{12})\\ C_{31}=A_{12}A_{23}-A_{22}A_{13} & C_{32}=-(A_{11}A_{23}-A_{21}A_{13})\\ C_{33}=A_{11}A_{22}-A_{21}A_{12} & \end{array}} \tag{A.13}$$

The rule of transposes for two $n\times n$ matrices is

$$([\boldsymbol{A}][\boldsymbol{B}])^T=[\boldsymbol{B}]^T[\boldsymbol{A}]^T. \tag{A.14}$$

The rule of inverses for two invertible $n\times n$ matrices is

$$([\boldsymbol{A}][\boldsymbol{B}])^{-1}=[\boldsymbol{B}]^{-1}[\boldsymbol{A}]^{-1} \quad\text{and}\quad [\boldsymbol{A}]^{-1}[\boldsymbol{A}]=[\boldsymbol{A}][\boldsymbol{A}]^{-1}=[\boldsymbol{I}], \tag{A.15}$$

where $[\boldsymbol{I}]$ is the identity matrix. Clearly, $[\boldsymbol{A}]^{-1}$ exists only when $det[\boldsymbol{A}]\neq 0$.

The mathematical definitions of an eigenvalue, a scalar denoted Λ and eigenvector, a vector denoted $\mathcal{E}$, of a matrix $[\boldsymbol{A}]$ are

$$[\boldsymbol{A}]\{\mathcal{E}\}=\Lambda\{\mathcal{E}\} \tag{A.16}$$

We note that for any given tensor $\boldsymbol{A}$ of order 2 (a 3×3 matrix), if we set the determinant $det[\boldsymbol{A} - \Lambda \boldsymbol{I}] = 0$, it can be shown that the so-called characteristic polynomial is

$$det(\boldsymbol{A} - \Lambda \boldsymbol{I}) = -\Lambda^3 + I_A \Lambda^2 - II_A \Lambda + III_A = 0, \tag{A.17}$$

where

$$\boxed{\begin{array}{l} I_A = tr(\boldsymbol{A}) = \Lambda_1 + \Lambda_2 + \Lambda_3 \\ II_A = \frac{1}{2}((tr(\boldsymbol{A}))^2 - tr(\boldsymbol{A}^2)) = \Lambda_1\Lambda_2 + \Lambda_2\Lambda_3 + \Lambda_1\Lambda_3 \\ III_A = det(\boldsymbol{A}) = \frac{1}{6}((tr\boldsymbol{A})^3 + 2tr\boldsymbol{A}^3 - 3(tr\boldsymbol{A}^2)(tr\boldsymbol{A})) = \Lambda_1\Lambda_2\Lambda_3. \end{array}} \tag{A.18}$$

Since I_A, II_A, and III_A can be written in terms of $tr\boldsymbol{A}$, which is invariant under frame rotation, they too are invariant under frame rotation. The main properties to remember about eigenvalues and eigenvectors are:

1. If $[\boldsymbol{A}]$ $(n \times n)$ has n linearly independent eigenvectors, then it is diagonalizable by a matrix formed by columns of the eigenvectors, for example, for a 3×3 matrix

$$\begin{bmatrix} \Lambda_1 & 0 & 0 \\ 0 & \Lambda_2 & 0 \\ 0 & 0 & \Lambda_3 \end{bmatrix} = \begin{bmatrix} \mathcal{E}_1^{(1)} & \mathcal{E}_1^{(2)} & \mathcal{E}_1^{(3)} \\ \mathcal{E}_2^{(1)} & \mathcal{E}_2^{(2)} & \mathcal{E}_2^{(3)} \\ \mathcal{E}_3^{(1)} & \mathcal{E}_3^{(2)} & \mathcal{E}_3^{(3)} \end{bmatrix}^{-1} \begin{bmatrix} A_{11} & A_{12} & A_{13} \\ A_{21} & A_{22} & A_{23} \\ A_{31} & A_{32} & A_{33} \end{bmatrix} \begin{bmatrix} \mathcal{E}_1^{(1)} & \mathcal{E}_1^{(2)} & \mathcal{E}_1^{(3)} \\ \mathcal{E}_2^{(1)} & \mathcal{E}_2^{(2)} & \mathcal{E}_2^{(3)} \\ \mathcal{E}_3^{(1)} & \mathcal{E}_3^{(2)} & \mathcal{E}_3^{(3)} \end{bmatrix} \tag{A.19}$$

2. If $[\boldsymbol{A}]$ $(n \times n)$ has n distinct eigenvalues, then the eigenvectors are linearly independent.
3. If $[\boldsymbol{A}]$ $(n \times n)$ is symmetric, then its eigenvalues are real. If the eigenvalues are distinct, then the eigenvectors are orthogonal.

A quadratic form is such that

$$\{\boldsymbol{x}\}^T[\boldsymbol{A}]\{\boldsymbol{x}\} \stackrel{\text{def}}{=} \begin{bmatrix} x_1 & x_2 & x_3 \end{bmatrix} \begin{bmatrix} A_{11} & A_{12} & A_{13} \\ A_{21} & A_{22} & A_{23} \\ A_{31} & A_{32} & A_{33} \end{bmatrix} \begin{bmatrix} x_1 \\ x_2 \\ x_3 \end{bmatrix}. \tag{A.20}$$

A matrix $[\boldsymbol{A}]$ is said to be positive definite if the quadratic form is positive for all nonzero vectors $\boldsymbol{x}$. Clearly, from Eq. A.19, a positive definite matrix must have positive eigenvalues.

To perform a coordinate transform for a 3×3 matrix $[\boldsymbol{A}]$ from one Cartesian coordinate system to another (denoted with a $\hat{(\cdot)}$), we apply a transformation matrix $[\boldsymbol{Q}]$:

$$[\hat{\boldsymbol{A}}] = [\boldsymbol{Q}][\boldsymbol{A}][\boldsymbol{Q}]^{-1}. \tag{A.21}$$

If $\boldsymbol{Q}$ is an orthogonal matrix, then $\boldsymbol{Q}^{-1} = \boldsymbol{Q}^T$ (denoted "unitary").

The standard axes rotators are (Fig. A.1), about the x_1-axis

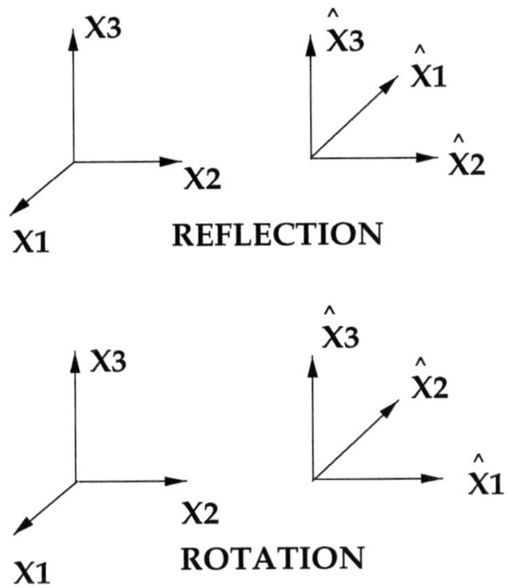

Fig. A.1 Top: reflection with respect to the x_2–x_3 plane. Bottom: rotation with respect to the x_3-axis

$$Rot(x_1) \stackrel{\text{def}}{=} \begin{bmatrix} 1 & 0 & 0 \\ 0 & cos\theta_1 & sin\theta_1 \\ 0 & -sin\theta_1 & cos\theta_1 \end{bmatrix}, \tag{A.22}$$

about the x_2-axis

$$Rot(x_2) \stackrel{\text{def}}{=} \begin{bmatrix} cos\theta_2 & 0 & -sin\theta_2 \\ 0 & 1 & 0 \\ sin\theta_2 & 0 & cos\theta_2 \end{bmatrix}, \tag{A.23}$$

about the x_3-axis

$$Rot(x_3) \stackrel{\text{def}}{=} \begin{bmatrix} cos\theta_3 & sin\theta_3 & 0 \\ -sin\theta_3 & cos\theta_3 & 0 \\ 0 & 0 & 1 \end{bmatrix}. \tag{A.24}$$

The standard axes reflectors are with respect to the x_2–x_3 plane

$$Ref(x_1) \stackrel{\text{def}}{=} \begin{bmatrix} -1 & 0 & 0 \\ 0 & 1 & 0 \\ 0 & 0 & 1 \end{bmatrix}, \tag{A.25}$$

with respect to the x_1–x_3 plane

$$Ref(x_2) \stackrel{\text{def}}{=} \begin{bmatrix} 1 & 0 & 0 \\ 0 & -1 & 0 \\ 0 & 0 & 1 \end{bmatrix}, \tag{A.26}$$

with respect to the x_1–x_2 plane

$$Ref(x_3) \stackrel{\text{def}}{=} \begin{bmatrix} 1 & 0 & 0 \\ 0 & 1 & 0 \\ 0 & 0 & -1 \end{bmatrix}. \tag{A.27}$$

A.3 Integral Transformations

The divergence of a vector-valued function (a contraction to a scalar-valued function) is defined by

$$\nabla_x \cdot \boldsymbol{u} = \sum_{i=1}^{N} u_{i,i}, \tag{A.28}$$

whereas for a second-order tensor (a contraction to a vector):

$$\nabla_x \cdot \boldsymbol{A} \text{ has components of } \sum_{j=1}^{N} A_{ij,j}. \tag{A.29}$$

The gradient of a vector (a dilation to a second-order tensor) is:

$$\nabla_x \boldsymbol{u} \text{ has components of } u_{i,j}, \tag{A.30}$$

whereas for a second-order tensor (a dilation to a third-order tensor):

$$\nabla_x \boldsymbol{A} \text{ has components of } A_{ij,k}. \tag{A.31}$$

The gradient of a scalar (a dilation to a vector) is:

$$\nabla_x \phi \text{ has components of } \phi_{,i}. \tag{A.32}$$

The scalar product of two second-order tensors, for example, the gradients of first-order vectors, is defined as

$$\nabla_x \boldsymbol{v} : \nabla_x \boldsymbol{u} = \underbrace{\frac{\partial v_i}{\partial x_j}\frac{\partial u_i}{\partial x_j}}_{\text{in Cartesian bases}} \stackrel{\text{def}}{=} v_{i,j} u_{i,j} \quad i, j = 1, 2, 3, \tag{A.33}$$

where $\partial u_i / \partial x_j$, $\partial v_i / \partial x_j$ are partial derivatives of u_i and v_i, and u_i, v_i are the Cartesian components of $\boldsymbol{u}$ and $\boldsymbol{v}$ and

$$\nabla_x \boldsymbol{u} \cdot \boldsymbol{n} \text{ has components of } \underbrace{u_{i,j} n_j}_{\text{in Cartesian bases}}, \quad i, j = 1, 2, 3. \tag{A.34}$$

For a scalar, we have

$$\int_{\Omega} \nabla_x \phi \, d\Omega = \int_{\partial\Omega} \phi \boldsymbol{n} \, dA \quad \int_{\Omega} \phi_{,i} \, d\Omega = \int_{\partial\Omega} \phi n_i \, dA \tag{A.35}$$

and for a vector

$$\int_{\Omega} \nabla_x \boldsymbol{u} \, d\Omega = \int_{\partial\Omega} \boldsymbol{u} \otimes \boldsymbol{n} \, dA \quad \int_{\Omega} u_{i,j} \, d\Omega = \int_{\partial\Omega} u_i n_j \, dA. \tag{A.36}$$

The divergence theorem for vectors is

$$\int_{\Omega} \nabla_x \cdot \boldsymbol{u} \, d\Omega = \int_{\partial\Omega} \boldsymbol{u} \cdot \boldsymbol{n} \, dA \quad \int_{\Omega} u_{i,i} \, d\Omega = \int_{\partial\Omega} u_i n_i \, dA \tag{A.37}$$

and analogously for a tensor $\boldsymbol{B}$

$$\int_{\Omega} \nabla_x \cdot \boldsymbol{B} \, d\Omega = \int_{\partial\Omega} \boldsymbol{B} \cdot \boldsymbol{n} \, dA \quad \int_{\Omega} B_{ij,j} \, d\Omega = \int_{\partial\Omega} B_{ij} n_j \, dA, \tag{A.38}$$

where $\boldsymbol{n}$ is the outward normal to the bounding surface. These standard operations arise throughout the analysis. A generalization of these last results is

$$\int_{\Omega} \nabla_x * \boldsymbol{B} \, d\Omega = \int_{\partial\Omega} \boldsymbol{n} * \boldsymbol{B} \, dA, \tag{A.39}$$

where, when $* = \cdot$, we have the divergence theorem and when $* = \times$ we have the "cross product" theorem.[1] For proofs, see Chandrasekharaiah and Debnath [1] or Malvern [2].

References

1. Chandrasekharaiah, D.S., Debnath, L.: Continuum Mechanics. Academic press (1994)
2. Malvern, L.: Introduction to the mechanics of a continuous medium. Prentice Hall (1968)

---

[1] Also, we have the point-wise product rule:

$$\frac{d}{dt}(\boldsymbol{a} * \boldsymbol{b}) = \frac{d\boldsymbol{a}}{dt} * \boldsymbol{b} + \boldsymbol{a} * \frac{d\boldsymbol{a}}{dt}. \tag{A.40}$$

Monograph Appendix B—CM Approaches: Effective Electrical Properties of Mixtures

A critical aspect of additive manufacturing, in particular printed electronics, is the estimation of the effective properties of particle-functionalized dielectric materials. One of the primary properties of interest is the overall "effective" electrical conductivity, defined via Ohm's law:

$$\langle \boldsymbol{J} \rangle_{\Omega} = \boldsymbol{\sigma}^* \langle \boldsymbol{E} \rangle_{\Omega}, \tag{B.1}$$

where $\boldsymbol{\sigma}^*$ is the effective conductivity for the mixture, $\langle \boldsymbol{E} \rangle_{\Omega}$ is the volume-averaged electric field, $\langle \boldsymbol{J} \rangle_{\Omega}$ is the volume-averaged current, the averaging operator is defined as $\langle \cdot \rangle_{\Omega} \stackrel{\text{def}}{=} \frac{1}{|\Omega|} \int_{\Omega} (\cdot) \, d\Omega$ over a statistically representative volume element with domain Ω. Other properties, although not needed immediately in the analysis, are:

- Overall electrical permittivity:

$$\langle \boldsymbol{D} \rangle_{\Omega} = \boldsymbol{\epsilon}^* \langle \boldsymbol{E} \rangle_{\Omega}, \tag{B.2}$$

  where $\boldsymbol{\epsilon}^*$ is the effective electrical permittivity for the mixture, $\langle \boldsymbol{E} \rangle_{\Omega}$ is the volume-averaged electric field, $\langle \boldsymbol{D} \rangle_{\Omega}$ is the volume-averaged electric field flux,
- Overall magnetic permeability:

$$\langle \boldsymbol{B} \rangle_{\Omega} = \boldsymbol{\mu}^* \langle \boldsymbol{H} \rangle_{\Omega}, \tag{B.3}$$

  where $\boldsymbol{\mu}^*$ is the effective magnetic permeability for the mixture, $\langle \boldsymbol{H} \rangle_{\Omega}$ is the volume-averaged magnetic field, $\langle \boldsymbol{B} \rangle_{\Omega}$ is the volume-averaged magnetic field flux,

Our objective is to estimate $\boldsymbol{\sigma}^*$ and then illustrate how to use it in conjunction with a model deposited electrical material problem. We refer the reader to Zohdi [1–15] for an in-depth discussion of the electromagnetic properties of materials.

T. I. Zohdi, *Modeling and Simulation of Functionalized Materials for Additive Manufacturing and 3D Printing: Continuous and Discrete Media*, Lecture Notes in Applied and Computational Mechanics 60, https://doi.org/10.1007/978-3-319-70079-3

B.1 Computing the Effective Electrical Conductivity

In order to make estimates of the overall properties of a mixture, we consider the widely used Hashin and Shtrikman bounds (see previous chapter) for isotropic materials with isotropic effective responses. These estimates provide one with upper and lower bounds on the overall response of the material. For two isotropic materials with an overall isotropic response, we utilize the following estimates,

$$\underbrace{\sigma_1 + \frac{v_2}{\frac{1}{\sigma_2-\sigma_1} + \frac{1-v_2}{3\sigma_1}}}_{\sigma^{*,-}} \leq \sigma^* \leq \underbrace{\sigma_2 + \frac{1-v_2}{\frac{1}{\sigma_1-\sigma_2} + \frac{v_2}{3\sigma_2}}}_{\sigma^{*,+}}, \tag{B.4}$$

where the conductivity of phase 2 (with volume fraction v_2) is larger than phase 1 ($\sigma_2 \geq \sigma_1$). Usually, v_2 corresponds to the particle material, although there can be applications where the matrix is more conductive than the particles. In that case, v_2 would correspond to the matrix material. Provided that the volume fractions and constituent conductivities are the only known information about the microstructure, the expressions are the tightest bounds for the overall isotropic effective responses for two-phase media, where the constituents are both isotropic. A critical observation is that the lower bound is more accurate when the material is composed of high-conductivity particles that are surrounded by a low-conductivity matrix (denoted case 1) and the upper bound is more accurate for a high-conductivity matrix surrounding low-conductivity particles (denoted case 2).

Remark As mentioned before, this can be explained by considering two cases of material combinations, one with 50% low-conductivity material and the other with 50% high-conductivity material. A material with a continuous low-conductivity (fine-scale powder) binder (50%) will isolate the high-conductivity particles (50%), and the overall system will not conduct electricity well (this is case 1, and the lower bound is more accurate), while a material formed by a continuous high-conductivity (fine-scale powder) binder (50%) surrounding low-conductivity particles (50%, case 2) will, in an overall sense, conduct electricity better than case 1. Thus, case 2 is more closely approximated by the upper bound and case 1 is closer to the lower bound. Since the true effective property lies between the upper and lower bounds, one can construct the following approximation

$$\sigma^* \approx \phi\sigma^{*,+} + (1-\phi)\sigma^{*,-}, \tag{B.5}$$

where $0 \leq \phi \leq 1$. ϕ *is a function of the microstructure and must be calibrated.* As mentioned, for high-conductivity spherical particles, at low volume fractions, for example, under 15%, where the particles are not making contact, the lower bound is more accurate. Thus, one would pick $\phi = \phi^s \leq 0.5$ to bias the estimate to the lower bound. However, if we take the same volume fraction of particles, but make the flat flakes, they will certainly touch and produce high-conductivity pathways. Their overall conductivity will be higher than those of sphere at the same volume fraction.

Thus, one would pick $\phi = \phi^f > \phi^s$. One can calibrate ϕ by comparing it to different experiments.

B.2 Concentration Tensors and Load-Sharing

The effective conductivity, $\boldsymbol{\sigma}^*$, defined via[2]

$$\langle \boldsymbol{J} \rangle_\Omega = \boldsymbol{\sigma}^* \cdot \langle \boldsymbol{E} \rangle_\Omega, \tag{B.6}$$

can be re-written in the following manner

$$\begin{aligned} \langle \boldsymbol{J} \rangle_\Omega &= v_1 \langle \boldsymbol{J} \rangle_{\Omega_1} + v_2 \langle \boldsymbol{J} \rangle_{\Omega_2} \\ &= v_1 \boldsymbol{\sigma}_1 \cdot \langle \boldsymbol{E} \rangle_{\Omega_1} + v_2 \boldsymbol{\sigma}_2 \cdot \langle \boldsymbol{E} \rangle_{\Omega_2} \\ &= \boldsymbol{\sigma}_1 \cdot (\langle \boldsymbol{E} \rangle_\Omega - v_2 \langle \boldsymbol{E} \rangle_{\Omega_2}) + v_2 \boldsymbol{\sigma}_2 \cdot \langle \boldsymbol{E} \rangle_{\Omega_2} \\ &= \underbrace{\left(\boldsymbol{\sigma}_1 + v_2 (\boldsymbol{\sigma}_2 - \boldsymbol{\sigma}_1) \cdot \boldsymbol{C}_{E,2}\right)}_{\boldsymbol{\sigma}^*} \cdot \langle \boldsymbol{E} \rangle_\Omega, \end{aligned} \tag{B.7}$$

where

$$\underbrace{\left(\frac{1}{v_2} (\boldsymbol{\sigma}_2 - \boldsymbol{\sigma}_1)^{-1} \cdot (\boldsymbol{\sigma}^* - \boldsymbol{\sigma}_1)\right)}_{\stackrel{\text{def}}{=} \boldsymbol{C}_{E,2}} \cdot \langle \boldsymbol{E} \rangle_\Omega = \langle \boldsymbol{E} \rangle_{\Omega_2}. \tag{B.8}$$

$\boldsymbol{C}_{E,2}$ is known as the electric field concentration tensor. Thus, the product of $\boldsymbol{C}_{E,2}$ with $\langle \boldsymbol{E} \rangle_\Omega$ yields $\langle \boldsymbol{E} \rangle_{\Omega_2}$. *It is important to realize that once either* $\boldsymbol{C}_{E,2}$ or $\boldsymbol{\sigma}^*$ *are known, the other can be computed.*

In order to determine the concentration tensor for phase 1, we have from Eq. B.69

$$\langle \boldsymbol{E} \rangle_{\Omega_1} = \frac{\langle \boldsymbol{E} \rangle_\Omega - v_2 \langle \boldsymbol{E} \rangle_{\Omega_2}}{v_1} = \frac{(\mathbf{1} - v_2 \boldsymbol{C}_{E,2}) \cdot \langle \boldsymbol{E} \rangle_\Omega}{v_1} \stackrel{\text{def}}{=} \boldsymbol{C}_{E,1} \cdot \langle \boldsymbol{E} \rangle_\Omega, \tag{B.9}$$

where

$$\boldsymbol{C}_{E,1} = \frac{1}{v_1} (\mathbf{1} - v_2 \boldsymbol{C}_{E,2}) = \frac{\mathbf{1} - v_2 \boldsymbol{C}_{E,2}}{1 - v_2}. \tag{B.10}$$

Note that Eq. B.10 implies

$$\underbrace{v_1 \boldsymbol{C}_{E,1}}_{\text{phase}-1 \text{ contribution}} + \underbrace{v_2 \boldsymbol{C}_{E,2}}_{\text{phase}-2 \text{ contribution}} = \mathbf{1}. \tag{B.11}$$

---

[2]Implicitly, we assume that (a) the contact between the phases is perfect and (b) the ergodicity hypothesis is satisfied (see Kröner [16] or Torquato [17]).

Similarly, for the current, we have

$$\langle \boldsymbol{J} \rangle_{\Omega} = \boldsymbol{\sigma}^{*} \cdot \langle \boldsymbol{E} \rangle_{\Omega} \Rightarrow \boldsymbol{\sigma}^{*-1} \cdot \langle \boldsymbol{J} \rangle_{\Omega} = \boldsymbol{C}_{E,2}^{-1} \cdot \langle \boldsymbol{E} \rangle_{\Omega_2} = \boldsymbol{C}_{E,2}^{-1} \cdot \boldsymbol{\sigma}_{2}^{-1} \cdot \langle \boldsymbol{J} \rangle_{\Omega_2}. \quad (B.12)$$

Thus,

$$\underbrace{\boldsymbol{\sigma}_2 \cdot \boldsymbol{C}_{E,2} \cdot \boldsymbol{\sigma}^{*-1}}_{\boldsymbol{C}_{J,2}} \cdot \langle \boldsymbol{J} \rangle_{\Omega} = \langle \boldsymbol{J} \rangle_{\Omega_2}, \quad (B.13)$$

and

$$\boldsymbol{C}_{J,1} \cdot \langle \boldsymbol{J} \rangle_{\Omega} = \langle \boldsymbol{J} \rangle_{\Omega_1} \quad (B.14)$$

where

$$\boldsymbol{C}_{J,1} = \frac{\boldsymbol{1} - v_2 \boldsymbol{C}_{J,2}}{1 - v_2} = \boldsymbol{\sigma}_1 \cdot \boldsymbol{C}_{E,1} \cdot \boldsymbol{\sigma}^{*-1}. \quad (B.15)$$

We remark that from Eq. B.81 implies

$$\underbrace{v_1 \boldsymbol{C}_{J,1}}_{\text{phase}-1 \text{ contribution}} + \underbrace{v_2 \boldsymbol{C}_{J,2}}_{\text{phase}-2 \text{ contribution}} = \boldsymbol{1}. \quad (B.16)$$

Summarizing, we have the following results:

- $\boldsymbol{C}_{E,1} \cdot \langle \boldsymbol{E} \rangle_{\Omega} = \langle \boldsymbol{E} \rangle_{\Omega_1}$ where $\boldsymbol{C}_{E,1} = \frac{1}{v_1}(\boldsymbol{1} - v_2 \boldsymbol{C}_{E,2}) = \frac{\boldsymbol{1} - v_2 \boldsymbol{C}_{E,2}}{1 - v_2}$,
- $\boldsymbol{C}_{E,2} \cdot \langle \boldsymbol{E} \rangle_{\Omega} = \langle \boldsymbol{E} \rangle_{\Omega_2}$ where $\boldsymbol{C}_{E,2} = \frac{1}{v_2}(\boldsymbol{\sigma}_2 - \boldsymbol{\sigma}_1)^{-1} \cdot (\boldsymbol{\sigma}^{*} - \boldsymbol{\sigma}_1)$,
- $\boldsymbol{C}_{J,1} \cdot \langle \boldsymbol{J} \rangle_{\Omega} = \langle \boldsymbol{J} \rangle_{\Omega_1}$ where $\boldsymbol{C}_{J,1} = \frac{\boldsymbol{1} - v_2 \boldsymbol{C}_{J,2}}{1 - v_2} = \boldsymbol{\sigma}_1 \cdot \boldsymbol{C}_{E,1} \cdot \boldsymbol{\sigma}^{*-1}$,
- $\boldsymbol{C}_{J,2} \cdot \langle \boldsymbol{J} \rangle_{\Omega} = \langle \boldsymbol{J} \rangle_{\Omega_2}$ where $\boldsymbol{C}_{J,2} = \boldsymbol{\sigma}_2 \cdot \boldsymbol{C}_{E,2} \cdot \boldsymbol{\sigma}^{*-1}$.

B.3 "Load-Sharing" Interpretation

One may write

$$\begin{aligned} \underbrace{v_1^{\sigma} \boldsymbol{C}_{E,1}}_{\text{phase}-1 \text{ contribution}} + \underbrace{v_2^{\sigma} \boldsymbol{C}_{E,2}}_{\text{phase}-2 \text{ contribution}} &= \boldsymbol{1} \\ \underbrace{v_1^{\sigma} \boldsymbol{C}_{J,1}}_{\text{phase}-1 \text{ contribution}} + \underbrace{v_2^{\sigma} \boldsymbol{C}_{J,2}}_{\text{phase}-2 \text{ contribution}} &= \boldsymbol{1}. \end{aligned} \quad (B.17)$$

Frequently, the first term in the above expressions is referred to as "phase-1's" share, while the second term is "phase-2's" share. Thus, the concentration tensor tells you how much of the overall load is carried by the particles and how much by the matrix. This is useful when we want to compute the Joule-heating in the system.

B.4 Joule-Heating

One of the key quantities of interest here is the amount of heat generated from running a current through a material, denoted H (a rate), which feeds into first law of thermodynamics,

$$\rho\dot{w} - \boldsymbol{T} : \nabla\dot{\boldsymbol{u}} + \nabla \cdot \boldsymbol{q} = aH. \tag{B.18}$$

In Eq. B.18, ρ is the mass density, w is the stored energy per unit mass, $\boldsymbol{T}$ is Cauchy stress, $\boldsymbol{u}$ is the displacement field, $\boldsymbol{q}$ is heat flux, and $H = (\boldsymbol{J} \cdot \boldsymbol{E})$ is the rate of electrical energy absorbed due to Joule-heating, where $\boldsymbol{J}$ is the current, $\boldsymbol{E}$ is the electric field, and $0 \leq a \leq 1$ is an absorption constant. Our objective of this chapter is to determine the phase-wise load-shares of the Joule-field, denoted $H = \boldsymbol{J} \cdot \boldsymbol{E}$, carried by the components in the heterogeneous mixture.

It is important to realize that heterogeneous mixtures (microstructures) distort the electrical and current fields within the material. For electrical flow to be properly controlled, in particular for heterogeneous mixtures, one needs accurate characterizations of the electrical loads carried by each of the phases in the system. In this phase of the research, as a model problem, we will consider a statistically representative volume element (RVE of volume $|\Omega|$) of a two-phase dielectric medium. *We assume that the material has been properly prepared so that there are no gaps between the phases (an idealization).* The microscale properties are characterized by a spatially variable electrical conductivity $\boldsymbol{\sigma}(\boldsymbol{x})$. For such a sample, one can decompose the electrical field carried by each phase in the material as follows:

$$\langle \boldsymbol{E} \rangle_\Omega = \frac{1}{|\Omega|}\left(\int_{\Omega_1} \boldsymbol{E}\, d\Omega + \int_{\Omega_2} \boldsymbol{E}\, d\Omega\right) = v_1 \langle \boldsymbol{E} \rangle_{\Omega_1} + v_2 \langle \boldsymbol{E} \rangle_{\Omega_2} \tag{B.19}$$

and the current can be decomposed as

$$\langle \boldsymbol{J} \rangle_\Omega = \frac{1}{|\Omega|}\left(\int_{\Omega_1} \boldsymbol{J}\, d\Omega + \int_{\Omega_2} \boldsymbol{J}\, d\Omega\right) = v_1 \langle \boldsymbol{J} \rangle_{\Omega_1} + v_2 \langle \boldsymbol{J} \rangle_{\Omega_2}, \tag{B.20}$$

and the Joule-heating field as

$$\langle H \rangle_\Omega = \frac{1}{|\Omega|}\left(\int_{\Omega_1} H\, d\Omega + \int_{\Omega_2} H\, d\Omega\right) = v_1 \langle H \rangle_{\Omega_1} + v_2 \langle H \rangle_{\Omega_2}, \tag{B.21}$$

where $\langle \cdot \rangle_\Omega \stackrel{\text{def}}{=} \frac{1}{|\Omega|}\int_\Omega \cdot\, d\Omega$ is a volume-averaging operator and v_1 and v_2 are the volume fractions of phases 1 and 2, respectively ($v_1 + v_2 = 1$). We denote $v_1 \frac{\langle H \rangle_{\Omega_1}}{\langle H \rangle_\Omega}$ and $v_2 \frac{\langle H \rangle_{\Omega_2}}{\langle H \rangle_\Omega}$ as the "load-shares," since

$$v_1 \frac{\langle H \rangle_{\Omega_1}}{\langle H \rangle_\Omega} + v_2 \frac{\langle H \rangle_{\Omega_2}}{\langle H \rangle_\Omega} = 1. \tag{B.22}$$

The objective is to determine the load-shares as functions of known (a priori) quantities

$$v_1 \frac{\langle H \rangle_{\Omega_1}}{\langle H \rangle_\Omega} = \mathcal{F}_1(v_1, \sigma_1, \sigma_2, \langle \boldsymbol{J} \rangle_\Omega, \langle \boldsymbol{E} \rangle_\Omega) \tag{B.23}$$

and

$$v_2 \frac{\langle H \rangle_{\Omega_2}}{\langle H \rangle_\Omega} = \mathcal{F}_2(v_2, \sigma_1, \sigma_2, \langle \boldsymbol{J} \rangle_\Omega, \langle \boldsymbol{E} \rangle_\Omega), \tag{B.24}$$

where σ_1 and σ_2 are the conductivities of phase 1 and phase 2, respectively. The overall volume averages, $\langle \boldsymbol{E} \rangle_\Omega$ and $\langle \boldsymbol{J} \rangle_\Omega$, are considered known, *since they can be determined by the boundary values from well-known results (discussed in the next section): (a) the Average Electric Field Theorem and (b) the Average Current Theorem.*

We will proceed as follows:

- Expressions are developed for the current field ($\boldsymbol{J}$) and electric field ($\boldsymbol{E}$) distribution for each component in the mixture,
- Expressions are developed for the Joule-heating field current distribution for each component in the materials ($\boldsymbol{J} \cdot \boldsymbol{E}$),
- Bounding principles are used to provide estimates of the overall response of the material,
- Asymptotic cases of extreme mixtures of insulators and superconductors are considered,
- Simple estimates for the time to heating are provided, and
- Extensions, involving numerical methods, are discussed.

Remark 1 The mathematical form for Joule-heating can be motivated by taking Faraday's law

$$\nabla \times \boldsymbol{E} = -\frac{\partial \boldsymbol{B}}{\partial t} \tag{B.25}$$

and Ampere's law

$$\nabla \times \boldsymbol{H} = \frac{\partial \boldsymbol{D}}{\partial t} + \boldsymbol{J}, \tag{B.26}$$

where $\boldsymbol{D}$ is the electric field flux, $\boldsymbol{H}$ is the magnetic field, $\boldsymbol{B}$ is the magnetic field flux, and forming the difference between the inner product of the electric field with Ampere's law and the inner product of the magnetic field with Faraday's law:

$$\underbrace{\boldsymbol{E} \cdot (\nabla \times \boldsymbol{H}) - \boldsymbol{H} \cdot (\nabla \times \boldsymbol{E})}_{-\nabla \cdot (\boldsymbol{E} \times \boldsymbol{H}) = -\nabla \cdot \boldsymbol{S}} = \boldsymbol{E} \cdot \boldsymbol{J} + \underbrace{\boldsymbol{E} \cdot \frac{\partial \boldsymbol{D}}{\partial t} + \boldsymbol{H} \cdot \frac{\partial \boldsymbol{B}}{\partial t}}_{= \frac{\partial W}{\partial t}}, \tag{B.27}$$

where $W = \frac{1}{2}(\boldsymbol{E} \cdot \boldsymbol{D} + \boldsymbol{H} \cdot \boldsymbol{B})$ is the electromagnetic energy and $\boldsymbol{S} = \boldsymbol{E} \times \boldsymbol{H}$ is the Poynting vector. This relation can be re-written as

$$\frac{\partial W}{\partial t} + \nabla \cdot \boldsymbol{S} = -\boldsymbol{J} \cdot \boldsymbol{E}. \tag{B.28}$$

Equation B.28 is usually referred to as Poynting's theorem. This can be interpreted, *for simple material laws, where the previous representation for W holds*, as stating that the rate of change of electromagnetic energy within a volume, plus the energy flowing out through a boundary, is equal to the negative of the total work done by the fields on the sources and conduction. This work is then converted into thermo-mechanical energy ("Joule-heating," H in Eq. B.18). Joule-heating stems from ions being pulled through a medium by electromagnetic fields, which generate heat when they collide with their surroundings.

Remark Generally, for detailed point-wise information, for example, localized effects in the matrix ligaments between particles ("hot spots"), one needs to solve boundary value problems posed over a statistically representative volume element (RVE) sample of heterogeneous media. This will be discussed toward the end of this chapter. However, the essential issue is that time-transient effects lead to coupling of electrical and magnetic fields, and the only viable approach is to employ direct numerical techniques to solve for Maxwell's equations. Generally, these equations are strongly coupled. Additionally, if the local material properties are thermally sensitive, and Joule-heating is significant, then the first law of thermodynamics must also be solved, simultaneously. Numerical techniques for the solution of coupled boundary value problems posed over heterogeneous electromagnetic media, undergoing thermo-mechano-chemical effects, can be found in Zohdi [2–15]. Later in the monograph, we will illustrate how such calculations proceed.

B.5 The Controllable Quantities: $\langle \boldsymbol{J} \rangle_{\Omega}$ and $\langle \boldsymbol{E} \rangle_{\Omega}$

For our model problem, two physically important test boundary ($\partial\Omega$) loading states are notable on a sample of heterogeneous material: (1) applied electric fields of the form: $\boldsymbol{E}|_{\partial\Omega} = \mathcal{E}$ and (2) applied current field of the form: $\boldsymbol{J}|_{\partial\Omega} = \mathcal{J}$, where $\mathcal{E}$ and $\mathcal{J}$ are constant electric field and current field vectors, respectively. Clearly, for these loading states to be satisfied within a macroscopic body under nonuniform external loading, the sample must be large enough to possess small boundary field fluctuations relative to its size. Therefore, applying (1)- or (2)-type boundary conditions to a large sample is a way of reproducing approximately what may be occurring in a statistically representative microscopic sample of material within a macroscopic body. The following two results render $\langle \boldsymbol{J} \rangle_{\Omega}$ and $\langle \boldsymbol{E} \rangle_{\Omega}$ as controllable quantities, via the boundary loading:

- *The Average Electric Field Theorem*: Consider a sample with boundary loading $\boldsymbol{E}|_{\partial\Omega} = \mathcal{E}$. We make use of the identity

$$\nabla \times (\boldsymbol{E} \otimes \boldsymbol{x}) = (\nabla \times \boldsymbol{E}) \otimes \boldsymbol{x} + \underbrace{\boldsymbol{E} \cdot \nabla \boldsymbol{x}}_{\boldsymbol{E}}, \tag{B.29}$$

and substitute this in the definition of the average electric field

$$\begin{aligned} \langle \boldsymbol{E} \rangle_\Omega &= \frac{1}{|\Omega|} \int_\Omega (\nabla \times (\boldsymbol{E} \otimes \boldsymbol{x}) - \underbrace{(\nabla \times \boldsymbol{E}) \otimes \boldsymbol{x}}_{=\boldsymbol{0}})\, d\Omega = \frac{1}{|\Omega|} \int_{\partial\Omega} \boldsymbol{n} \times (\boldsymbol{E} \otimes \boldsymbol{x})\, dA \\ &= \frac{1}{|\Omega|} \int_{\partial\Omega} \boldsymbol{n} \times (\mathcal{E} \otimes \boldsymbol{x})\, dA = \frac{1}{|\Omega|} \left(\int_{\partial\Omega} (\nabla \times \mathcal{E}) \otimes \boldsymbol{x}\, d\Omega + \int_{\partial\Omega} \mathcal{E} \cdot \nabla \boldsymbol{x}\, d\Omega \right). \end{aligned} \tag{B.30}$$

Thus, if $\nabla \times \boldsymbol{E} = \boldsymbol{0}$, then $\langle \boldsymbol{E} \rangle_\Omega = \mathcal{E}$, when $\boldsymbol{E}|_{\partial\Omega} = \mathcal{E}$.

- *The Average Current Field Theorem*: Consider a sample with boundary loading $\boldsymbol{J}|_{\partial\Omega} = \mathcal{J}$. We make use of the identity

$$\nabla \cdot (\boldsymbol{J} \otimes \boldsymbol{x}) = (\nabla \cdot \boldsymbol{J})\boldsymbol{x} + \underbrace{\boldsymbol{J} \cdot \nabla \boldsymbol{x}}_{\boldsymbol{J}}, \tag{B.31}$$

and substitute this in the definition of the average current

$$\begin{aligned} \langle \boldsymbol{J} \rangle_\Omega &= \frac{1}{|\Omega|} \int_\Omega (\nabla \cdot (\boldsymbol{J} \otimes \boldsymbol{x}) - \underbrace{(\nabla \cdot \boldsymbol{J})\boldsymbol{x}}_{=0})\, d\Omega = \frac{1}{|\Omega|} \int_{\partial\Omega} (\boldsymbol{J} \otimes \boldsymbol{x}) \cdot \boldsymbol{n}\, dA \\ &= \frac{1}{|\Omega|} \int_{\partial\Omega} \boldsymbol{n} \cdot (\mathcal{J} \otimes \boldsymbol{x})\, dA = \frac{1}{|\Omega|} \left(\int_{\partial\Omega} (\nabla \cdot \mathcal{J}) \otimes \boldsymbol{x}\, d\Omega + \int_{\partial\Omega} \mathcal{J} \cdot \nabla \boldsymbol{x}\, d\Omega \right). \end{aligned} \tag{B.32}$$

Thus, if $\nabla \cdot \boldsymbol{J} = 0$, then $\langle \boldsymbol{J} \rangle_\Omega = \mathcal{J}$, when $\boldsymbol{J}|_{\partial\Omega} = \mathcal{J}$.

Remark 1 The importance of the *Average Electric Field Theorem* and the *Average Current Field Theorem* is that we can consider $\langle \boldsymbol{E} \rangle_\Omega$ and $\langle \boldsymbol{J} \rangle_\Omega$ to be controllable quantities, via $\mathcal{E}$ or $\mathcal{J}$ on the boundary. Applying these boundary conditions should be made with the understanding that these idealizations reproduce what a representative volume element (which is much smaller than the structural component of intended use) would experience within the system of intended use. Uniform loading is an idealization and would be present within a vanishingly small microstructure relative to a finite-sized engineering (macro)structure. These types of loadings are somewhat standard in computational analyses of samples of heterogeneous materials (see Ghosh [18], Ghosh and Dimiduk [19], Zohdi [2–15]).

Remark 2 In the analysis that follows, we will utilize the following energy–power relation

$$\langle H \rangle_\Omega = \langle \boldsymbol{J} \cdot \boldsymbol{E} \rangle_\Omega = \langle \boldsymbol{J} \rangle_\Omega \cdot \langle \boldsymbol{E} \rangle_\Omega, \tag{B.33}$$

which is referred to as an ergodicity condition in statistical mechanics (Kröner [16], Torquato [17]) and as a Hill-type condition in the solid mechanics literature (Hill [20]). This is essentially a statement that the microenergy (power) must equal the macro-energy (power). Equation B.33 is developed by first splitting the current and electric fields into mean (average) and purely fluctuating (zero mean) parts. For the current field, one has $\boldsymbol{J} = \langle \boldsymbol{J} \rangle_\Omega + \tilde{\boldsymbol{J}}$, where $\langle \tilde{\boldsymbol{J}} \rangle_\Omega = \boldsymbol{0}$, and for the electric field $\boldsymbol{E} = \langle \boldsymbol{E} \rangle_\Omega + \tilde{\boldsymbol{E}}$, where $\langle \tilde{\boldsymbol{E}} \rangle_\Omega = \boldsymbol{0}$. The product yields

$$\langle (\langle \boldsymbol{J} \rangle_\Omega + \tilde{\boldsymbol{J}}) \cdot (\langle \boldsymbol{E} \rangle_\Omega + \tilde{\boldsymbol{E}}) \rangle_\Omega = \langle \boldsymbol{J} \rangle_\Omega \cdot \langle \boldsymbol{E} \rangle_\Omega + \langle \tilde{\boldsymbol{J}} \cdot \tilde{\boldsymbol{E}} \rangle_\Omega, \tag{B.34}$$

since $\langle \tilde{\boldsymbol{J}} \rangle_\Omega = \boldsymbol{0}$ and $\langle \tilde{\boldsymbol{E}} \rangle_\Omega = \boldsymbol{0}$. The ergodicity assumption is that $\langle \tilde{\boldsymbol{J}} \cdot \tilde{\boldsymbol{E}} \rangle_\Omega \to 0$, as the volume, $|\Omega| \to \infty$ (relative to the inherent length scales in the microstructure). The implication is that as the sample becomes infinitely large, $\tilde{\boldsymbol{J}} \cdot \tilde{\boldsymbol{E}}$ is purely fluctuating, and hence, $\langle \tilde{\boldsymbol{J}} \cdot \tilde{\boldsymbol{E}} \rangle_\Omega = 0$. In other words, the product of two purely fluctuating random fields is also purely fluctuating. These results are consistent with the use of the uniform boundary loadings introduced earlier, since they can be shown to satisfy Eq. B.33.

Remark As a consequence of previous results on concentration tensors, the Joule-fields can be written in a variety of useful forms:

$$\begin{aligned}
0 \le \langle H \rangle_{\Omega_i} \stackrel{\text{def}}{=} \langle \boldsymbol{J} \rangle_{\Omega_i} \cdot \langle \boldsymbol{E} \rangle_{\Omega_i} &= \underbrace{\boldsymbol{\sigma}_i^{-1} \cdot \langle \boldsymbol{J} \rangle_{\Omega_i} \cdot \langle \boldsymbol{J} \rangle_{\Omega_i}}_{\text{in terms of phase averages of } \boldsymbol{J}} \\
&= \underbrace{\boldsymbol{\sigma}_i \cdot \langle \boldsymbol{E} \rangle_{\Omega_i} \cdot \langle \boldsymbol{E} \rangle_{\Omega_i}}_{\text{in terms of phase averages of } \boldsymbol{E}} \\
&= \underbrace{(\boldsymbol{C}_{J,i} \cdot \langle \boldsymbol{J} \rangle_\Omega) \cdot (\boldsymbol{C}_{E_i} \cdot \langle \boldsymbol{E} \rangle_\Omega)}_{\text{in terms of overall averages of } \boldsymbol{J} \text{ and } \boldsymbol{E}} \\
&= \underbrace{\boldsymbol{\sigma}_i^{-1} \cdot (\boldsymbol{C}_{J,i} \cdot \langle \boldsymbol{J} \rangle_\Omega) \cdot (\boldsymbol{C}_{J_i} \cdot \langle \boldsymbol{J} \rangle_\Omega)}_{\text{in terms of overall averages of } \boldsymbol{J}} \\
&= \underbrace{\boldsymbol{\sigma}_i \cdot (\boldsymbol{C}_{E,i} \cdot \langle \boldsymbol{E} \rangle_\Omega) \cdot (\boldsymbol{C}_{E_i} \cdot \langle \boldsymbol{E} \rangle_\Omega)}_{\text{in terms of overall averages of } \boldsymbol{E}}.
\end{aligned} \tag{B.35}$$

B.6 Joule-Heating Load-Shares

Using Eq. B.35, the Joule-fields can be bounded as follows (using the Cauchy–Schwartz inequality)

$$\langle H \rangle_{\Omega_i} = (\boldsymbol{C}_{J,i} \cdot \langle \boldsymbol{J} \rangle_\Omega) \cdot (\boldsymbol{C}_{E_i} \cdot \langle \boldsymbol{E} \rangle_\Omega) \le ||\boldsymbol{C}_{E,i}||||\boldsymbol{C}_{J,i}||\langle H \rangle_\Omega \Rightarrow ||\boldsymbol{C}_{E,i}||||\boldsymbol{C}_{J,i}|| \ge \frac{\langle H \rangle_{\Omega_i}}{\langle H \rangle_\Omega}. \tag{B.36}$$

If the overall property is isotropic, and each of the constituents is isotropic (e.g., a microstructure comprised of a continuous isotropic binder embedded with randomly distributed isotropic particles), then we have the following, $\boldsymbol{C}_{E,i} = C_{E,i}\boldsymbol{1}$ where for a two-phase material

$$C_{E,1} = \frac{1}{1-v_2}\frac{\sigma_2-\sigma^*}{\sigma_2-\sigma_1} \text{ and } C_{E,2} = \frac{1}{v_2}\frac{\sigma^*-\sigma_1}{\sigma_2-\sigma_1}, \tag{B.37}$$

and $\boldsymbol{C}_{J,i} = C_{J,i}\mathbf{1}$, leading to

$$C_{J,1} = \frac{\sigma_1}{\sigma^*(1-v_2)}\left(\frac{\sigma_2-\sigma^*}{\sigma_2-\sigma_1}\right) \text{ and } C_{J,2} = \frac{\sigma_2}{\sigma^* v_2}\left(\frac{\sigma^*-\sigma_1}{\sigma_2-\sigma_1}\right). \tag{B.38}$$

Thus, in the case of isotropy, Eq. B.36 asserts

$$C_{E,1}C_{J,1} \geq \frac{\langle H\rangle_{\Omega_1}}{\langle H\rangle_{\Omega}} \text{ and } C_{E,2}C_{J,2} \geq \frac{\langle H\rangle_{\Omega_2}}{\langle H\rangle_{\Omega}}. \tag{B.39}$$

The product of the concentration functions takes the following form:

$$C_{E,1}C_{J,1} = \frac{\sigma_1}{\sigma^*}\left(\frac{1}{(1-v_2)}\left(\frac{\sigma_2-\sigma^*}{\sigma_2-\sigma_1}\right)\right)^2 \tag{B.40}$$

and

$$C_{E,2}C_{J,2} = \frac{\sigma_2}{\sigma^*}\left(\frac{1}{v_2}\left(\frac{\sigma^*-\sigma_1}{\sigma_2-\sigma_1}\right)\right)^2. \tag{B.41}$$

Because the concentration functions depend on σ^*, which in turn depend on σ_1, σ_2, v_2, and the microstructure, we need to employ estimates for σ^*.

Remark Recall, as introduced earlier, one class of estimates are the Hashin–Shtrikman bounds (Hashin [21]) for two isotropic materials with an overall isotropic response

$$\underbrace{\sigma_1 + \frac{v_2}{\frac{1}{\sigma_2-\sigma_1}+\frac{1-v_2}{3\sigma_1}}}_{\sigma^{*,-}} \leq \sigma^* \leq \underbrace{\sigma_2 + \frac{1-v_2}{\frac{1}{\sigma_1-\sigma_2}+\frac{v_2}{3\sigma_2}}}_{\sigma^{*,+}}, \tag{B.42}$$

where the conductivity of phase 2 (with volume fraction v_2) is larger than phase 1 ($\sigma_2 \geq \sigma_1$). Provided that the volume fractions and constituent conductivities are the only known information about the microstructure, the expressions in Eq. C.37 are the tightest bounds for the overall isotropic effective responses for two-phase media, where the constituents are both isotropic. A critical observation is that the lower bound is more accurate when the material is composed of high-conductivity particles that are surrounded by a low-conductivity matrix (denoted case 1) and the upper bound is more accurate for a high-conductivity matrix surrounding low-conductivity particles (denoted case 2). Since the true effective property lies between the upper and lower bounds, one can construct the following approximation

$$\sigma^* \approx \phi\sigma^{*,+} + (1-\phi)\sigma^{*,-}, \tag{B.43}$$

where $0 \leq \phi \leq 1$. ϕ *is an unknown function of the microstructure.* However, the general trends are (a) for cases where the upper bound is more accurate, $\phi > \frac{1}{2}$ and (b) for cases when the lower bound is more accurate, $\phi < \frac{1}{2}$. Explicitly, for the product of concentration functions, embedding the effective property estimates, we have

$$C_{E,1}C_{J,1} \approx \frac{\sigma_1}{(\phi\sigma^{*,+} + (1-\phi)\sigma^{*,-})}\left(\frac{1}{(1-v_2)}\left(\frac{\sigma_2 - (\phi\sigma^{*,+} + (1-\phi)\sigma^{*,-})}{\sigma_2 - \sigma_1}\right)\right)^2 \tag{B.44}$$

and

$$C_{E,2}C_{J,2} \approx \frac{\sigma_2}{(\phi\sigma^{*,+} + (1-\phi)\sigma^{*,-})}\left(\frac{1}{v_2}\left(\frac{(\phi\sigma^{*,+} + (1-\phi)\sigma^{*,-}) - \sigma_1}{\sigma_2 - \sigma_1}\right)\right)^2. \tag{B.45}$$

Remark 1 There are a vast literature of methods, dating back to Maxwell [22, 23] and Lord Rayleigh [24], to estimate the overall macroscopic properties of heterogeneous materials. For an authoritative review of (a) the general theory of random heterogeneous media, see Torquato [17], (b) for more mathematical homogenization aspects, see Jikov et al. [25], (c) for solid mechanics inclined accounts of the subject, see Hashin [21], Mura [26], Nemat-Nasser and Hori [27], (d) for analyses of cracked media, see Sevostianov and Kachanov [28], and (e) for computational aspects, see Ghosh [18], Ghosh and Dimiduk [19], and Zohdi and Wriggers [2–15]. Tighter estimates, including generalized N-phase bounds, can be found in Torquato [17].[3]

Remark 2 The governing equation used in developing effective conductivity bounds is $\nabla \cdot \boldsymbol{J} = 0$, which stems from taking the divergence of Ampere's law: $\nabla \cdot \left(\nabla \times \boldsymbol{H} - \frac{\partial \boldsymbol{D}}{\partial t} - \boldsymbol{J}\right) = 0$; one obtains, since $\nabla \cdot (\nabla \times \boldsymbol{H}) = 0$,

$$\nabla \cdot \left(\frac{\partial \boldsymbol{D}}{\partial t} + \boldsymbol{J}\right) = \frac{\partial}{\partial t}\underbrace{\nabla \cdot \boldsymbol{D}}_{\mathcal{P}} + \nabla \cdot \boldsymbol{J} = \frac{\partial \mathcal{P}}{\partial t} + \nabla \cdot \boldsymbol{J} = 0, \tag{B.46}$$

where $\mathcal{P}$ is the charge per unit volume. Thus, if $\mathcal{P} = 0$, $\nabla \cdot \boldsymbol{J} = 0$. If one employs the constitutive relation $\boldsymbol{J} = \boldsymbol{\sigma} \cdot \boldsymbol{E}$, then this allows for Hashin–Shtrikman-type estimates to be used for the effective conductivity, as does $\nabla \cdot \boldsymbol{D} = 0$ (which is valid only when $\mathcal{P} = 0$) for estimates of the effective permittivity, $\langle \boldsymbol{D} \rangle_\Omega = \boldsymbol{\epsilon}^* \cdot \langle \boldsymbol{E} \rangle_\Omega$, when $\boldsymbol{D} = \boldsymbol{\epsilon} \cdot \boldsymbol{E}$. For example, one case when these two physical situations are compatible is when $\boldsymbol{E} = \boldsymbol{\sigma}^{-1} \cdot \boldsymbol{J} = \boldsymbol{\epsilon}^{-1} \cdot \boldsymbol{D} \Rightarrow \boldsymbol{J} = (\boldsymbol{\sigma} \cdot \boldsymbol{\epsilon}^{-1}) \cdot \boldsymbol{D}$.

[3]Such N-phase bounds go well beyond the simple Wiener bounds (Wiener [29]), $\left(\sum_{i=1}^{N} v_i \boldsymbol{\sigma}_i^{-1}\right)^{-1} \leq \boldsymbol{\sigma}^* \leq \sum_{i=1}^{N} v_i \boldsymbol{\sigma}_i$.

B.7 Examples of Joule-Heating Load-Sharing

B.7.1 A General Dielectric Mixture

Figures B.1 and B.2 illustrate a surface (using $\phi = \frac{1}{2}$) in parameter space $(\frac{\sigma_1}{\sigma_2}, v_2)$ for the normalized Joule-heating load-share, $v_i \frac{\langle H \rangle_{\Omega_i}}{\langle H \rangle_{\Omega}}$, of each component, $i = 1, 2$. The plots illustrate the proportion of the Joule-heating that will be delivered to each phase in the system. Directly from Eqs. B.44 and B.45, the load-share quantities of interest are

$$v_1 \frac{\langle H \rangle_{\Omega_1}}{\langle H \rangle_{\Omega}} \approx \frac{\sigma_1}{v_1(\phi\sigma^{*,+} + (1-\phi)\sigma^{*,-})} \left(\frac{\sigma_2 - (\phi\sigma^{*,+} + (1-\phi)\sigma^{*,-})}{\sigma_2 - \sigma_1} \right)^2, \quad \text{(B.47)}$$

and

$$v_2 \frac{\langle H \rangle_{\Omega_2}}{\langle H \rangle_{\Omega}} \approx \frac{\sigma_2}{v_2(\phi\sigma^{*,+} + (1-\phi)\sigma^{*,-})} \left(\frac{(\phi\sigma^{*,+} + (1-\phi)\sigma^{*,-}) - \sigma_1}{\sigma_2 - \sigma_1} \right)^2. \quad \text{(B.48)}$$

The trends are:

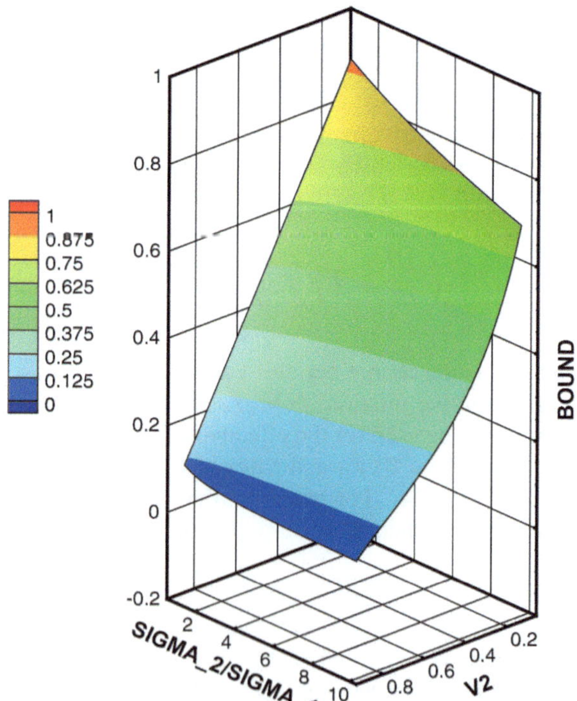

Fig. B.1 A load-share surface in parameter space $(\frac{\sigma_2}{\sigma_1}, v_2)$ for the normalized Joule-heating, $v_1 \frac{\langle H \rangle_{\Omega_1}}{\langle H \rangle_{\Omega}}$, for phase 1 (using $\phi = 1/2$)

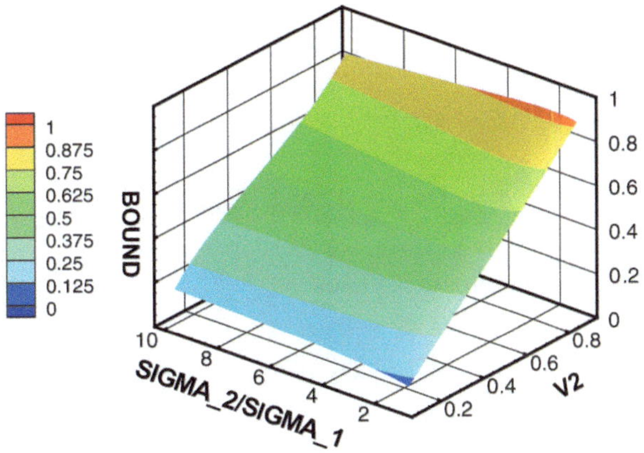

Fig. B.2 A load-share surface in parameter space ($\frac{\sigma_2}{\sigma_1}$, v_2) for the normalized Joule-heating, $v_2 \frac{\langle H \rangle_{\Omega_2}}{\langle H \rangle_{\Omega}}$, for phase 2 (using $\phi = 1/2$).

- **For phase 1**: Decreasing the volume fraction of phase 2 (v_2), for fixed σ_2/σ_1, leads to a larger load-share for phase 1, while decreasing the mismatch σ_2/σ_1, for a fixed v_2, leads to an increased load-share for phase 1, for a fixed volume fraction v_2.
- **For phase 2**: Increasing the volume fraction of phase 2 (v_2), for fixed σ_2/σ_1, leads to a larger load-share for phase 2, while increasing the mismatch σ_2/σ_1, for a fixed v_2, leads to an extremely slight change in the load-share of phase 2 (it is virtually flat).

B.7.2 An Extreme Mixture: High-Conductivity ("Superconducting") Particles in a Low-Conductivity Matrix

For the case of high-conductivity particles (phase 2) in a lower-conductivity matrix (phase 1), we have

$$1 << \frac{\sigma_2}{\sigma_1} \stackrel{\text{def}}{=} \alpha. \tag{B.49}$$

Inserting this expression into the Hashin–Shtrikman bounds and taking the limit as $\alpha \to \infty$ yield (σ_2 tending to infinity),

$$\sigma_1 \left(\frac{1 + 2v_2}{1 - v_2} \right) \stackrel{\text{def}}{=} \sigma_1 \zeta \leq \sigma^* \leq \infty, \tag{B.50}$$

where the lower (Hashin-Shtrikman) bound is more accurate ($\phi \to 0$). Correspondingly, for the concentration tensors for phase 1 (assuming isotropy)[4]

[4]These expressions are asymptotically consistent with the identities

$$v_1 C_{E,1} + v_2 C_{E,2} = 1 \quad \text{and} \quad v_1 C_{J,1} + v_2 C_{J,2} = 1. \tag{B.51}$$

$$C_{E,1} = \frac{1}{1 - v_2} \quad \text{and} \quad C_{J,1} = \frac{1}{\zeta(1 - v_2)} = \frac{1}{1 + 2v_2} \tag{B.52}$$

and for phase 2 (particle)

$$C_{E,2} = 0 \quad \text{and} \quad C_{J,2} = \frac{1}{v_2}\left(1 - \frac{1}{\zeta}\right) = \frac{3}{1 + 2v_2}. \tag{B.53}$$

Forming the products yields

$$C_{E,1}C_{J,1} = \left(\frac{1}{1 - v_2}\right)\left(\frac{1}{1 + 2v_2}\right) \tag{B.54}$$

and

$$C_{E,2}C_{J,2} = 0. \tag{B.55}$$

The expressions are appropriate for small v_2 (superconducting particles in a binding matrix). Thus, we have for the load-share

$$\frac{1}{1 + 2v_2} \geq v_1 \frac{\langle H \rangle_{\Omega_1}}{\langle H \rangle_{\Omega}}, \tag{B.56}$$

while for phase 2 (particle superconductor, no Joule-field)

$$v_2 \frac{\langle H \rangle_{\Omega_2}}{\langle H \rangle_{\Omega}} = 0. \tag{B.57}$$

Remark As $v_2 \rightarrow 0$ (no particle (phase 2) material), the expressions collapse to restrictions on the pure matrix (here, phase 1) material.

B.7.3 An Extreme Mixture: Low-Conductivity ("Insulator") Particles in a High-Conductivity Matrix

For the case of low-conductivity particles (phase 1) in a higher-conductivity matrix (phase 2), we have

$$1 >> \frac{\sigma_1}{\sigma_2} \stackrel{\text{def}}{=} \gamma. \tag{B.58}$$

Inserting this expression into the Hashin–Shtrikman bounds and taking the limit as $\gamma \rightarrow 0$ (σ_1 tending to zero) yield,

$$0 \leq \sigma^* \leq \sigma_2 \left(\frac{2v_2}{3 - v_2}\right) \stackrel{\text{def}}{=} \sigma_2 \Lambda, \tag{B.59}$$

where the upper (Hashin-Shtrikman) bound is more accurate ($\phi \to 1$). Correspondingly, for the concentration tensors (as $\gamma \to 0$), for phase 1 (particle)

$$C_{E,1} = \frac{1-\Lambda}{1-v_2} = \frac{3}{3-v_2} \quad \text{and} \quad C_{J,1} = 0 \tag{B.60}$$

and for phase 2 (matrix)

$$C_{E,2} = \frac{\Lambda}{v_2} = \frac{2}{3-v_2} \quad \text{and} \quad C_{J,2} = \frac{1}{v_2}. \tag{B.61}$$

The expressions are appropriate for large v_2 (insulating particles in a binding matrix).

$$C_{E,1}C_{J,1} = 0 \tag{B.62}$$

and

$$C_{E,2}C_{J,2} = \left(\frac{2}{3-v_2}\right)\left(\frac{1}{v_2}\right). \tag{B.63}$$

Thus, we have for the load-shares, for phase 1 (particle insulator, no Joule-field)

$$v_1 \frac{\langle H \rangle_{\Omega_1}}{\langle H \rangle_{\Omega}} = 0, \tag{B.64}$$

and for phase 2 (matrix)

$$\frac{2}{3-v_2} \geq v_2 \frac{\langle H \rangle_{\Omega_2}}{\langle H \rangle_{\Omega}}. \tag{B.65}$$

Remark As $v_2 \to 1$ (no particle (here, phase 1) material), the expressions collapse to restrictions on the pure matrix (here, phase 2) material.

B.8 Optimization Example: Dielectric Properties Using Genetic Algorithms

In order to utilize the previous results, we now provide examples of how to optimize the material properties. In particular, we will optimize the overall electrical conductivity of a mixture of material, subject to Joule-heating constraints on each phase. The effective conductivity is given by σ^* using convex combinations of the Hashin–Shtrikman bounds as approximations for the effective conductivity $\sigma^* \approx \phi\sigma^{*,+} + (1-\phi)\sigma^{*,-}$, where $0 \leq \theta \leq 1$. The micro–macro objective function is

$$\Pi = w_1 | \frac{\sigma^*}{\sigma^{*,D}} - 1|^2 + \hat{w}_2 \left((C_{J1}C_{E1} - 1) - H^{tol}\right)^2 + \hat{w}_3 \left((C_{J2}C_{E2} - 1) - H_2^{tol}\right)^2, \tag{B.66}$$

where the constraints are unilaterally activated in the following manner:

- $C_{J1}C_{E1} - 1 \leq H_1^{tol}$ then $\hat{w}_2 = 0$,
- $C_{J1}C_{E1} - 1 > H_1^{tol}$ then $\hat{w}_2 = w_2$,
- $C_{J2}C_{E2} - 1 \leq H_2^{tol}$ then $\hat{w}_3 = 0$,
- $C_{J2}C_{E2} - 1 > H_2^{tol}$ then $\hat{w}_3 = w_3$,

Here, the design variables are $\boldsymbol{\Lambda} = \{\sigma_1, \sigma_2, v_2\}$, and their constrained ranges are

- $\sigma_1^{(-)} \leq \sigma_1 \leq \sigma_1^{(+)}$,
- $\sigma_2^{(-)} \leq \sigma_2 \leq \sigma_2^{(+)}$, and
- $v_2^{(-)} \leq v_2 \leq v_2^{(+)}$.

As for the previous optimization formulation for mechanical properties, two characteristics of such a formulation make the application of standard gradient-type minimization schemes, such as Newton's method, difficult:

- (I) The incorporation of limits on the microfield behavior, as well as design search space restrictions, renders the objective function not continuously differentiable in design space and
- (II) The objective function is nonconvex, i.e., the system Hessian is not positive definite (invertible) throughout the design space.

Thus, as before, we apply a genetic algorithm:

- **STEP 1:** Randomly generate a population of S starting genetic strings, $\boldsymbol{\Lambda}^i$, $(i = 1, \ldots, S)$:
  $\boldsymbol{\Lambda}^i \stackrel{\text{def}}{=} \{\Lambda_1^i, \Lambda_2^i, \Lambda_3^i, \Lambda_4^i, \ldots, \Lambda_N^i\} = \{\sigma_2^i, \sigma_1^i, v_2^i, \ldots\}$
- **STEP 2:** Compute fitness of each string $\Pi(\boldsymbol{\Lambda}^i)$, (i=1, ..., S)
- **STEP 3:** Rank genetic strings: $\boldsymbol{\Lambda}^i$, (i=1, ..., S)
- **STEP 4:** Mate nearest pairs and produce two offspring, (i=1, ..., S)
  $\boldsymbol{\lambda}^i \stackrel{\text{def}}{=} \Phi^{(I)}\boldsymbol{\Lambda}^i + (1 - \Phi^{(I)})\boldsymbol{\Lambda}^{i+1}, \quad \boldsymbol{\lambda}^{i+1} \stackrel{\text{def}}{=} \Phi^{(II)}\boldsymbol{\Lambda}^i + (1 - \Phi^{(II)})\boldsymbol{\Lambda}^{i+1}$
- **NOTE:** $\Phi^{(I)}$ and $\Phi^{(II)}$ are random numbers, such that $0 \leq \Phi^{(I)}, \Phi^{(II)} \leq 1$, which are different for each component of each genetic string
- **STEP 5:** Kill off bottom $M < S$ strings and keep top $K < N$ parents and top K offspring (K offspring $+$ K parents $+$ $M = S$)
- **STEP 6:** Repeat STEPS 1-6 with top gene pool (K offspring and K parents), plus M new, randomly generated, strings
- **Option:** Rescale and restart search around best performing parameter set every few generations

Remark After application of such a global search algorithm, one can apply a gradient-based method, if the objective function is sufficiently smooth in that region of the design parameter space.

B.9 Additional Dielectric Properties: Electrical Permittivity and Magnetic Permeability

Other overall dielectric properties may be needed for other applications and can be generated with similar formulas as in the beginning of this analysis. For the overall electric permittivity

$$\langle \epsilon^{-1}(\boldsymbol{x})\rangle_{\Omega}^{-1} \leq \underbrace{\epsilon_1 + \frac{v_2^{\epsilon}}{\frac{1}{\epsilon_2-\epsilon_1} + \frac{1-v_2^{\epsilon}}{3\epsilon_1}}}_{\epsilon^{*,-}} \leq \epsilon^* \leq \underbrace{\epsilon_2 + \frac{1-v_2^{\epsilon}}{\frac{1}{\epsilon_1-\epsilon_2} + \frac{v_2^{\epsilon}}{3\epsilon_2}}}_{\epsilon^{*,+}} \leq \langle \epsilon(\boldsymbol{x})\rangle_{\Omega} \tag{B.67}$$

and the overall magnetic permeability

$$\langle \mu^{-1}(\boldsymbol{x})\rangle_{\Omega}^{-1} \leq \underbrace{\mu_1 + \frac{v_2^{\mu}}{\frac{1}{\mu_2-\mu_1} + \frac{1-v_2^{\mu}}{3\mu_1}}}_{\mu^{*,-}} \leq \mu^* \leq \underbrace{\mu_2 + \frac{1-v_2^{\mu}}{\frac{1}{\mu_1-\mu_2} + \frac{v_2^{\mu}}{3\mu_2}}}_{\mu^{*,+}} \leq \langle \mu(\boldsymbol{x})\rangle_{\Omega} \tag{B.68}$$

where $\epsilon_2 \geq \epsilon_1$, $\mu_2 \geq \mu_1$, v_2^{ϵ} is the volume fraction of ("higher") phase 2 for the permittivity mismatch, and v_2^{μ} is the volume fraction of ("higher") phase 2 for the permeability mismatch.[5]

Remark The speed of electromagnetic wave propagation is $c = \frac{1}{\sqrt{\epsilon_o \mu_o}} \approx 2.997924562 \times 10^8 \pm 1.1\,\text{m/s}$ in vacuum. The free space electric permittivity is $\epsilon_o = \frac{1}{c^2\mu_o} = 8.8542 \times 10^{-12}\,\text{CN}^{-1}\text{m}^{-1}$, and the free space magnetic permeability is $\mu_o = 4\pi \times 10^{-7}\,\text{WbA}^{-1}\text{m}^{-1} = 1.2566 \times 10^{-6}\,\text{WbA}^{-1}\text{m}^{-1}$. Usually, for electromagnetic materials, we write $\epsilon = \epsilon_o \epsilon_r$, where $\epsilon_o = 8.854 \times 10^{-12}$ F/m is the free space permittivity and ϵ_r is the relative permittivity or "dielectric" constant and $\mu_o = 4\pi \times 10^{-7}\text{Ns}^2/\text{C}^2$ and $\mu = \mu_r \mu_o$, where μ_r is the relative magnetic permeability.

B.10 The Concentration Tensor

In order to estimate how much of the load is carried by each material in the mixture, we use the concept of "load-sharing," introduced earlier. For example, for the relation between the electrical field and the electrical field flux, consider the following identities, relating to $\langle \boldsymbol{D}\rangle_{\Omega} = \epsilon^* \cdot \langle \boldsymbol{E}\rangle_{\Omega}$,

$$\langle \boldsymbol{E}\rangle_{\Omega} = \frac{1}{|\Omega|}\left(\int_{\Omega_1} \boldsymbol{E}\, d\Omega + \int_{\Omega_2} \boldsymbol{E}\, d\Omega\right) = v_1^{\epsilon}\langle \boldsymbol{E}\rangle_{\Omega_1} + v_2^{\epsilon}\langle \boldsymbol{E}\rangle_{\Omega_2} \tag{B.69}$$

[5] For either case, the volume fraction of the other phase is v_1, where $v_1 + v_2 = 1$.

and

$$\langle \boldsymbol{D} \rangle_{\Omega} = \frac{1}{|\Omega|}\left(\int_{\Omega_1} \boldsymbol{D}\, d\Omega + \int_{\Omega_2} \boldsymbol{D}\, d\Omega\right) = v_1^{\epsilon}\langle \boldsymbol{D} \rangle_{\Omega_1} + v_2^{\epsilon}\langle \boldsymbol{D} \rangle_{\Omega_2}. \tag{B.70}$$

By direct manipulation, we obtain

$$\begin{aligned}\langle \boldsymbol{D} \rangle_{\Omega} &= v_1^{\epsilon}\langle \boldsymbol{D} \rangle_{\Omega_1} + v_2^{\epsilon}\langle \boldsymbol{D} \rangle_{\Omega_2} \\ &= v_1^{\epsilon}\boldsymbol{\epsilon}_1 \cdot \langle \boldsymbol{E} \rangle_{\Omega_1} + v_2^{\epsilon}\boldsymbol{\epsilon}_2 \cdot \langle \boldsymbol{E} \rangle_{\Omega_2} \\ &= \boldsymbol{\epsilon}_1 \cdot (\langle \boldsymbol{E} \rangle_{\Omega} - v_2^{\epsilon}\langle \boldsymbol{E} \rangle_{\Omega_2}) + v_2^{\epsilon}\boldsymbol{\epsilon}_2 \cdot \langle \boldsymbol{E} \rangle_{\Omega_2} \\ &= \underbrace{\left(\boldsymbol{\epsilon}_1 + v_2^{\epsilon}(\boldsymbol{\epsilon}_2 - \boldsymbol{\epsilon}_1) \cdot \boldsymbol{C}_{E,2}\right)}_{\boldsymbol{\epsilon}^*} \cdot \langle \boldsymbol{E} \rangle_{\Omega}\end{aligned} \tag{B.71}$$

where

$$\underbrace{\left(\frac{1}{v_2^{\epsilon}}(\boldsymbol{\epsilon}_2 - \boldsymbol{\epsilon}_1)^{-1} \cdot (\boldsymbol{\epsilon}^* - \boldsymbol{\epsilon}_1)\right)}_{\stackrel{\text{def}}{=} \boldsymbol{C}_{E,2}} \cdot \langle \boldsymbol{E} \rangle_{\Omega} = \langle \boldsymbol{E} \rangle_{\Omega_2}. \tag{B.72}$$

In the special case of isotropy

$$C_{E,2} \stackrel{\text{def}}{=} \frac{1}{v_2^{\epsilon}} \frac{\epsilon^* - \epsilon_1}{\epsilon_2 - \epsilon_1} \tag{B.73}$$

Thereafter, we may write, for the variation in the electric field flux

$$\boldsymbol{C}_{E,2} \cdot \boldsymbol{\epsilon}^{*-1} \cdot \langle \boldsymbol{D} \rangle_{\Omega} = \boldsymbol{\epsilon}_2^{-1} \cdot \langle \boldsymbol{D} \rangle_{\Omega_2} \tag{B.74}$$

which reduces to

$$\boldsymbol{\epsilon}_2 \cdot \boldsymbol{C}_{E,2} \cdot \boldsymbol{\epsilon}^{*-1} \cdot \langle \boldsymbol{D} \rangle_{\Omega} \stackrel{\text{def}}{=} \boldsymbol{C}_{D,2} \cdot \langle \boldsymbol{D} \rangle_{\Omega} = \langle \boldsymbol{D} \rangle_{\Omega_2} \tag{B.75}$$

$\boldsymbol{C}_{D,2}$ is known as the electric field flux concentration tensor. Therefore, once either $\boldsymbol{C}_{D,2}$ or $\boldsymbol{\epsilon}^*$ are known, the other can be determined. In the case of isotropy, we may write

$$C_{D,2} \stackrel{\text{def}}{=} \frac{1}{v_2^{\epsilon}} \frac{\epsilon_2}{\epsilon^*} \frac{\epsilon^* - \epsilon_1}{\epsilon_2 - \epsilon_1}. \tag{B.76}$$

For the matrix

$$\langle \boldsymbol{E} \rangle_{\Omega_1} = \frac{\langle \boldsymbol{E} \rangle_{\Omega} - v_2^{\epsilon}\langle \boldsymbol{E} \rangle_{\Omega_2}}{v_1^{\epsilon}} = \frac{(\boldsymbol{1} - v_2^{\epsilon}\boldsymbol{C}_{E,2}) \cdot \langle \boldsymbol{E} \rangle_{\Omega}}{v_1^{\epsilon}} = \boldsymbol{C}_{E,1} \cdot \langle \boldsymbol{E} \rangle_{\Omega}. \tag{B.77}$$

where

$$\boldsymbol{C}_{E,1} \stackrel{\text{def}}{=} \frac{1}{v_1^{\epsilon}}(\boldsymbol{1} - v_2^{\epsilon}\boldsymbol{C}_{E,2}) = \frac{\boldsymbol{1} - v_2^{\epsilon}\boldsymbol{C}_{E,2}}{1 - v_2^{\epsilon}}, \tag{B.78}$$

where in the case of isotropy

$$C_{E,1} = \frac{1 - v_2^{\epsilon} C_{E,2}}{1 - v_2^{\epsilon}}. \tag{B.79}$$

Similarly, for the electric field flux:

$$\boldsymbol{C}_{D,1} \cdot \langle \boldsymbol{D} \rangle_{\Omega} = \langle \boldsymbol{D} \rangle_{\Omega_1}, \tag{B.80}$$

where

$$\boldsymbol{C}_{D,1} = \frac{\boldsymbol{1} - v_2^{\epsilon} \boldsymbol{C}_{D,2}}{1 - v_2^{\epsilon}} \tag{B.81}$$

where in the case of isotropy

$$C_{D,1} = \frac{1 - v_2^{\epsilon} C_{D,2}}{1 - v_2^{\epsilon}}. \tag{B.82}$$

The exact procedure holds for the magnetic fields, with $\boldsymbol{H}$ replacing $\boldsymbol{E}$, $\boldsymbol{B}$ replacing $\boldsymbol{D}$, $\boldsymbol{\mu}$ replacing ϵ, v_1^{μ} replacing v_1^{ϵ}, and v_2^{μ} replacing v_2^{ϵ}. Similarly, the exact procedure holds for the electrical current fields, with $\boldsymbol{J}$ replacing $\boldsymbol{D}$, $\boldsymbol{\sigma}$ replacing ϵ, v_1^{σ} replacing v_1^{ϵ}, and v_2^{σ} replacing v_2^{ϵ}.

In summary, we have the following concentration tensors:

- $\boldsymbol{C}_{E,2} \cdot \langle \boldsymbol{E} \rangle_{\Omega} = \langle \boldsymbol{E} \rangle_{\Omega_2}$ where $\boldsymbol{C}_{E,2} = \frac{1}{v_2^{\epsilon}}(\epsilon_2 - \epsilon_1)^{-1} \cdot (\epsilon^* - \epsilon_1)$
- $\boldsymbol{C}_{E,1} \cdot \langle \boldsymbol{E} \rangle_{\Omega} = \langle \boldsymbol{E} \rangle_{\Omega_1}$ where $\boldsymbol{C}_{E,1} = \frac{1}{v_1^{\epsilon}}(\boldsymbol{1} - v_2^{\epsilon} \boldsymbol{C}_{E,2}) = \frac{\boldsymbol{1} - v_2^{\epsilon} \boldsymbol{C}_{E,2}}{1 - v_2^{\epsilon}}$
- $\boldsymbol{C}_{D,2} \cdot \langle \boldsymbol{D} \rangle_{\Omega} = \langle \boldsymbol{D} \rangle_{\Omega_2}$ where $\boldsymbol{C}_{D,2} = \epsilon_2 \cdot \boldsymbol{C}_{E,2} \cdot \epsilon^{*-1}$
- $\boldsymbol{C}_{D,1} \cdot \langle \boldsymbol{D} \rangle_{\Omega} = \langle \boldsymbol{D} \rangle_{\Omega_1}$ where $\boldsymbol{C}_{D,1} = \frac{\boldsymbol{1} - v_2^{\epsilon} \boldsymbol{C}_{D,2}}{1 - v_2^{\epsilon}}$
- $\boldsymbol{C}_{H,2} \cdot \langle \boldsymbol{H} \rangle_{\Omega} = \langle \boldsymbol{H} \rangle_{\Omega_2}$ where $\boldsymbol{C}_{H,2} = \frac{1}{v_2^{\mu}}(\boldsymbol{\mu}_2 - \boldsymbol{\mu}_1)^{-1} \cdot (\boldsymbol{\mu}^* - \boldsymbol{\mu}_1)$
- $\boldsymbol{C}_{H,1} \cdot \langle \boldsymbol{H} \rangle_{\Omega} = \langle \boldsymbol{H} \rangle_{\Omega_1}$ where $\boldsymbol{C}_{H,1} = \frac{1}{v_1^{\mu}}(\boldsymbol{1} - v_2^{\mu} \boldsymbol{C}_{H,2}) = \frac{\boldsymbol{1} - v_2^{\mu} \boldsymbol{C}_{H,2}}{1 - v_2^{\mu}}$
- $\boldsymbol{C}_{B,2} \cdot \langle \boldsymbol{B} \rangle_{\Omega} = \langle \boldsymbol{B} \rangle_{\Omega_2}$ where $\boldsymbol{C}_{B,2} = \boldsymbol{\mu}_2 \cdot \boldsymbol{C}_{H,2} \cdot \boldsymbol{\mu}^{*-1}$
- $\boldsymbol{C}_{B,1} \cdot \langle \boldsymbol{B} \rangle_{\Omega} = \langle \boldsymbol{B} \rangle_{\Omega_1}$ where $\boldsymbol{C}_{B,1} = \frac{\boldsymbol{1} - v_2^{\mu} \boldsymbol{C}_{B,2}}{1 - v_2^{\mu}}$
- $\boldsymbol{C}_{E,2} \cdot \langle \boldsymbol{E} \rangle_{\Omega} = \langle \boldsymbol{E} \rangle_{\Omega_2}$ where $\boldsymbol{C}_{E,2} = \frac{1}{v_2^{\sigma}}(\boldsymbol{\sigma}_2 - \boldsymbol{\sigma}_1)^{-1} \cdot (\boldsymbol{\sigma}^* - \boldsymbol{\sigma}_1)$
- $\boldsymbol{C}_{E,1} \cdot \langle \boldsymbol{E} \rangle_{\Omega} = \langle \boldsymbol{E} \rangle_{\Omega_1}$ where $\boldsymbol{C}_{E,1} = \frac{1}{v_1^{\sigma}}(\boldsymbol{1} - v_2^{\sigma} \boldsymbol{C}_{E,2}) = \frac{\boldsymbol{1} - v_2^{\sigma} \boldsymbol{C}_{E,2}}{1 - v_2^{\sigma}}$
- $\boldsymbol{C}_{J,2} \cdot \langle \boldsymbol{J} \rangle_{\Omega} = \langle \boldsymbol{J} \rangle_{\Omega_2}$ where $\boldsymbol{C}_{J,2} = \boldsymbol{\sigma}_2 \cdot \boldsymbol{C}_{E,2} \cdot \boldsymbol{\sigma}^{*-1}$
- $\boldsymbol{C}_{J,1} \cdot \langle \boldsymbol{J} \rangle_{\Omega} = \langle \boldsymbol{J} \rangle_{\Omega_1}$ where $\boldsymbol{C}_{J,1} = \frac{\boldsymbol{1} - v_2^{\sigma} \boldsymbol{C}_{J,2}}{1 - v_2^{\sigma}}$

Remark The concentration tensors indicate the amplification of the field within the particle relative to the average of the field. Clearly, the microelectric flux fields are minimally distorted when the $\boldsymbol{C}$'s equal $\boldsymbol{1}$. *There has been no approximation yet.* The "burden" in the computations has shifted to the determination of the $\boldsymbol{C}$s. Classical methods approximate them. For example, the simplest approximation is $\boldsymbol{C}_{E,2} = \boldsymbol{1}$, which is the Voigt approximation (a constant $\boldsymbol{E}$-field throughout the microstructure), while the Reuss approximation is $\boldsymbol{C}_{D,2} = \boldsymbol{1}$ (a constant $\boldsymbol{D}$-field throughout the microstructure). The same relations can be derived for the electrical conduction and the magnetic field flow.

B.11 "Load-Sharing" Interpretation

Directly from Eqs. B.69 and B.70, one may write

$$
\begin{aligned}
\underbrace{v_1^{\epsilon}\boldsymbol{C}_{E,1}}_{\text{phase}-1\text{ contribution}} &+ \underbrace{v_2^{\epsilon}\boldsymbol{C}_{E,2}}_{\text{phase}-2\text{ contribution}} &= \boldsymbol{1}\\
\underbrace{v_1^{\epsilon}\boldsymbol{C}_{D,1}}_{\text{phase}-1\text{ contribution}} &+ \underbrace{v_2^{\epsilon}\boldsymbol{C}_{D,2}}_{\text{phase}-2\text{ contribution}} &= \boldsymbol{1}\\
\underbrace{v_1^{\mu}\boldsymbol{C}_{H,1}}_{\text{phase}-1\text{ contribution}} &+ \underbrace{v_2^{\mu}\boldsymbol{C}_{H,2}}_{\text{phase}-2\text{ contribution}} &= \boldsymbol{1}\\
\underbrace{v_1^{\mu}\boldsymbol{C}_{B,1}}_{\text{phase}-1\text{ contribution}} &+ \underbrace{v_2^{\mu}\boldsymbol{C}_{B,2}}_{\text{phase}-2\text{ contribution}} &= \boldsymbol{1}\\
\underbrace{v_1^{\sigma}\boldsymbol{C}_{E,1}}_{\text{phase}-1\text{ contribution}} &+ \underbrace{v_2^{\sigma}\boldsymbol{C}_{E,2}}_{\text{phase}-2\text{ contribution}} &= \boldsymbol{1}\\
\underbrace{v_1^{\sigma}\boldsymbol{C}_{J,1}}_{\text{phase}-1\text{ contribution}} &+ \underbrace{v_2^{\sigma}\boldsymbol{C}_{J,2}}_{\text{phase}-2\text{ contribution}} &= \boldsymbol{1}
\end{aligned}
\tag{B.83}
$$

Frequently, the first term in the above expressions is referred to as "phase-1's" share, while the second term is "phase-2's" share.

B.12 Thermal Conductivity

The process in the chapter can easily be used to compute the thermal conductivity, since the transport equation (conservation law) has the same form as steady-state electrical conduction. In order to make estimates of the overall properties of a mixture, we consider the widely used Hashin and Shtrikman bounds (see previous chapter) for isotropic materials with isotropic effective responses. These estimates provide one with upper and lower bounds on the overall response of the material. For two isotropic materials with an overall isotropic response, we utilize the following estimates.

$$
\underbrace{I\!K_1 + \frac{v_2}{\frac{1}{I\!K_2 - I\!K_1} + \frac{1-v_2}{3I\!K_1}}}_{I\!K^{*,-}} \le I\!K^* \le \underbrace{I\!K_2 + \frac{1-v_2}{\frac{1}{I\!K_1 - I\!K_2} + \frac{v_2}{3I\!K_2}}}_{I\!K^{*,+}},
\tag{B.84}
$$

where the conductivity of phase 2 (with volume fraction v_2) is larger than phase 1 ($I\!K_2 \geq I\!K_1$). Usually, v_2 corresponds to the particle material, although there can be applications where the matrix is more conductive than the particles. In that case, v_2 would correspond to the matrix material. As before, provided that the volume fractions and constituent conductivities are the only known information about the microstructure, the expressions are the tightest bounds for the overall isotropic effective responses for two-phase media, where the constituents are both isotropic. A critical observation is that the lower bound is more accurate when the material is composed of high-conductivity particles that are surrounded by a low-conductivity matrix (denoted case 1) and the upper bound is more accurate for a high-conductivity matrix surrounding low-conductivity particles (denoted case 2). All of the concentration tensor calculations proceed as before.

To obtain the concentration of the thermal fields in each phase, the exact procedure holds with $\nabla\theta$ replacing $\boldsymbol{E}$, $\boldsymbol{q}$ replacing $\boldsymbol{D}$, $I\!K$ replacing ϵ, v_1^K replacing v_1^ϵ, and v_2^K replacing v_2^ϵ. In summary, we have the following concentration tensors:

- $\boldsymbol{C}_{\theta,2} \cdot \langle\nabla\theta\rangle_\Omega = \langle\nabla\theta\rangle_{\Omega_2}$ where $\boldsymbol{C}_{\theta,2} = \frac{1}{v_2^K}(I\!K_2 - I\!K_1)^{-1} \cdot (I\!K^* - I\!K_1)$
- $\boldsymbol{C}_{\theta,1} \cdot \langle\nabla\theta\rangle_\Omega = \langle\nabla\theta\rangle_{\Omega_1}$ where $\boldsymbol{C}_{\theta,1} = \frac{1}{v_1^K}(\boldsymbol{1} - v_2^K \boldsymbol{C}_{\theta,2}) = \frac{\boldsymbol{1} - v_2^K \boldsymbol{C}_{\theta,2}}{1 - v_2^\theta}$
- $\boldsymbol{C}_{q,2} \cdot \langle\boldsymbol{q}\rangle_\Omega = \langle\boldsymbol{q}\rangle_{\Omega_2}$ where $\boldsymbol{C}_{q,2} = I\!K_2 \cdot \boldsymbol{C}_{\theta,2} \cdot I\!K^{*-1}$
- $\boldsymbol{C}_{q,1} \cdot \langle\boldsymbol{q}\rangle_\Omega = \langle\boldsymbol{q}\rangle_{\Omega_1}$ where $\boldsymbol{C}_{q,1} = \frac{\boldsymbol{1} - v_2^K \boldsymbol{C}_{q,2}}{1 - v_2^K}$

As mentioned before, the concentration tensors indicate the amplification of the field within the particle relative to the average of the field. Clearly, the microelectric field flux fields are minimally distorted when the $\boldsymbol{C}$'s equal 1. *There has been no approximation yet.* The "burden" in the computations has shifted to the determination of the $\boldsymbol{C}$s. Classical methods approximate them. For example, the simplest approximation is $\boldsymbol{C}_{\theta,2} = \boldsymbol{1}$, which is the Voigt approximation (a constant $\nabla\theta$-field throughout the microstructure, while the Reuss approximation is $\boldsymbol{C}_{q,2} = \boldsymbol{1}$ (a constant $\boldsymbol{q}$-field throughout the microstructure). Directly from Eqs. B.69 and B.70, one may write

$$
\begin{aligned}
\underbrace{v_1^K \boldsymbol{C}_{\theta,1}}_{\text{phase}-1\text{ contribution}} + \underbrace{v_2^K \boldsymbol{C}_{\theta,2}}_{\text{phase}-2\text{ contribution}} &= \boldsymbol{1} \\
\underbrace{v_1^K \boldsymbol{C}_{q,1}}_{\text{phase}-1\text{ contribution}} + \underbrace{v_2^K \boldsymbol{C}_{q,2}}_{\text{phase}-2\text{ contribution}} &= \boldsymbol{1}
\end{aligned}
\tag{B.85}
$$

Frequently, the first term in the above expressions is referred to as "phase-1's" share, while the second term is "phase-2's" share.

References

1. Zohdi, T.I., Monteiro, P.J.M., Lamour, V.: Extraction of elastic moduli from granular compacts. The International Journal of Fracture/Letters in Micromechanics. **115**, L49–L54 (2002)

2. Zohdi, T.I.: Genetic design of solids possessing a random-particulate microstructure. Philos. Trans. R. Soc. Math. Phys. Eng. Sci. **361**(1806), 1021–1043 (2003)
3. Zohdi, T.I.: On the compaction of cohesive hyperelastic granules at finite strains. Proc. R. Soc. **454**(2034), 1395–1401 (2003)
4. Zohdi, T.I.: Constrained inverse formulations in random material design. Comput. Methods Appl. Mech. Eng. **192**(28–30), 3179–3194 (2003)
5. Zohdi, T.I.: Staggering error control for a class of inelastic processes in random microheterogeneous solids. Int. J. Nonlinear Mech. **39**, 281–297 (2004)
6. Zohdi, T.I.: Modeling and simulation of a class of coupled thermo-chemo-mechanical processes in multiphase solids. Comput. Methods Appl. Mech. Eng. **193**(6–8), 679–699 (2004)
7. Zohdi, T.I.: Statistical ensemble error bounds for homogenized microheterogeneous solids. J. Appl. Math. Phys. (Z. Angew. Math. Phys.) **56**(3), 497–515 (2005)
8. Zohdi, T.I., Kachanov, M.: A note on the micromechanics of plastic yield of porous solids. Int. J. Fract. Lett. Micromech. **133**, L31–L35 (2005)
9. Zohdi, T.I.: On the computation of the coupled thermo-electromagnetic response of continua with particulate microstructure. Int. J. Numer. Methods Eng. **76**, 1250–1279 (2008)
10. Zohdi, T.I., Kuypers, F.A., Lee, W.C.: Estimation of red blood cell volume fraction from overall permittivity measurement. Int. J. Eng. Sci. **48**, 1681–1691 (2010)
11. Zohdi, T.I.: Simulation of coupled microscale multiphysical-fields in particulate-doped dielectrics with staggered adaptive FDTD. Comput. Methods Appl. Mech. Eng. **199**, 79–101 (2010)
12. Zohdi, T.I.: Joule-heating field phase-amplification in particulate-doped dielectrics. Int. J. Eng. Sci. **49**, 30–40 (2011)
13. Zohdi, T.I.: Estimation of electrical-heating load-shares for sintering of powder mixtures. Proc. R. Soc. **468**, 2174–2190 (2012)
14. Zohdi, T.I.: On cross-correlation between thermal gradients and electric fields. Int. J. Eng. Sci. **74**, 143–150 (2014)
15. Zohdi, T.I.: Modeling and simulation of cooling-induced residual stresses in heated particulate mixture depositions. Comput. Mech. **56**, 613–630 (2015)
16. Kröner, E.: Statistical Continuum Mechanics. CISM Lecture Notes, vol. 92. Springer-Verlag (1972)
17. Torquato, S.: Random Heterogeneous Materials: Microstructure and Macroscopic Properties. Springer-Verlag, New York (2002)
18. Ghosh, S.: Micromechanical Analysis and Multi-Scale Modeling Using the Voronoi Cell Finite Element Method. CRC Press/Taylor & Francis (2011)
19. Ghosh, S., Dimiduk, D.: Computational Methods for Microstructure-Property Relations. Springer, NY (2011)
20. Hill, R.: The elastic behaviour of a crystalline aggregate. Proc. Phys. Soc. (Lond.) **A65**, 349–354 (1952)

21. Hashin, Z.: Analysis of composite materials: a survey. ASME J. Appl. Mech. **50**, 481–505 (1983)
22. Maxwell, J.C.: On the dynamical theory of gases. Philos. Trans. Soc. London. **157**, 49 (1867)
23. Maxwell, J.C.: A Treatise on Electricity and Magnetism, 3rd edn. Clarendon Press, Oxford (1873)
24. Rayleigh, J.W.: On the influence of obstacles arranged in rectangular order upon properties of a medium. Phil. Mag. **32**, 481–491 (1892)
25. Jikov, V.V., Kozlov, S.M., Olenik, O.A.: Homogenization of Differential Operators and Integral Functionals. Springer-Verlag (1994)
26. Mura, T.: Micromechanics of Defects in Solids, 2nd edn. Kluwer Academic Publishers (1993)
27. Nemat-Nasser, S., Hori, M.: Micromechanics: Overall Properties of Heterogeneous Solids, 2nd edn. Elsevier, Amsterdam (1999)
28. Sevostianov, I., Kachanov, M.: Effective properties of heterogeneous materials: proper application of the non-interaction and the "dilute limit" approximations. Int. J. Eng. Sci. **58**, 124–128 (2012)
29. Wiener, O.: Zur Theorie der Refraktionskonstanten. Berichte über die Verhandlungen der Königlich-Sächsischen Gesellschaft der Wissenshaften zu Leipzig, vol. Math.-phys Klasses, Band 62, pp. 256–277 (1910)

Monograph Appendix C—CM Approaches: Extensions to Multiphase Materials

In order to make estimates of the overall properties of a mixture of more than two materials mixed together (Fig. C.1), we consider extensions to the Hashin–Shtrikman bounds (Hashin and Shtrikman [1]). We consider each major property.

C.1 Electrical Conductivity

Consider a material with N different phases, σ_1, σ_2,...,σ_N, where σ_N has the highest conductivity and σ_1 has the lowest. The corresponding volume fractions are $v_1, v_2, \ldots, v_N$. There are two possible approximation formulas: (1) the Hill–Reuss–Voigt–Weiner bounds and (2) the multiphase Hashin–Shtrikman bounds.

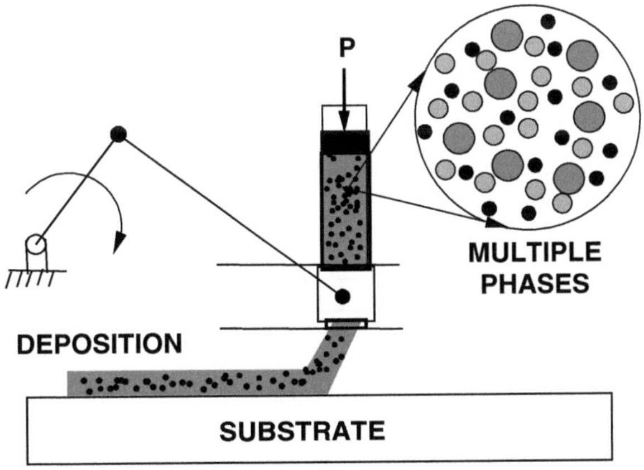

Fig. C.1 A multiphase material

T. I. Zohdi, *Modeling and Simulation of Functionalized Materials for Additive Manufacturing and 3D Printing: Continuous and Discrete Media*, Lecture Notes in Applied and Computational Mechanics 60, https://doi.org/10.1007/978-3-319-70079-3

C.1.1 The Hill–Reuss–Voigt–Weiner (HRVW) Bounds

The HRVW bounds are

$$\left(\sum_{i=1}^{N} \frac{v_i}{\sigma_i}\right)^{-1} \leq \sigma^* \leq \sum_{i=1}^{N} v_i \sigma_i. \tag{C.1}$$

C.1.2 The Hashin–Shtrikman (HS) Bounds

The HS bounds are

$$\sigma^{*,-} = \sigma_1 + \frac{A_1}{1 - \alpha_1 A_1} \leq \sigma^* \leq \sigma_N + \frac{A_N}{1 - \alpha_N A_N} \leq \sigma^{*,+}, \tag{C.2}$$

where

$$\alpha_1 = \frac{1}{3\sigma_1}, \tag{C.3}$$

and

$$\alpha_N = \frac{1}{3\sigma_N} \tag{C.4}$$

and

$$A_1 = \sum_{i=2}^{N} \frac{v_i}{(\sigma_i - \sigma_1)^{-1} + \alpha_1} \tag{C.5}$$

and

$$A_N = \sum_{i=1}^{N-1} \frac{v_i}{(\sigma_i - \sigma_N)^{-1} + \alpha_N} \tag{C.6}$$

For two isotropic materials with an overall isotropic response, this collapses to

$$\underbrace{\sigma_1 + \frac{v_2}{\frac{1}{\sigma_2 - \sigma_1} + \frac{1 - v_2}{3\sigma_1}}}_{\sigma^{*,-}} \leq \sigma^* \leq \underbrace{\sigma_2 + \frac{1 - v_2}{\frac{1}{\sigma_1 - \sigma_2} + \frac{v_2}{3\sigma_2}}}_{\sigma^{*,+}}, \tag{C.7}$$

where the conductivity of phase 2 (with volume fraction v_2) is larger than phase 1 ($\sigma_2 \geq \sigma_1$). Usually, v_2 corresponds to the particle material, although there can be applications where the matrix is more conductive that the particles. In that case, v_2 would correspond to the matrix material.

C.2 Electrical Permittivity

Consider a material with N different phases, $\epsilon_1, \epsilon_2, \ldots, \epsilon_N$, where ϵ_N has the highest permittivity and ϵ_1 has the lowest. The corresponding volume fractions are $v_1, v_2, \ldots, v_N$. There are two possible approximation formulas: (1) the Hill–Reuss–Voigt–Weiner bounds and (2) the multiphase Hashin–Shtrikman bounds.

C.2.1 The Hill–Reuss–Voigt–Weiner Bounds

The HRVW bounds are

$$\left(\sum_{i=1}^{N} \frac{v_i}{\epsilon_i}\right)^{-1} \leq \epsilon^* \leq \sum_{i=1}^{N} v_i \epsilon_i. \tag{C.8}$$

C.2.2 The Hashin–Shtrikman Bounds

The HS bounds are

$$\epsilon^{*,-} = \epsilon_1 + \frac{A_1}{1 - \alpha_1 A_1} \leq \epsilon^* \leq \epsilon_N + \frac{A_N}{1 - \alpha_N A_N} \leq \epsilon^{*,+}, \tag{C.9}$$

where

$$\alpha_1 = \frac{1}{3\epsilon_1}, \tag{C.10}$$

and

$$\alpha_N = \frac{1}{3\epsilon_N} \tag{C.11}$$

and

$$A_1 = \sum_{i=2}^{N} \frac{v_i}{(\epsilon_i - \epsilon_1)^{-1} + \alpha_1} \tag{C.12}$$

and

$$A_N = \sum_{i=1}^{N-1} \frac{v_i}{(\epsilon_i - \epsilon_N)^{-1} + \alpha_N} \tag{C.13}$$

For two isotropic materials with an overall isotropic response, this collapses to

$$\underbrace{\epsilon_1 + \frac{v_2}{\frac{1}{\epsilon_2-\epsilon_1} + \frac{1-v_2}{3\epsilon_1}}}_{\epsilon^{*,-}} \leq \epsilon^* \leq \underbrace{\epsilon_2 + \frac{1-v_2}{\frac{1}{\epsilon_1-\epsilon_2} + \frac{v_2}{3\epsilon_2}}}_{\epsilon^{*,+}}. \tag{C.14}$$

C.3 Magnetic Permeability

Consider a material with N different phases, $\mu_1, \mu_2, \ldots, \mu_N$, where μ_N has the highest permeability and μ_1 has the lowest. The corresponding volume fractions are $v_1, v_2, \ldots, v_N$. There are two possible approximation formulas: (1) the Hill–Reuss–Voigt–Weiner bounds and (2) the multiphase Hashin–Shtrikman bounds.

C.3.1 The Hill–Reuss–Voigt–Weiner Bounds

The HRVW bounds are

$$\left(\sum_{i=1}^{N} \frac{v_i}{\mu_i}\right)^{-1} \leq \mu^* \leq \sum_{i=1}^{N} v_i \mu_i. \tag{C.15}$$

C.3.2 The Hashin–Shtrikman Bounds

The HS bounds are

$$\mu^{*,-} - \mu_1 + \frac{A_1}{1-\alpha_1 A_1} \leq \mu^* \leq \mu_N + \frac{A_N}{1-\alpha_N A_N} \leq \mu^{*,+}, \tag{C.16}$$

where

$$\alpha_1 = \frac{1}{3\mu_1}, \tag{C.17}$$

and

$$\alpha_N = \frac{1}{3\mu_N} \tag{C.18}$$

and

$$A_1 = \sum_{i=2}^{N} \frac{v_i}{(\mu_i - \mu_1)^{-1} + \alpha_1} \tag{C.19}$$

and

$$A_N = \sum_{i=1}^{N-1} \frac{v_i}{(\mu_i - \mu_N)^{-1} + \alpha_N} \tag{C.20}$$

For two isotropic materials with an overall isotropic response, this collapses to

$$\underbrace{\mu_1 + \frac{v_2}{\frac{1}{\mu_2-\mu_1} + \frac{1-v_2}{3\mu_1}}}_{\mu^{*,-}} \leq \mu^* \leq \underbrace{\mu_2 + \frac{1-v_2}{\frac{1}{\mu_1-\mu_2} + \frac{v_2}{3\mu_2}}}_{\mu^{*,+}}. \tag{C.21}$$

C.4 Thermal Conductivity

Consider a material with N different phases, $\mathbb{K}_1, \mathbb{K}_2, \ldots, \mathbb{K}_N$, where $\mathbb{K}_N$ has the highest thermal conductivity and $\mathbb{K}_1$ has the lowest. The corresponding volume fractions are $v_1, v_2, \ldots, v_N$. There are two possible approximation formulas: (1) the Hill–Reuss–Voigt–Weiner bounds and (2) the multiphase Hashin–Shtrikman bounds.

C.4.1 The Hill–Reuss–Voigt–Weiner Bounds

The HRVW bounds are

$$\left(\sum_{i=1}^{N} \frac{v_i}{\mathbb{K}_i}\right)^{-1} \leq \mathbb{K}^* \leq \sum_{i=1}^{N} v_i \mathbb{K}_i. \tag{C.22}$$

C.4.2 The Hashin–Shtrikman Bounds

The HS bounds are

$$\mathbb{K}^{*,-} = \mathbb{K}_1 + \frac{A_1}{1 - \alpha_1 A_1} \leq \mathbb{K}^* \leq \mathbb{K}_N + \frac{A_N}{1 - \alpha_N A_N} \leq \mathbb{K}^{*,+}, \tag{C.23}$$

where

$$\alpha_1 = \frac{1}{3\mathbb{K}_1}, \tag{C.24}$$

and

$$\alpha_N = \frac{1}{3\mathbb{K}_N} \tag{C.25}$$

and

$$A_1 = \sum_{i=2}^{N} \frac{v_i}{(\mathbb{K}_i - \mathbb{K}_1)^{-1} + \alpha_1} \tag{C.26}$$

and

$$A_N = \sum_{i=1}^{N-1} \frac{v_i}{(\mathbb{K}_i - \mathbb{K}_N)^{-1} + \alpha_N} \tag{C.27}$$

For two isotropic materials with an overall isotropic response, this collapses to

$$\underbrace{\mathbb{K}_1 + \frac{v_2}{\frac{1}{\mathbb{K}_2 - \mathbb{K}_1} + \frac{1-v_2}{3\mathbb{K}_1}}}_{\mathbb{K}^{*,-}} \leq \mathbb{K}^* \leq \underbrace{\mathbb{K}_2 + \frac{1-v_2}{\frac{1}{\mathbb{K}_1 - \mathbb{K}_2} + \frac{v_2}{3\mathbb{K}_2}}}_{\mathbb{K}^{*,+}}. \tag{C.28}$$

C.5 Elastic Moduli

C.5.1 Bulk Modulus

Consider a material with N different phases, $\kappa_1, \kappa_2, \ldots, \kappa_N$, where κ_N has the highest bulk modulus and κ_1 has the lowest. The corresponding volume fractions are $v_1, v_2, \ldots, v_N$. There are two possible approximation formulas: (1) the Hill–Reuss–Voigt–Weiner bounds and (2) the multiphase Hashin–Shtrikman bounds.

C.5.1.1 The Hill–Reuss–Voigt–Weiner Bounds

The HRVW bounds are

$$\left(\sum_{i=1}^{N} \frac{v_i}{\kappa_i}\right)^{-1} \leq \kappa^* \leq \sum_{i=1}^{N} v_i \kappa_i. \tag{C.29}$$

C.5.1.2 The Hashin–Shtrikman Bounds

The HS bounds are

$$\kappa^{*,-} = \kappa_1 + \frac{A_1}{1 - \alpha_1 A_1} \leq \kappa^* \leq \kappa_N + \frac{A_N}{1 - \alpha_N A_N} \leq \kappa^{*,+}, \tag{C.30}$$

where

$$\alpha_1 = \frac{3}{3\kappa_1 + 4\mu_1}, \tag{C.31}$$

and

$$\alpha_N = \frac{3}{3\kappa_N + 4\mu_N}, \tag{C.32}$$

and

$$A_1 = \sum_{i=2}^{N} \frac{v_i}{(\kappa_i - \kappa_1)^{-1} + \alpha_1} \tag{C.33}$$

and

$$A_N = \sum_{i=1}^{N-1} \frac{v_i}{(\kappa_i - \kappa_N)^{-1} + \alpha_N} \tag{C.34}$$

For a two-phase microstructure, this collapses to

$$\underbrace{\kappa_1 + \frac{v_2}{\frac{1}{\kappa_2-\kappa_1} + \frac{3(1-v_2)}{3\kappa_1+4\mu_1}}}_{\text{bulk modulus H–S lower bound} \stackrel{\text{def}}{=} \kappa^{*,-}} \leq \kappa^* \leq \underbrace{\kappa_2 + \frac{1-v_2}{\frac{1}{\kappa_1-\kappa_2} + \frac{3v_2}{3\kappa_2+4\mu_2}}}_{\text{bulk modulus H–S upper bound} \stackrel{\text{def}}{=} \kappa^{*,+}}, \tag{C.35}$$

where κ_1, μ_1 and κ_2, μ_2 are the bulk and shear moduli for the phases, while v_2 is the phase 2 volume fraction.

C.5.2 Shear Modulus

Consider a material with N different phases, $\mu_1, \mu_2, \ldots, \mu_N$, where μ_N has the highest shear modulus and μ_1 has the lowest. The corresponding volume fractions are $v_1, v_2, \ldots, v_N$. There are two possible approximation formulas: (1) the Hill–Reuss–Voigt–Weiner bounds and (2) the multiphase Hashin–Shtrikman bounds.

C.5.2.1 The Hill–Reuss–Voigt–Weiner Bounds

The HRVW bounds are

$$\left(\sum_{i=1}^{N} \frac{v_i}{\mu_i}\right)^{-1} \leq \mu^* \leq \sum_{i=1}^{N} v_i \mu_i. \tag{C.36}$$

C.5.2.2 The Hashin–Shtrikman Bounds

The HS bounds are

$$\mu^{*,-} = \mu_1 + \frac{B_1}{1-\gamma_1 B_1} \leq \mu^* \leq \mu_N + \frac{B_N}{1-\gamma_N B_N} \leq \mu^{*,+}, \tag{C.37}$$

where

$$\gamma_1 = \frac{3(\kappa_1+\mu_1)}{5\mu_1(3\kappa_1+4\mu_1)}, \tag{C.38}$$

and

$$\gamma_N = \frac{3(\kappa_N+\mu_N)}{5\mu_N(3\kappa_N+4\mu_N)}, \tag{C.39}$$

and

$$B_1 = \sum_{i=2}^{N} \frac{v_i}{(\mu_i-\mu_1)^{-1}+\gamma_1} \tag{C.40}$$

and

$$B_N = \sum_{i=1}^{N-1} \frac{v_i}{(\mu_i-\mu_N)^{-1}+\gamma_N} \tag{C.41}$$

For a two-phase microstructure, this collapses to

$$\underbrace{\mu_1 + \frac{v_2}{\frac{1}{\mu_2-\mu_1}+\frac{6(1-v_2)(\kappa_1+2\mu_1)}{5\mu_1(3\kappa_1+4\mu_1)}}}_{\text{shear modulus H}-\text{S lower bound}\stackrel{\text{def}}{=}-*,-} \leq \mu^* \leq \underbrace{\mu_2 + \frac{(1-v_2)}{\frac{1}{\mu_1-\mu_2}+\frac{6v_2(\kappa_2+2\mu_2)}{5\mu_2(3\kappa_2+4\mu_2)}}}_{\text{shear modulus H}-\text{S upper bound}\stackrel{\text{def}}{=}-*,+}. \tag{C.42}$$

C.6 Concentration Tensors for Multiphase materials

As for a two-phase material, the load carried by each phase in the microstructure is characterized via stress and strain concentration tensors, which we now discuss. These provide a measure of the deviation away from the mean fields throughout the material. One can decompose averages of an arbitrary quantity over Ω into averages over each of the phases in the following manner, recall:

$$\begin{aligned}\langle \boldsymbol{A}\rangle_{\Omega} &= (1/|\Omega|)\left(\int_{\Omega_1}\boldsymbol{A}\,d\Omega + \int_{\Omega_2}\boldsymbol{A}\,d\Omega + \cdots + \int_{\Omega_N}\boldsymbol{A}\,d\Omega\right) \\ &= v_1\langle \boldsymbol{A}\rangle_{\Omega_1} + v_2\langle \boldsymbol{A}\rangle_{\Omega_2} + \cdots = \sum_{i=1}^{N} v_i\langle \boldsymbol{A}\rangle_{\Omega_i}.\end{aligned} \tag{C.43}$$

If we make use of this decomposition, we have

$$
\begin{aligned}
\langle \boldsymbol{\sigma} \rangle_{\Omega} = \sum_{i=1}^{N} v_i \langle \boldsymbol{\sigma} \rangle_{\Omega_i} &= \sum_{i=1}^{N} v_i \boldsymbol{I\!E}_i : \langle \boldsymbol{\epsilon} \rangle_{\Omega_i} \\
&= \boldsymbol{I\!E}_1 : (\langle \boldsymbol{\epsilon} \rangle_{\Omega} - \sum_{j=2}^{N} v_j \langle \boldsymbol{\epsilon} \rangle_{\Omega_j}) + \sum_{j=2}^{N} v_j \boldsymbol{I\!E}_j : \langle \boldsymbol{\epsilon} \rangle_{\Omega_j} \\
&= \left(\boldsymbol{I\!E}_1 + \sum_{j=2}^{N} v_j (\boldsymbol{I\!E}_j - \boldsymbol{I\!E}_1) : \boldsymbol{C}^{\epsilon,j} \right) : \langle \boldsymbol{\epsilon} \rangle_{\Omega}
\end{aligned}
\tag{C.44}
$$

where $\boldsymbol{C}^{\epsilon,j} \stackrel{\text{def}}{=} \left(\frac{1}{v_j} (\boldsymbol{I\!E}_j - \boldsymbol{I\!E}_1)^{-1} : (\boldsymbol{I\!E}^* - \boldsymbol{I\!E}_j) \right)$ with $\boldsymbol{C}^{\epsilon,j} : \langle \boldsymbol{\epsilon} \rangle_{\Omega} = \langle \boldsymbol{\epsilon} \rangle_{\Omega_j}$. The strain concentration tensor $\boldsymbol{C}^{\epsilon,j}$ relates the average strain over the phase (j) to the average strain over all phases. Similarly, for the variation in the stress, we have $\boldsymbol{C}^{\epsilon,j} : \boldsymbol{I\!E}^{*-1} : \langle \boldsymbol{\sigma} \rangle_{\Omega} = \boldsymbol{I\!E}_j^{-1} : \langle \boldsymbol{\sigma} \rangle_{\Omega_j}$, which reduces to $\boldsymbol{I\!E}_j : \boldsymbol{C}^{\epsilon,j} : \boldsymbol{I\!E}^{*-1} : \langle \boldsymbol{\sigma} \rangle_{\Omega} \stackrel{\text{def}}{=} \boldsymbol{C}^{\sigma,j} : \langle \boldsymbol{\sigma} \rangle_{\Omega} = \langle \boldsymbol{\sigma} \rangle_{\Omega_j}$. $\boldsymbol{C}^{\sigma,j}$ is known as the stress concentration tensor; it relates the average stress in the particle phase to that in the whole RVE. Note that once either the $\boldsymbol{C}^{\sigma,j}$s or $\boldsymbol{I\!E}^*$ are known, the other can be determined. In the case of isotropy, we may write ($j \neq 1$)

$$
C_{\kappa}^{\sigma,j} \stackrel{\text{def}}{=} \frac{1}{v_j} \frac{\kappa_j}{\kappa^*} \frac{\kappa^* - \kappa_1}{\kappa_j - \kappa_1} \quad \text{and} \quad C_{\mu}^{\sigma,j} \stackrel{\text{def}}{=} \frac{1}{v_j} \frac{\mu_j}{\mu^*} \frac{\mu^* - \mu_1}{\mu_j - \mu_1}
\tag{C.45}
$$

where $C_{\kappa}^{\sigma,j} \langle \frac{tr\boldsymbol{\sigma}}{3} \rangle_{\Omega} = \langle \frac{tr\boldsymbol{\sigma}}{3} \rangle_{\Omega_j}$ and $C_{\mu}^{\sigma,j} \langle \boldsymbol{\sigma}' \rangle_{\Omega} = \langle \boldsymbol{\sigma}' \rangle_{\Omega_j}$. Clearly, the microstress fields are minimally distorted when $C_{\kappa}^{\sigma,j} = C_{j,\mu}^{\sigma,2} = 1$; there are no stress concentrations in a homogeneous material. For the matrix,

$$
\begin{aligned}
\langle \boldsymbol{\sigma} \rangle_{\Omega_1} &= \frac{\langle \boldsymbol{\sigma} \rangle_{\Omega} - \sum_{j=2}^{N} v_j \langle \boldsymbol{\sigma} \rangle_{\Omega_j}}{v_1} = \frac{\langle \boldsymbol{\sigma} \rangle_{\Omega} - \sum_{j=2}^{N} v_j \boldsymbol{C}^{\sigma,j} : \langle \boldsymbol{\sigma} \rangle_{\Omega}}{v_1} \\
&= \frac{(\boldsymbol{1} - \sum_{j=2}^{N} v_j \boldsymbol{C}^{\sigma,j}) : \langle \boldsymbol{\sigma} \rangle_{\Omega}}{v_1} \stackrel{\text{def}}{=} \boldsymbol{C}^{\sigma,1} : \langle \boldsymbol{\sigma} \rangle_{\Omega}.
\end{aligned}
\tag{C.46}
$$

Therefore, in the case of isotropy,

$$
C_{\kappa}^{\sigma,1} \stackrel{\text{def}}{=} \frac{1}{v_1} (1 - \sum_{j=2}^{N} v_j C_{\kappa}^{\sigma,j}) \quad \text{and} \quad C_{\mu}^{\sigma,1} \stackrel{\text{def}}{=} \frac{1}{v_1} (1 - \sum_{j=2}^{N} v_j C_{\mu}^{\sigma,j}).
\tag{C.47}
$$

The fraction of the total stress carried by each phase can be determined by multiplying the concentration factors by the corresponding volume fractions

$$\langle \boldsymbol{\sigma} \rangle_{\Omega} = v_1 \langle \boldsymbol{\sigma} \rangle_{\Omega_1} + \sum_{j=2}^{N} v_j \langle \boldsymbol{\sigma} \rangle_{\Omega_j}$$

$$= v_1 \boldsymbol{C}^{\sigma,1} : \langle \boldsymbol{\sigma} \rangle_{\Omega} + \sum_{j=2}^{N} v_j \boldsymbol{C}^{\sigma,j} : \langle \boldsymbol{\sigma} \rangle_{\Omega}. \tag{C.48}$$

Remark Virtually, the identical process can be used to generate the concentration tensors for the electrical conductivity, electrical permittivity, magnetic permeability, and thermal conductivity.

References

1. Hashin, Z., Shtrikman, S.: A variational approach to the theory of the elastic behaviour of multiphase materials. J. Mech. Phys. Solids **11**, 127–140 (1963)

Monograph Appendix D—Pumping of Fluidized Particle-Laden Materials

D.1 Introduction

In addition to additive manufacturing and 3D printing, in a variety of industries, ranging from next-generation engines, turbo machinery, food processing, etc, new types of heterogeneous materials, comprised of particulates in a solvent or binding surrounding matrix, are being developed and utilized. The macroscopic material characteristics are dictated by the aggregate response of an assemblage of particles suspended in a binding matrix material. In the fabrication of such materials, the basic philosophy is to select material combinations to produce desired aggregate responses. For example, in structural engineering applications, the classical choice is a harder particulate phase that serves as a stiffening agent for a ductile, easy to form, base matrix material. Oftentimes, such materials start in particulate form and are then mixed with a binder and delivered as a flowing slurry to be cast into their final shape.[6] Thus, because of the increasing demands for faster and faster manufacturing of new complex particle-laden materials, the determination of pumping pressures needed to move such fluids through channels is critical (Fig. D.1).

For particle-laden fluids delivered through channels, the increase in viscosity can lead to system malfunction, due to an inability to supply necessary pressures to pump the more viscous material properly. The derived expression provides the pressure gradient needed to maintain a given flow rate, explicitly as a function of the volume fraction of particles present in the fluid. The expression is general and easy to apply for the analysis of pumping particle-laden fluids. Furthermore, it is crucial to control voids in the resulting casted products, which are correlated to air entrainment, spurious internal reactions, de-wetting, etc. These effects are correlated to high Reynolds numbers. Accordingly, an expression for the resulting Reynolds number as a function of the particle volume fraction and flow rate is also developed. Numerical examples are provided to illustrate the practical use of the derived relations

[6]Over 50% (by mass) of man-made materials start in granulated form.

T. I. Zohdi, *Modeling and Simulation of Functionalized Materials for Additive Manufacturing and 3D Printing: Continuous and Discrete Media*, Lecture Notes in Applied and Computational Mechanics 60, https://doi.org/10.1007/978-3-319-70079-3

to characterize the necessary pumping pressures for process-driven particle-laden fluid flows. Because resulting voids may be impossible to avoid, we can use the expressions derived in the beginning of this monograph that provide the overall effective properties of a heterogeneous two-phase slurry consisting of particles and a binding interstitial material based on embedded, double application of the Hashin–Shtrikman bounds, whereby, on the first level, the effective properties due to voids are computed, and on the second level, the smaller-scale heterogeneous material is taken into account. This research is also quite relevant to the development of high-resolution electrohydrodynamic jet printing processes. For overviews, see Wei and Dong [1], who also develop specialized processes employing phase-change inks. Such processes are capable of producing micron-level footprints for high-resolution additive manufacturing.

D.2 Channel Flow

As indicated in the introduction, the presence of secondary particles in fluids, particularly within channels, is wide-ranging and their presence can dramatically increase the effective overall viscosity, thus requiring increased applied pressure to maintain nominal flow rates (Fig. D.1). The primary objective of the first part of this analysis is to derive a relatively easy to use expression for the pressure gradient required to maintain a given flow rate in a channel, as a function of the volume fraction of secondary particles present in the fluid.

Accordingly, consider an idealized channel with a circular cross section of area $A = \pi R^2$, with a velocity profile given by a classical channel flow of the form:

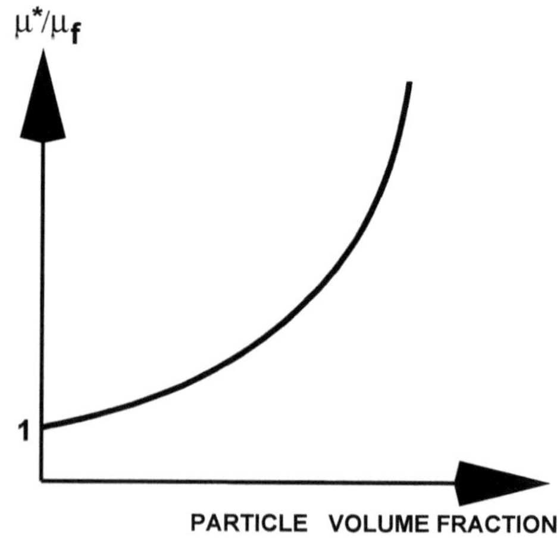

Fig. D.1 Increase in the ratio of effective viscosity to baseline fluid viscosity (μ^*/μ_f) as a function of secondary particle volume fraction (v_p)

$$v = v_{\max}\left(1 - \left(\frac{r}{R}\right)^q\right), \tag{D.1}$$

where $v_{\max}$ is the centerline velocity and r is the radial coordinate from the centerline of the channel. For fully developed laminar flow, $q = 2$, while for increasing q, one characterizes, phenomenologically, progressively turbulent flow ($q \geq 2$). The shear stress is given by

$$\tau = \mu^* \frac{\partial v}{\partial r} = -\frac{\mu^* v_{\max} q}{R}\left(\frac{r}{R}\right)^{q-1}, \tag{D.2}$$

where μ^* is the effective viscosity of the particle-laden fluid. We assume that the overall flow rate is assumed constant; thus,

$$Q = \int_A v\, dA = Q_o. \tag{D.3}$$

One can show that

$$v_{\max} = \frac{Q_o(q+2)}{Aq} = \frac{Q_o(q+2)}{\pi R^2 q}. \tag{D.4}$$

The stress at the wall becomes

$$\tau_w = -\tau(r = R) = \frac{\mu^* v_{\max} q}{R} = \frac{\mu^* Q_o(q+2)}{\pi R^3}. \tag{D.5}$$

We have the following observations:

- Increasing μ^*, Q_o, or q increases the stress at the wall (τ_w),
- Increasing q leads to an increasingly more blunted flow profile, and
- Decreasing R increases the stress at the wall (τ_w).

Remark In the remaining analysis, we will assume steady flow, that the particles are not elongated and that they are well distributed within the base fluid.[7] Furthermore, we will adopt a generalization of the classical Poiseuille solution for fully developed flow in a pipe (assuming the velocity depends on some undetermined power q instead of the standard parabolic dependence for laminar single-phase flow).

D.3 Pressure Gradients

The previous expressions allow us to correlate the pressure applied to a volume of particle-laden to allow it to move at a constant flow rate. By performing a force balance, we have in the positive x-direction (assuming steady flow, no acceleration)

[7]In long channels, elongated particles can tend to align themselves in a particular direction that could also affect their viscosity. The assumptions made eliminate this possibility for the problems under consideration.

$$(-(P + \Delta P) + P)\pi R^2 - \tau_w 2\pi R \Delta x = 0, \tag{D.6}$$

where x is the coordinate along the length of the channel and Δx is the differential length, leading to

$$-\Delta P = \mu^* \frac{Q_o(q+2)}{\pi^2 R^5} 2\pi R \Delta x = \frac{2\mu^* Q_o(q+2)\Delta x}{\pi R^4}, \tag{D.7}$$

where we used the expression for v_{max} and the effective viscosity is a function of the volume fraction of particles, $\mu^* = \mu^*(v_p)$. An explicit relation for $\mu^*(v_p)$ will be given shortly. Solving for the pressure gradient yields

$$-\frac{\Delta P}{\Delta x} = \underbrace{\frac{2\mu^*(q+2)}{\pi R^4}}_{C} Q_o \stackrel{\text{def}}{=} C Q_o. \tag{D.8}$$

If we fix the flow rate Q_o, the multiplier C identifies the pressure gradient needed to achieve the flow rate Q_o. For a fixed value of q, the expression directly indicates that an increase in viscosity will require an increase in the pressure gradient. For small channels, this can be a problem, as indicated by the R^4 term in the denominator. However, in general, q is a function of the Reynolds number. This case will be considered next.

D.4 Velocity Profile Characteristics

As the Reynolds number increases, the velocity profile will change from a quadratic ($q = 2$) to a more blunted profile ($q >> 2$), which represents, phenomenologically, turbulent (inertia-dominated) behavior (Fig. D.2). The effect of a changing profile is described by representing q by a linear function of the centerline Reynolds number ($\mathcal{R}_{ec}$)

$$q = q(\mathcal{R}_{ec}) = c_1 \mathcal{R}_{ec} + c_2, \tag{D.9}$$

where $\mathcal{R}_{ec} = \frac{\rho^* v_{max} 2R}{\mu^*}$ and c_1 and c_2 are constants. Models of this type, linking the profile exponent (q) to the centerline Reynolds number ($\mathcal{R}_{ec}$), are quite well-established, for example, see Hinze [2]. Usually, $0 \leq c_1 << 1$ and $c_2 \approx 2$, and in

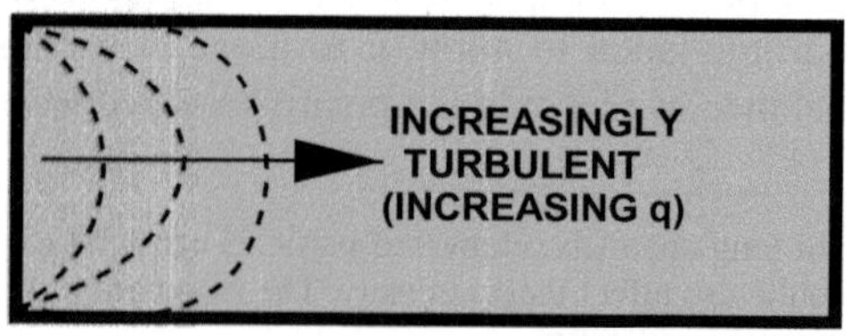

Fig. D.2 Progressive blunting of the velocity profile with increasing Reynolds number

the limit, we have, for $c_1 = 0$ and $c_2 = 2$, laminar flow ($q = 2$). For the general case, combining Eq. D.4 with Eq. D.9 and the definition of the centerline Reynolds number, we obtain a quadratic relationship for q,

$$q^2 - (\gamma^* + c_2)q - 2\gamma^* = 0, \tag{D.10}$$

where $\gamma^* = \frac{2c_1 Q_o \rho^*}{\pi R \mu^*}$, where ρ^* is the effective density and μ^* is the effective viscosity. This quadratic relationship can be solved in closed form for q to yield

$$q(\mathcal{R}_{ec}) = \frac{1}{2}\left((\gamma^* + c_2) \pm \sqrt{(\gamma^* + c_2)^2 + 8\gamma^*}\right). \tag{D.11}$$

The larger root is the physically correct choice (since the smaller root can become negative)[8]. We further observe that $q(\mathcal{R}_{ec})$ is a function of R^{-1} and decreasing R increases q, for fixed Q_o.

D.5 Models for Effective Properties of Particle-Laden Fluids

D.5.1 Effective Density

It is important to be able to characterize the effective properties of a particle-laden fluid as a function of the volume fraction of particles and the baseline (interstitial) fluid properties. The density of the particle-laden fluid is actually an "effective density," since it actually is a mixture of materials (particles and interstitial fluid). Effective properties are defined through volume averages. For example, the effective density of the mixture is

$$\rho^* \stackrel{\text{def}}{=} \langle \rho(\boldsymbol{x}) \rangle_V \stackrel{\text{def}}{=} \frac{1}{V}\int_V \rho(\boldsymbol{x})\, dV = \frac{1}{V}\left(\int_{V_f} \rho_f\, dV + \int_{V_p} \rho_p\, dV\right) = v_f \rho_f + v_p \rho_p \tag{D.12}$$

where v_f and v_p are the volume fractions of the fluid and particles, respectively. The volume fractions have to sum to unity:

$$v_f + v_p = 1 \Rightarrow v_f = 1 - v_p \tag{D.13}$$

[8]In the special case of laminar flow ($c_1 = 0$ and $c_2 = 2$), there are two roots to Eq. D.16, $q = 2$ and $q = 0$.

D.5.2 *Ancillary Effective Viscosities*

Similar approaches can be used to calculate various types of properties, such as the effective viscosity (a transport property). However, to calculate them is a bit more complicated, since they require one to estimate the types of interaction between the constituents. There are a number of models which provide expressions for the effective viscosity of the fluid containing particles. For the purposes of this flow analysis, the particles are considered to be rigid, relative to the surrounding fluid. For example, in 1906, Einstein [3] developed an approximation which is quite simple, but only valid at extremely low volume fractions of particles (under one percent). It reads as

$$\mu^* = \mu_f(1 + 2.5v_p), \tag{D.14}$$

where μ_f is the viscosity of the surrounding (incompressible) fluid and the particles are assumed rigid. At even quite moderate volume fractions, this approximation is inaccurate. A better approximation, which is in fact a rigorous lower bound on the effective viscosity, can be derived from the well-known Hashin and Shtrikman [4–6] bounds. To derive effective fluid viscosities, one can take the limit of the particle phase becoming rigid, i.e., the bulk and shear moduli tending toward infinity, $\kappa_p \rightarrow \infty$ and $\mu_p \rightarrow \infty$, signifying that the particles are much stiffer than the interstitial fluid, while simultaneously specifying that the interstitial fluid is incompressible, i.e., $\kappa_f/\mu_f \rightarrow \infty$ with μ_f being finite. This yields

$$\mu^* = \mu_f(1 + 2.5\frac{v_p}{1 - v_p}). \tag{D.15}$$

One can then assign μ_f the value of μ_f to obtain Eq. D.15. See, for example, Abedian and Kachanov [7] and Sevostianov and Kachanov [8] for more details. The expression in Eq. D.15 is the tightest known lower bound on the effective viscosity of a two-phase material comprised of rigid particles in a surrounding incompressible fluid. This expression remains quite accurate up to about $v_p = 0.25$, which is sufficient for most applications and allows us to directly correlate the pressure gradient to the volume fraction of the particles.

D.6 Correlation of Pressure Gradient to Particle Volume Fraction

Using the effective properties, we have an expression for the velocity profile exponent

$$q(\mathcal{R}_{ec}(\mu^*, \rho^*), \gamma^*) = \frac{1}{2}\left((\gamma^* + c_2) \pm \sqrt{(\gamma^* + c_2)^2 + 8\gamma^*}\right). \tag{D.16}$$

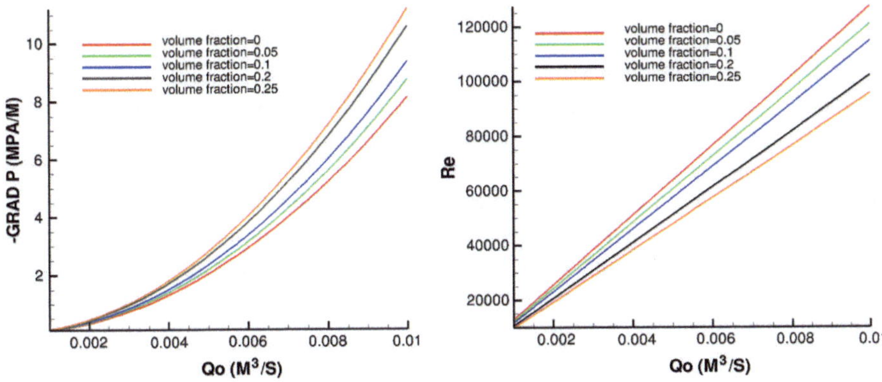

Fig. D.3 Trends: Left: Pressure gradient needed ($-\frac{\Delta P}{\Delta x}$) as a function of the desired volumetric flow rate (Q_o) for various volume fractions of v_p. Right: Resulting Reynolds number as a function of the volumetric flow rate (Q_o)

Consequently, the pressure gradient's dependency on the volume fraction of particles can be written as

$$-\frac{\Delta P}{\Delta x} = \frac{2(\mu_f(1 + 2.5\frac{v_p}{1-v_p}))(q(\mathcal{R}_{ec}(\mu^*, \rho^*), \gamma^*) + 2)}{\pi R^4} Q_o \stackrel{\text{def}}{=} C^* Q_o, \quad \text{(D.17)}$$

where $C^* = C^*(Q_o)$. For a fixed flow rate, Q_o, increasing the volume fraction of particles (v_p) requires a corresponding increase in the pressure differential. Explicitly, the Reynolds number is

$$R_e = \frac{v_{max} D \rho^*}{\mu^*} = \frac{2Q_o(q+2)}{\pi R q} \frac{((1 - v_p)\rho_f + v_p \rho_p)}{\mu_f(1 + 2.5\frac{v_p}{1-v_p})}. \quad \text{(D.18)}$$

D.7 Trends

To illustrate the trends, we varied Q_o from 10^{-3} m^3/s to 10^{-2} m^3/s and utilized the expression in Eq. D.17. We plotted the pressure gradient and Reynolds number as a function of the volumetric flow rate (Q_o) in Fig. D.3 for various values of v_p, with the following parameters used:[9]

- Viscosity: $\mu_f = 0.01$ Pa-s,
- Fluid density: $\rho_f = 2000$ kg/m^3,
- Particle density: $\rho_p = 5000$ kg/m^3,
- Channel radius: $R = 0.01$ m, and

[9]For reference, the viscosity of water is $\mu_f = 0.001$ Pa–s and for honey is $\mu_f = 1$ Pa–s.

- Profile constants: $c_1 = 0.01$ and $c_2 = 2$.

Generally, the trends are that a steady increase in the pressure gradient (approximately 40% more) is needed to maintain a fixed Q_o, for increasing volume fraction of particles. Due to the increase in the particle volume fraction, the viscosity increases, thus decreasing the Reynolds number. High Reynolds numbers, and consequential turbulence, can lead to aspiration (air entrainment), spurious internal reactions, de-wetting, etc., which can lead to voids. The point of this example was not to illustrate an all-encompassing parameter set, but simply to show the explicit dependency of the pressure gradient and Reynolds number on the presence of secondary particles. Other parameter sets can be easily simulated.

D.8 Summary

The presence of particle-laden fluids is widespread. Because the presence of particles increases the overall viscosity of the fluid, the pressure gradients needed to pump such fluids through channels at a nominal flow rate can increase dramatically. The present analysis and model can provide a useful guide to designing systems that pump particle-laden flows, with the purpose to be able to cast materials. This chapter derived the pressure gradient needed to maintain a given flow rate, as a function volume fraction of particles present in the fluid. The expression explicitly correlates the dependency of the pressure gradient to the particle volume fraction and is hopefully easy to use by researchers in the field. Furthermore, the developed expressions also provide estimates on the Reynolds numbers that arise for given flow rates. The tracking of the Reynolds number is important, since turbulence can lead to improper processing due to the resulting voids.

A fluidized particle-laden material is often referred to as a "slurry." In general terms, a slurry is comprised of an interstitial fluid (solvent/lubricant/heated binder) with viscosity μ_f and rigid particles (effectively rigid compared to the surrounding fluidized solvent/lubricant/heated binder material). Generally, there are several options for improving a fluidized particle-laden delivery process:

- The use of particle coatings (surfactants) to make them slippery and to reduce the overall viscosity of the particle-laden slurry.
- The decreasing of the viscosity, either by higher heating or adding solvents/lubricants. A slurry viscosity decreases *exponentially* with temperature

$$\mu_f = \mu_o e^{\frac{U}{R\theta}}, \tag{D.19}$$

where μ_o is the base viscosity, U is the activation energy, R is the universal gas constant, and θ is the absolute temperature. *Thus, any change in temperature can dramatically change the viscosity.*

- The use of specially designed funnels that allow air to escape, which is designed to keep the flow rate constant to reduce aspiration. An option is the adaptive choke mechanism and controlled pouring.
- The addition of degassing chemicals to the process: There are of course many issues (potential increased cured shrinkage, possible reduction of strength).
- The use of degassing propellers and direct feed lines into the system (this is often done in other industries that use slurries).
- Evacuating the print area of all air a priori to eliminate air in the deposition.
- Carefully controlled cooling of the base of the substrate: Thick sections will cure at different rates than thin sections, due to different temperature gradients, thus leading to inhomogeneous, possibly dendritic, curing. This could be achieved by heating the substrate during curing and also modifying the conductivity.

While these options are not explicitly discussed in this chapter, they should be considered by a practitioner. Furthermore, we mention that oftentimes the detrimental growth of channel walls (thus clogging feed lines) starts with the adhesion of particles to the surfaces. This is a complex process, which is likely to involve low fluid-induced shear stress (allowing particles stick to the walls, Zohdi [9–13]).

In summary, the present analysis and model can provide a useful guide to designing and interpreting experiments. However, while the model can provide qualitative information, extensions are almost certainly going to require complex spatiotemporal discretization resolving multiparticle particle–fluid interaction. Such particle/fluid systems are strongly coupled, due to the drag forces induced by the fluid onto the particles and vice versa. For example, in Zohdi [9–13], a flexible and robust solution strategy was developed to resolve coupled systems comprised of large groups of flowing particles embedded within a continuous flowing fluid. The focus of that work was to develop adaptive time-stepping schemes which properly resolve the coupling, via a staggered recursive time-stepping process. The approach can be used in conjunction with computational fluid mechanics codes based on finite difference, finite element, finite volume, or discrete element discretization, such as those developed in Onate et al. [14, 15], Rojek et al. [16], Carbonell et al. [17], Labra and Onate [18], Leonardi et al. [19], Cante et al. [20], Rojek [21], Onate et al. [22], Bolintineanu et al. [23], Avci and Wriggers [24], and Zohdi [9–13]. Clearly, one could approach the problem with a large-scale CFD analysis. However, for direct numerical simulation of particle-laden continua, spatiotemporal discretization grids must be extremely fine, with several thousand numerical unknowns needed per particle length scale for numerically accurate results. Thus, for several hundred thousand particles in a system, a proper discretization would require several billion numerical unknowns. Although such simulations are possible in high-performance computing centers, their usefulness for rapid daily design analysis is debatable. Direct numerical approaches are discussed in Monograph Appendix E.

References

1 Wei, C., Dong, J.: Development and Modeling of Melt Electrohydrodynamic-Jet Printing of Phase-Change Inks for High-Resolution Additive Manufacturing. J. Manuf. Sci. Eng. **136**, 061010 (2014). Paper no: MANU-14-1179. https://doi.org/10.1115/1.4028483
2. Hinze, J.O.: Turbulence. McGraw-Hill, New York (1975)
3. Einstein, A.: A new determination of molecular dimensions. Ann. Phys. **19**(4), 289–306 (1906)
4. Hashin, Z., Shtrikman, S.: On some variational principles in anisotropic and nonhomogeneous elasticity. J. Mech. Phys. Solids **10**, 335–342 (1962)
5. Hashin, Z., Shtrikman, S.: A variational approach to the theory of the elastic behaviour of multiphase materials. J. Mech. Phys. Solids **11**, 127–140 (1963)
6. Hashin, Z.: Analysis of composite materials: a survey. ASME J. Appl. Mech. **50**, 481–505 (1983)
7. Abedian, B., Kachanov, M.: On the effective viscosity of suspensions. Int. J. Eng. Sci. **48**(11), 962–965 (2010). https://doi.org/10.1016/j.ijengsci.2010.08.012
8. Sevostianov, I., Kachanov, M.: Effective properties of heterogeneous materials: proper application of the non-interaction and the "dilute limit" approximations. Int. J. Eng. Sci. **58**, 124–128 (2012)
9. Zohdi, T.I.: On the reduction of heat generation in lubricants using microscale additives. Int. J. Eng. Sci. **62**, 84–89 (2013)
10. Zohdi, T.I.: Mechanically-driven accumulation of microscale material at coupled solid-fluid interfaces in biological channels. Proc. R. Soc. Interface **11**, 20130922 (2014)
11. Zohdi, T.I.: Additive particle deposition and selective laser processing-a computational manufacturing framework. Comput. Mech. **54**, 171–191 (2014)
12. Zohdi, T.I.: Embedded electromagnetically sensitive particle motion in functionalized fluids. Comput. Part. Mech. **1**, 27–45 (2014)
13. Zohdi, T.I.: On necessary pumping pressures for industrial process-driven particle-laden fluid flows. J. Manuf. Sci. Eng. ASME (2015). https://doi.org/10.1115/1.4030620
14. Onate, E., Idelsohn, S.R., Celigueta, M.A., Rossi, R.: Advances in the particle finite element method for the analysis of fluid-multibody interaction and bed erosion in free surface flows. Comput. Methods Appl. Mech. Eng. **197**(19–20), 1777–1800 (2008)
15. Onate, E., Celigueta, M.A., Idelsohn, S.R., Salazar, F., Surez, B.: Possibilities of the particle finite element method for fluid-soil-structure interaction problems. Comput. Mech. **48**, 307–318 (2011)
16. Rojek, J., Labra, C., Su, O., Onate, E.: Comparative study of different discrete element models and evaluation of equivalent micromechanical parameters. Int. J. Solids Struct. **49**, 1497–1517 (2012). https://doi.org/10.1016/j.ijsolstr.2012.02.032

17 Carbonell, J.M., Onate, E., Suarez, B.: Modeling of ground excavation with the particle finite element method. J. Eng. Mech. ASCE **136**, 455–463 (2010)
18 Labra, C., Onate, E.: High-density sphere packing for discrete element method simulations. Commun. Numer. Methods Eng. **25**(7), 837–849 (2009)
19. Leonardi, A., Wittel, F.K., Mendoza, M., Herrmann, H.J.: Coupled DEM-LBM method for the free-surface simulation of heterogeneous suspensions. Comput. Part. Mech. **1**(1), 3–13 (2014)
20 Cante, J., Davalos, C., Hernandez, J.A., Oliver, J., Jonsen, P., Gustafsson, G., Haggblad, H.A.: PFEM-based modeling of industrial granular flows. Comput. Part. Mech. **1**(1), 47–70 (2014)
21 Rojek, J.: Discrete element thermomechanical modelling of rock cutting with valuation of tool wear. Comput. Part. Mech. **1**(1), 71–84 (2014)
22 Onate, E., Celigueta, M.A., Latorre, S., Casas, G., Rossi, R., Rojek, J.: Lagrangian analysis of multiscale particulate flows with the particle finite element method. Comput. Part. Mech. **1**(1), 85–102 (2014)
23 Bolintineanu, D.S., Grest, G.S., Lechman, J.B., Pierce, F., Plimpton, S.J., Schunk, P.R.: Particle dynamics modeling methods for colloid suspensions. Comput. Part. Mech. **1**(3), 321–356 (2014)
24 Avci, B., Wriggers, P.: A DEM-FEM coupling approach for the direct numerical simulation of 3D particulate flows. J. Appl. Mech. **79**, 010901-1–7 (2012)

Monograph Appendix E—Hybrid DEM-CM Approaches for Particle-Functionalized Fluids

There are applications where Discrete Element Method (DEM) modeling alone is not adequate (because too many very small particles would be needed to describe the interstitial solvent) nor are pure continuum formulations (CF) based descriptions alone sufficient (because the mesh sizes have to be extremely fine to describe the particles). In these cases, hybrid methods, combining aspects of DEM and CF, are advantageous. As an example, we discuss one of the most common additive manufacturing materials, particle-functionalized inks, which are constructed by adding particles to a solvent. Oftentimes, the intended purpose is to add electromagnetically sensitive particles to a solvent to create an electromagnetically sensitive ink.

E.1 Applications

Electromagnetically sensitive fluids are typically constructed ("functionalized") by embedding charged or magnetic particles in a neutral fluid. Such fluids date back, at least, to Winslow [1, 2] in 1947. The most widely used class of such fluids are electrorheological fluids, which are comprised of extremely fine suspensions of charged particles (on the order of 50 microns) in an electrically neutral fluid.[10] They have been historically used in specialized niche applications involving hydraulic valves, actuators, flexible screens, brakes, liquid seals and recently in electromagnetically sensitive inks (so-called e-inks), for large-scale printing applications and coatings. The first step in achieving this overall goal is the analysis of the movement of seeded charged particles. In such a system, there are two primary effects that play a role in the charged particle dynamics: (1) drag forces due to the fluid flow and (2) electromagnetic forces due to an external source. The goal of this chapter is to formulate such models.[11] Specifically, the goal is to characterize the motion of electromagnetically sensitive

[10] Later in the chapter, we also discuss ferrofluids.

[11] More details on this specific application in mind are given in the conclusions.

T. I. Zohdi, *Modeling and Simulation of Functionalized Materials for Additive Manufacturing and 3D Printing: Continuous and Discrete Media*, Lecture Notes in Applied and Computational Mechanics 60, https://doi.org/10.1007/978-3-319-70079-3

small-scale particles embedded in a flowing neutral fluid. Initially, the dynamics of a single particle is studied. Thereafter, a three-dimensional representative volume element of a flowing particle-laden fluid, under the action of external electromagnetic fields, is investigated. A fully implicit finite difference discretization of the Navier–Stokes equations is used for the fluid, and a direct particle dynamics discretization is performed for the particles. Because of the large computational difficulty and expense of a conforming spatial discretization needed for large numbers of embedded particles, simplifying assumptions are made for the coupling, based on semianalytical computation of drag coefficients, which allows for the use of coarser meshes. Even after these simplifications, the particle–fluid system is strongly coupled. The approach taken in the present work is to construct a submodel for each primary physical process. In order to resolve the coupling, a recursive staggering scheme is constructed, which builds on works found in Zohdi [3–43]. The procedure is as follows (at a given time increment): (1) Each submodel equation (fluid system or particle system) is solved individually, "freezing" the other (coupled) fields in the system, allowing only the primary field to be active, (2) after the solution of each submodel, the associated field variable is updated, and the next submodel is solved, and (3) the process is then repeated, until convergence. The time-steps are adjusted to control the rates of convergence, which is dictated by changes in the overall physics. Specifically, the approach is a staggered implicit time-stepping scheme, with an internal recursion that automatically adapts the time-step sizes to control the rates of convergence within a time-step. If the process does not converge (below an error tolerance) within a preset number of iterations, the time-step is adapted (reduced) by utilizing an estimate of the spectral radius of the coupled system. The modular approach is designed for easy replacement of submodels for the fluid, particles, and their interaction. The approach allows researchers to rapidly compute such systems with laptop/desktop resources. Numerical examples are provided to illustrate the model and numerical solution scheme, and limitations and extensions of the approach are discussed.

Remark 1 From this point forth, we add a subscript p for particle velocities, to delineate between particles and fluids. We will consider a drag force which is given by:

$$\boldsymbol{\Psi}^{drag} = C_D \frac{1}{2} A \rho_f ||\boldsymbol{v}_f - \boldsymbol{v}_p||^2 \boldsymbol{\tau}, \tag{E.1}$$

where

$$\boldsymbol{\tau} \stackrel{\text{def}}{=} \frac{\boldsymbol{v}_f - \boldsymbol{v}_p}{||\boldsymbol{v}_f - \boldsymbol{v}_p||}, \tag{E.2}$$

and $A = \pi R^2$, where R is the particle radius.

Remark 2 Qualitatively, one can expect a drag-induced drift in the particle trajectory. For example, consider an isolated particle, moving in one dimension, in a surrounding fluid in the Stokesian regime, governed by

$$m\dot{v}_p = \underbrace{c(v_f - v_p)}_{drag}, \tag{E.3}$$

which yields

$$v_p(t) = (v_p(0) - v_f)e^{-ct/m} + v_f. \tag{E.4}$$

In the special case where $m = \rho_p \frac{4}{3}\pi R^3$ and the Stokesian drag is $c = c_o 6\pi R$ we have:

$$\dot{v}_p = \underbrace{\frac{9c_o}{2\rho_p \mathcal{R}^2}}_{drag \stackrel{\text{def}}{=} \alpha}(v_f - v_p), \tag{E.5}$$

and the velocity can be determined to be

$$v_p(t) = (v_p(0) - v_f)e^{-\alpha t} + v_f. \tag{E.6}$$

Thus, qualitatively, we expect such a fluid-induced drift to occur.

Remark 3 A special class or functionalized fluids are ferrofluids, which are colloidal liquids of ferromagnetic nanoscale (10 nanometers or less) particles, often from a iron-based compound, such as magnetite or hematite, usually in an organic solvent. The volume fraction of the particles is typically under 5 percent by volume. The particles are usually coated with a surfactant (typically oleic acid, citric acid, soy lecithin, tetramethylammonium hydroxide) to avoid agglomeration. The surfactant is strong enough to counteract near-field interaction effects between particles. The particles usually do not retain magnetization and can be considered as paramagnetic. Furthermore, they lose their magnetic properties at sufficiently high (Curie) temperatures.[12] We refer the reader to Albrecht et al. [43], Andelman and Rosensweig [44], and Berger et al. [45] for reviews. If the particles exhibit magnetization, an additional force is exerted, independent of the electrodynamically induced Lorentz forces. A simple model to characterize this force is given by

$$\boldsymbol{\Psi}^{mag} = -\nabla(\gamma \boldsymbol{B}^{ext} \cdot \boldsymbol{B}^{ext}), \tag{E.7}$$

where γ is a material parameter that is related to the magnetic dipole properties, the magnetization of the particle, and which is dependent on the magnetic susceptibility, the magnetic permeability, and the internal magnetic moment density of the material (see Feynman et al. [46], Cullity and Graham [47], Boyer [48], or Jackson [49]. We note that while we will not simulate purely magnetic (or magnetizable) particles in numerical examples presented later, we include the relevant terms in the formulation. An implementation of this model can be found in Zohdi [3–42].

[12] Another class of electromagnetically sensitive particle-laden fluid are magnetorheological fluids which are seeded with micron-sized particles, and consequently, the particles can sediment over time.

E.2 A Quick Review of General Governing Fluid Equations

Recall, for a hydrostatic fluid, the stress can be written as

$$\boldsymbol{\sigma} = -P_o \mathbf{1}, \tag{E.8}$$

where $P_o = \frac{tr\boldsymbol{\sigma}}{3}$ is the hydrostatic pressure. In other words, there are no shear stresses in a fluid at rest. In the dynamic case, the pressure, denoted the "thermodynamic pressure," is related to the temperature and the fluid density by an equation of state

$$\mathcal{Z}(P, \rho_f, \theta) = 0. \tag{E.9}$$

For a fluid in motion

$$\boldsymbol{\sigma} = -P\mathbf{1} + \boldsymbol{\tau}^{vs} \tag{E.10}$$

where $\boldsymbol{\tau}^{vs}$ is a so-called viscous stress tensor, needed in a balance of linear momentum:[13]

$$\nabla_x \cdot \boldsymbol{\sigma} + \boldsymbol{f} = \rho_f \frac{d\boldsymbol{v}_f}{dt}, \tag{E.11}$$

where $\boldsymbol{v}_f$ is the fluid velocity at point $\boldsymbol{x}$ and $\boldsymbol{f}$ are the body forces. Thus, for a compressible fluid in motion:

$$\frac{tr\boldsymbol{\sigma}}{3} = -P + \frac{tr\boldsymbol{\tau}^{vs}}{3}. \tag{E.12}$$

In general, for a fluid, we have

$$\boldsymbol{\tau}^{vs} = \mathcal{G}(\boldsymbol{D}) \quad \text{and} \quad \boldsymbol{D} \stackrel{\text{def}}{=} \frac{1}{2}(\nabla_x \boldsymbol{v}_f + (\nabla_x \boldsymbol{v}_f)^T), \tag{E.13}$$

where $\boldsymbol{v}_f = \dot{\boldsymbol{u}}$ is the velocity and $\boldsymbol{D}$ is the symmetric part of the velocity gradient. For a Newtonian fluid, where a linear relation exists between the viscous stresses ($\boldsymbol{\tau}^{vs}$) and $\boldsymbol{D}$

$$\boldsymbol{\tau}^{vs} = \mathcal{G}(\boldsymbol{D}) = \boldsymbol{C} : \boldsymbol{D} \tag{E.14}$$

where $\boldsymbol{C}$ is a symmetric positive definite (fourth-order) viscosity tensor. For an isotropic (standard) Newtonian fluid, we have

$$\boldsymbol{\sigma} = -P\mathbf{1} + \lambda tr\boldsymbol{D}\mathbf{1} + 2\mu\boldsymbol{D} = -P\mathbf{1} + 3\kappa\frac{tr\boldsymbol{D}}{3}\mathbf{1} + 2\mu\boldsymbol{D}', \tag{E.15}$$

where κ is called the bulk viscosity, λ is a viscosity constant, μ the shear viscosity, and $\boldsymbol{D}' = \boldsymbol{D} - \frac{tr\boldsymbol{D}}{3}\mathbf{1}$. Explicitly, with an (x_1, x_2, x_3) Cartesian triad

[13] An inviscid or "perfect" fluid is one where $\boldsymbol{\tau}^{vs}$ is taken to be zero, even when motion is present.

$$\underbrace{\begin{Bmatrix} \sigma_{11} \\ \sigma_{22} \\ \sigma_{33} \\ \sigma_{12} \\ \sigma_{23} \\ \sigma_{31} \end{Bmatrix}}_{\stackrel{\text{def}}{=}\{\boldsymbol{\sigma}\}} = \underbrace{\begin{Bmatrix} -P \\ -P \\ -P \\ 0 \\ 0 \\ 0 \end{Bmatrix}}_{\stackrel{\text{def}}{=}\{-\boldsymbol{P}\}} + \underbrace{\begin{bmatrix} c_1 & c_2 & c_2 & 0 & 0 & 0 \\ c_2 & c_1 & c_2 & 0 & 0 & 0 \\ c_2 & c_2 & c_1 & 0 & 0 & 0 \\ 0 & 0 & 0 & \mu & 0 & 0 \\ 0 & 0 & 0 & 0 & \mu & 0 \\ 0 & 0 & 0 & 0 & 0 & \mu \end{bmatrix}}_{\stackrel{\text{def}}{=}[\boldsymbol{C}]} \underbrace{\begin{Bmatrix} D_{11} \\ D_{22} \\ D_{33} \\ 2D_{12} \\ 2D_{23} \\ 2D_{31} \end{Bmatrix}}_{\stackrel{\text{def}}{=}\{\boldsymbol{D}\}}, \tag{E.16}$$

where $c_1 = \kappa + \frac{4}{3}\mu$ and $c_2 = \kappa - \frac{2}{3}\mu$, where $D_{ij} = \frac{1}{2}\left(\frac{\partial v_i}{\partial x_j} + \frac{\partial v_j}{\partial x_i}\right)$. The so-called Stokes' condition attempts to force the thermodynamic pressure to collapse to the classical definition of mechanical pressure, i.e.,

$$\frac{tr\boldsymbol{\sigma}}{3} = -P + 3\kappa\frac{tr\boldsymbol{D}}{3} = -P, \tag{E.17}$$

leading to the conclusion that $\kappa = 0$ or $\lambda = -\frac{2}{3}\mu$. Thus, a Newtonian fluid obeying the Stokes' condition has the following constitutive law:

$$\boldsymbol{\sigma} = -P\mathbf{1} - \frac{2}{3}\mu tr\boldsymbol{D}\mathbf{1} + 2\mu\boldsymbol{D} = -P\mathbf{1} + 2\mu\boldsymbol{D}'. \tag{E.18}$$

Note that

$$\dot{J} = \frac{d}{dt}det\boldsymbol{F} = (det\boldsymbol{F})tr(\dot{\boldsymbol{F}}\cdot\boldsymbol{F}^{-1}) = Jtr\boldsymbol{L} = J\nabla_x\cdot\boldsymbol{v}_f, \tag{E.19}$$

where $\boldsymbol{L} = \nabla_x\boldsymbol{v}_f$ is the velocity gradient. Note that $\nabla_x\cdot\boldsymbol{v}_f = tr\boldsymbol{L} = tr\boldsymbol{D}$. Therefore, if the fluid is incompressible, $\dot{J} = 0$, then $\nabla_x\cdot\boldsymbol{v}_f = 0 = tr\boldsymbol{L} = tr\boldsymbol{D}$. Therefore,

$$\boldsymbol{\sigma} = -P\mathbf{1} + 2\mu\boldsymbol{D}. \tag{E.20}$$

A conservation of mass dictates

$$\frac{d}{dt}(\rho_{fo}) = \frac{d}{dt}(\rho_f J) = J\frac{d\rho_f}{dt} + \rho_f\frac{dJ}{dt} = 0, \tag{E.21}$$

which leads to

$$\frac{d\rho_f}{dt} + \frac{\rho_f}{J}\frac{dJ}{dt} = 0. \tag{E.22}$$

Using Eqs. E.19 and E.21 becomes

$$\frac{d\rho_f}{dt} + \rho_f\nabla_x\cdot\boldsymbol{v}_f = 0. \tag{E.23}$$

Now write the total temporal ("material") derivative in convective form:

$$\frac{d\rho_f}{dt} = \frac{\partial \rho_f}{\partial t} + (\nabla_x \rho_f) \cdot \frac{d\boldsymbol{x}}{dt} = \frac{\partial \rho_f}{\partial t} + \nabla_x \rho_f \cdot \boldsymbol{v}_f. \tag{E.24}$$

Thus, Eq. E.19 becomes

$$\frac{\partial \rho_f}{\partial t} + \nabla_x \rho_f \cdot \boldsymbol{v}_f + \rho_f \nabla_x \cdot \boldsymbol{v}_f = \frac{\partial \rho_f}{\partial t} + \nabla_x \cdot (\rho_f \boldsymbol{v}_f) = 0. \tag{E.25}$$

Thus, writing the total time derivatives appearing previously as

$$\frac{d\boldsymbol{v}_f}{dt} = \frac{\partial \boldsymbol{v}_f}{\partial t}|_x + (\nabla_x \boldsymbol{v}_f)|_t \cdot \frac{d\boldsymbol{x}}{dt}, \tag{E.26}$$

the coupled governing equations are (ignoring thermal effects)

$$\boxed{\begin{aligned} &\frac{\partial \rho_f}{\partial t} = -\nabla_x \rho_f \cdot \boldsymbol{v}_f - \rho_f \nabla_x \cdot \boldsymbol{v}_f, \\ &\rho_f(\frac{\partial \boldsymbol{v}_f}{\partial t} + (\nabla_x \boldsymbol{v}_f) \cdot \boldsymbol{v}_f) = \nabla_x \cdot \boldsymbol{\sigma} + \boldsymbol{f}, \\ &\boldsymbol{\sigma} = -P\mathbf{1} + \lambda tr\boldsymbol{D}\mathbf{1} + 2\mu\boldsymbol{D} = -P\mathbf{1} + 3\kappa\tfrac{tr\boldsymbol{D}}{3}\mathbf{1} + 2\mu\boldsymbol{D}', \end{aligned}} \tag{E.27}$$

where, for example, P is given by an equation of state. Collectively, we refer to these equations as the "Navier–Stokes" equations. There are a total of three variables: ρ_f, $\boldsymbol{v}_f$, and P. It is customary to specify $\boldsymbol{v}_f$ and P on the boundary and to determine ρ_f on the boundary through the equation of state. It is important to emphasize that physically compatible boundary data must be applied, and this is not a trivial matter for compressible flow.

Remark There are a variety of possible equations of state that connect the density to the pressure, such as a Boussinesq-like relation, which is adequate to describe dense gases and fluids, derived from[14]

$$\rho_f \approx \rho_{fo}(P_o) + \frac{\partial \rho_f}{\partial P}\Delta P, \tag{E.28}$$

where ρ_{fo} and P_o are reference values and $\Delta P = P - P_o$. We define the bulk (compressibility) modulus by $\zeta \stackrel{\text{def}}{=} \rho_f \frac{\partial P}{\partial \rho_f}$, yielding

$$\rho_f \approx \rho_{fo}\left(1 + \frac{1}{\zeta}\Delta P\right) \Rightarrow P \approx P_o + \zeta\left(\frac{\rho_f}{\rho_{fo}} - 1\right). \tag{E.29}$$

For a constant density case, $\rho_f = \rho_{fo}$, and utilizing the Boussinesq-like relation, $P = P_o$.

[14] We have ignored thermal effects in this representation.

E.3 Numerical Simulation of Coupled Fluid–Multiparticle Systems

There are three components to the numerical discretization of the coupled fluid–particle system:

- Temporal discretization of the particle dynamics model,
- Spatiotemporal discretization of the fluid continuum model, and
- Iterative staggering scheme to resolve the coupling between the particles and fluid continuum.

E.4 The Overall Approach

The present subsection develops a flexible and robust solution strategy to resolve two-way coupled systems comprised of large groups of flowing particles embedded within a fluid. Such particle/fluid systems are strongly coupled due to the forces induced by the fluid onto the particles and vice versa. Specifically, a staggered approach is developed whereby at each time-step:

- A pure particle-only system is solved where the fluid's influence is accounted for by drag forces on the particles.
- A pure fluid-only system is then solved where the particles' influence is accounted for by reverse drag forces on the fluid continuum.
- The interactions between the particles and fluid are then updated. Within a time-step, this process is repeated until the system converges. If the system does not converge within a preset number of iterations, the time-steps are reduced by a prescribed amount, dictated by an estimate of the spectral radius of the coupled system.

Because the coupling of the various particles and fluid fields can dramatically change over the course of a flow process, a primary focus is the development of a recursive "staggering" solution scheme, whereby the time-steps are adaptively adjusted to control the error associated with the incomplete resolution of the coupled interaction between the various solid particulates and continuum fluid fields.

E.5 Simplifying Assumptions

We make the following simplifying assumptions:

- We assume that each particle is small enough that their rotation with respect to their mass centers is deemed insignificant. The particles are considered to be spheres.
- We assume that any pre-applied particle coatings/surfactants are thin enough to be considered as part of the particle.

- We assume that the external electromagnetic fields are given and that action of the particle-laden fluid does not affect the fields, i.e., they are considered control variables.
- We assume that the fluid is incompressible.

Other simplifying assumptions will be made later in the presentation wherever appropriate.

E.6 Modeling and Simulation of the Particle Dynamics Problem

As in the previous chapter on particle dynamics, we consider a group of nonintersecting particles ($i = 1, 2, \ldots, N_p$). The equation of motion for the ith particle in the system is

$$m_i \dot{\boldsymbol{v}}_{pi} = m_i \ddot{\boldsymbol{r}}_{pi} = \boldsymbol{\Psi}_i^{tot}(\boldsymbol{r}_{p1}, \boldsymbol{r}_{p2}, \ldots, \boldsymbol{r}_{pN_p}) = \boldsymbol{\Psi}_i^{con} + \boldsymbol{\Psi}_i^{drag} + \boldsymbol{\Psi}_i^{e+m}, \quad \text{(E.30)}$$

where $\boldsymbol{r}_{pi}$ is the position vector of the ith particle and $\boldsymbol{\Psi}_i^{tot}$ represents all forces acting on particle i, which is decomposed into the sum of forces due to the normal and frictional contact forces ($\boldsymbol{\Psi}_i^{con}$), the drag forces arising from the surrounding fluid environment ($\boldsymbol{\Psi}_i^{drag}$), and electromagnetic forces ($\boldsymbol{\Psi}_i^{e+m}$). In summary, we have the following forces acting on each particle ($i = 1, 2, \ldots, N_p$)[15]

$$\boldsymbol{\Psi}_i^{tot} \stackrel{\text{def}}{=} \boldsymbol{\Psi}_i^{con} + \boldsymbol{\Psi}_i^{drag} + \boldsymbol{\Psi}_i^{e+m}. \quad \text{(E.31)}$$

Integrating Eq. E.30 leads to (using a trapezoidal rule with variable integration metric, $0 \leq \phi \leq 1$)

$$\begin{aligned} \boldsymbol{v}_{pi}(t + \Delta t) &= \boldsymbol{v}_{pi}(t) + \frac{1}{m_i} \int_t^{t+\Delta t} \boldsymbol{\Psi}_i^{tot}\, dt \\ &\approx \boldsymbol{v}_{pi}(t) + \frac{\Delta t}{m_i} \left(\phi \boldsymbol{\Psi}_i^{tot}(t + \Delta t) + (1 - \phi)\boldsymbol{\Psi}_i^{tot}(t)\right). \end{aligned} \quad \text{(E.32)}$$

The position can be computed via application of the trapezoidal rule again:

$$\boldsymbol{r}_{pi}(t + \Delta t) \approx \boldsymbol{r}_{pi}(t) + \Delta t(\phi \boldsymbol{v}_{pi}(t + \Delta t) + (1 - \phi)\boldsymbol{v}_{pi}(t)), \quad \text{(E.33)}$$

which can be consolidated into

$$\boldsymbol{r}_{pi}(t + \Delta t) = \boldsymbol{r}_{pi}(t) + \boldsymbol{v}_{pi}(t)\Delta t + \frac{\phi(\Delta t)^2}{m_i} \left(\phi \boldsymbol{\Psi}_i^{tot}(t + \Delta t) + (1 - \phi)\boldsymbol{\Psi}_i^{tot}(t)\right). \quad \text{(E.34)}$$

[15] For completeness, we have included an additional force which can be exerted on magnetic particles.

As introduced previously, this leads to a coupled system of equations, which are solved using an adaptive iterative scheme, building on approaches found in various forms in Zohdi [3–42]. This is dealt with after introducing the governing fluid mechanics equations.

E.7 Characterization of Particle/Fluid Interaction

We first consider drag force interactions between the fluid and particles. The drag force acting on an object in a fluid flow (occupying domain Ω and outward surface normal $\boldsymbol{n}$) which is defined as

$$\boldsymbol{\Psi}^{drag} = \int_{\partial\Omega} \boldsymbol{\sigma} \cdot \boldsymbol{n}\, dA, \tag{E.35}$$

where $\boldsymbol{\sigma}$ is the Cauchy stress. For a Newtonian fluid, obeying the Stokes' condition has the following constitutive law (see appendix):

$$\boldsymbol{\sigma} = -P\mathbf{1} - \frac{2}{3}\mu tr\boldsymbol{D}\mathbf{1} + 2\mu\boldsymbol{D} = -P\mathbf{1} + 2\mu\boldsymbol{D}'. \tag{E.36}$$

where P is the thermodynamic pressure, μ is the absolute viscosity, $\boldsymbol{D} = \frac{1}{2}(\nabla_x \boldsymbol{v}_f + (\nabla_x \boldsymbol{v}_f)^T)$ is the symmetric part of the velocity gradient, $tr\boldsymbol{D}$ is the trace of $\boldsymbol{D}$, and $\boldsymbol{D}' = \boldsymbol{D} - \frac{tr\boldsymbol{D}}{3}\mathbf{1}$ is the deviatoric part of $\boldsymbol{D}$. The stress is determined by solving the balance of linear momentum (Navier–Stokes)

$$\rho_f \left(\frac{\partial \boldsymbol{v}_f}{\partial t} + (\nabla_x \boldsymbol{v}_f) \cdot \boldsymbol{v}_f \right) = \nabla_x \cdot \boldsymbol{\sigma} + \boldsymbol{f}_f, \tag{E.37}$$

where, at a point, ρ_f is the fluid density, $\boldsymbol{v}_f$ is the fluid velocity, and $\boldsymbol{f}_f$ represents body forces per unit volume.

The fluid domain will require spatial discretization with some type of mesh, for example, using a finite difference, finite volume, or finite element method. Usually, it is extremely difficult to resolve the flow in the immediate neighborhood of the particles, in particular if there are several particles. However, if the primary interest is in the dynamics of the particles, *as it is in this work*, an appropriate approach, which permits coarser discretization of the fluid continuum, is to employ effective drag coefficients, for example, as introduced earlier[16]

---

[16] In the general compressible case, one would use

$$\langle \rho_{fi} \rangle_{\Omega_{pi}} \stackrel{\text{def}}{=} \frac{1}{\Omega_{pi}} \int_{\Omega_{pi}} \rho(\boldsymbol{x})\, d\Omega, \tag{E.38}$$

instead of ρ_f.

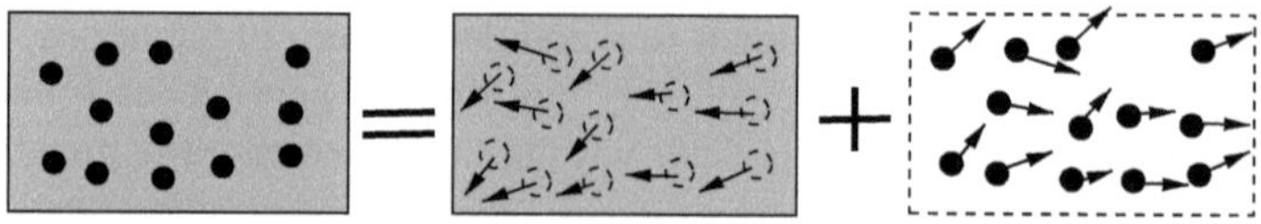

Fig. E.1 Decomposition of the coupled fluid system

$$C_D \stackrel{\text{def}}{=} \frac{||\boldsymbol{\Psi}_i^{drag}||}{\frac{1}{2}\rho_f ||\langle \boldsymbol{v}_f \rangle_{\Omega_i} - \boldsymbol{v}_{pi}||^2 A_i}, \tag{E.39}$$

where $\langle(\cdot)\rangle_{\Omega_i} \stackrel{\text{def}}{=} \frac{1}{|\Omega_i|}\int_{\Omega_i}(\cdot)\, d\Omega_i$ is the volumetric average of the argument over the domain occupied by the ith particle, $\langle \boldsymbol{v}_f\rangle_{\Omega_i}$ is the volumetric average of the fluid velocity, and A_i is the cross-sectional area of the ith (solid) particle. The use of this simple concept is relatively straightforward to account for the presence of the solid particles in the fluid by augmenting the flow calculations with drag forces (Fig. E.1). Algorithmically speaking, one must compute the fluid flow with reaction forces due to the presence of the particles. To this end, one can use the volumetric forces ($\boldsymbol{f}_f$) within the fluid domain for this purpose by writing

$$\boldsymbol{f}_f = -\frac{\boldsymbol{\Psi}_{pi}^{drag}}{|\Omega_{pi}^*|} = -\frac{C_D \frac{1}{2}\rho_f ||\langle \boldsymbol{v}_f\rangle_{\Omega_{pi}^*} - \boldsymbol{v}_{pi}||^2 A_i}{|\Omega_{pi}^*|}\hat{\boldsymbol{\tau}} \quad \left(\hat{\boldsymbol{\tau}} = \frac{\langle \boldsymbol{v}_f\rangle_{\Omega_{pi}^*} - \boldsymbol{v}_{pi}}{||\langle \boldsymbol{v}_f\rangle_{\Omega_{pi}^*} - \boldsymbol{v}_{pi}||}\right), \tag{E.40}$$

where $|\Omega_{pi}^*|$ is the equivalent volume of the number of nodes (each node with assigned volume $\Delta x_1 \Delta x_2 \Delta x_3$) that fall within the particle domain and $\boldsymbol{f}_f$ (per unit volume) is the drag force on the fluid, which is nonzero if a node falls within the particle domain and is zero otherwise.[17] *This drag-based approach is designed to account for particles in the fluid using a coarse mesh.* In summary, the drag force is computed by calculating the difference between the average fluid velocity in the ith particle domain $\langle \boldsymbol{v}_{fi}\rangle_{\Omega_{pi}}$ and the particle velocity (no particle rotation is assumed), $\boldsymbol{v}_{pi}$, to yield the drag force $\boldsymbol{\Psi}_i^{drag} \stackrel{\text{def}}{=} \frac{1}{2}C_D\rho_f ||\langle \boldsymbol{v}_{fi}\rangle_{\Omega_{pi}} - \boldsymbol{v}_{pi}||^2 A_i \hat{\boldsymbol{\tau}}$ (Fig. E.2).

Remark 1 From this point forward, we will consider incompressible cases (ρ_f is constant) and that P is given (known).

Remark 2 There are a variety of empirical relations for the drag coefficient. One possible way to represent the drag coefficient is with a piecewise definition, as a function of the Reynolds number (Chow [60]):

- For $0 < Re \leq 1$, $C_D = \frac{24}{Re}$,
- For $1 < Re \leq 400$, $C_D = \frac{24}{Re^{0.646}}$,

[17] If the particles are significantly smaller than the mesh spacing, then the drag forces associated with the particles are computed from the nearest node/particle center pair.

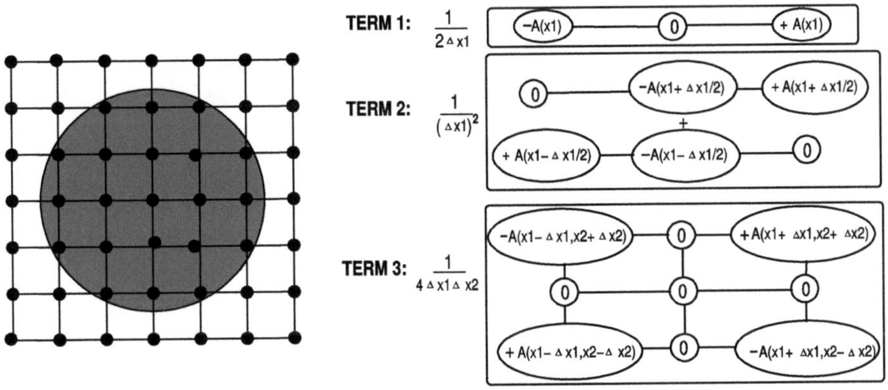

Fig. E.2 Left: grid points where the average values of drag are computed. Right: various finite difference stencils in "computational molecule" form (centered at (x_1, x_2, x_3)), where (1) $A(x)\frac{\partial v}{\partial x_1}$, (2) $\frac{\partial}{\partial x_1}\left(A\frac{\partial v}{\partial x_1}\right)$, and (3) $\frac{\partial}{\partial x_2}\left(A\frac{\partial v}{\partial x_1}\right)$

- For $400 < Re \leq 3 \times 10^5$, $C_D = 0.5$,
- For $3 \times 10^5 < Re \leq 2 \times 10^6$, $C_D = 0.000366 Re^{0.4275}$,
- For $2 \times 10^6 < Re < \infty$, $C_D = 0.18$,

where the local Reynolds number for a particle is $Re \stackrel{\text{def}}{=} \frac{2R_i \rho_f ||\langle \boldsymbol{v}_f \rangle_{\Omega_i} - \boldsymbol{v}_{pi}||}{\mu_f}$ and $\mathcal{R}_i$ is the radius of the ith particle.

E.8 Discretization of the Fluid

E.8.1 Temporal Discretization

For the fluid, we write

$$\frac{d\boldsymbol{v}_f}{dt} = \frac{\partial \boldsymbol{v}_f}{\partial t} + \nabla_x \boldsymbol{v}_f \cdot \boldsymbol{v}_f = \frac{1}{\rho_f}\left(\nabla_x \cdot \boldsymbol{\sigma} + \boldsymbol{f}\right), \tag{E.41}$$

leading to

$$\frac{\partial \boldsymbol{v}_f}{\partial t} = \frac{1}{\rho_f}\left(\nabla_x \cdot \boldsymbol{\sigma} + \boldsymbol{f}\right) - \nabla_x \boldsymbol{v}_f \cdot \boldsymbol{v}_f \stackrel{\text{def}}{=} \boldsymbol{L}. \tag{E.42}$$

We discretize for time $= t + \phi \Delta t$, and using a trapezoidal "$\phi - scheme$" $(0 \leq \phi \leq 1)$

$$\frac{\partial \boldsymbol{v}_f}{\partial t} \approx \frac{\boldsymbol{v}_f(t + \Delta t) - \boldsymbol{v}_f(t)}{\Delta t} \approx \boldsymbol{L}(t + \phi \Delta t) \approx \phi \boldsymbol{L}(t + \Delta t) + (1 - \phi)\boldsymbol{L}(t). \tag{E.43}$$

Rearranging yields

$$\boldsymbol{v}_f(t + \Delta t) \approx \boldsymbol{v}_f(t) + \Delta t \left(\phi \boldsymbol{L}(t + \Delta t) + (1 - \phi)\boldsymbol{L}(t)\right) \tag{E.44}$$

where the previously introduced spatial discretization is applied to the derivative terms (such as $\nabla_x \cdot \boldsymbol{\sigma}$) in $\boldsymbol{L}$. The discretized system is formulated next as an implicit time-stepping scheme within each time-step L.

E.8.2 Spatial Discretization: Spatial Finite Difference Stencils

The following standard approximations are used:

1. For the first derivative of a primal variable v at (x_1, x_2, x_3):

$$\frac{\partial v_f}{\partial x_1} \approx \frac{v_f(x_1 + \Delta x_1, x_2, x_3) - v_f(x_1 - \Delta x_1, x_2, x_3)}{2\Delta x_1} \tag{E.45}$$

2. For the derivative of a flux at (x_1, x_2, x_3):

$$\begin{aligned}\frac{\partial}{\partial x_1}\left(A\frac{\partial v_f}{\partial x_1}\right) &\approx \frac{\left(A\frac{\partial v_f}{\partial x_1}\right)\Big|_{x_1+\frac{\Delta x_1}{2},x_2,x_3} - \left(A\frac{\partial v_f}{\partial x_1}\right)\Big|_{x_1-\frac{\Delta x_1}{2},x_2,x_3}}{\Delta x_1} \\ &= \frac{1}{\Delta x_1}\left[A(x_1 + \frac{\Delta x_1}{2}, x_2, x_3)\left(\frac{v_f(x_1 + \Delta x_1, x_2, x_3) - v_f(x_1, x_2, x_3)}{\Delta x_1}\right)\right] \\ &- \frac{1}{\Delta x_1}\left[A(x_1 - \frac{\Delta x_1}{2}, x_2, x_3)\left(\frac{v_f(x_1, x_2, x_3) - v_f(x_1 - \Delta x_1, x_2, x_3)}{\Delta x_1}\right)\right],\end{aligned} \tag{E.46}$$

where we have used

$$A(x_1 + \frac{\Delta x_1}{2}, x_2, x_3) \approx \frac{1}{2}\left(A(x_1 + \Delta x_1, x_2, x_3) + A(x_1, x_2, x_3)\right) \tag{E.47}$$

and

$$A(x_1 - \frac{\Delta x_1}{2}, x_2, x_3) \approx \frac{1}{2}\left(A(x_1, x_2, x_3) + A(x_1 - \Delta x_1, x_2, x_3)\right) \tag{E.48}$$

3. For the cross-derivative of a flux at (x_1, x_2):

$$\begin{aligned}\frac{\partial}{\partial x_2}\left(A\frac{\partial v_f}{\partial x_1}\right) &\approx \frac{\partial}{\partial x_2}\left(A(x_1,x_2,x_3)\left(\frac{v_f(x_1+\Delta x_1,x_2,x_3)-v_f(x_1-\Delta x_1,x_2,x_3)}{2\Delta x_1}\right)\right)\\ &\approx \frac{1}{4\Delta x_1\Delta x_2}\big(A(x_1,x_2+\Delta x_2,x_3)\left[v_f(x_1+\Delta x_1,x_2+\Delta x_2,x_3)\right.\\ &\qquad \left.-v_f(x_1-\Delta x_1,x_2+\Delta x_2,x_3)\right]-A(x_1,x_2-\Delta x_2,x_3)\\ &\qquad \left[v_f(x_1+\Delta x_1,x_2-\Delta x_2,x_3)\ -v_f(x_1-\Delta x_1,x_2-\Delta x_2,x_3)\right]\big).\end{aligned} \tag{E.49}$$

E.9 Overall Iterative (Implicit) Solution Method

As in the body of the monograph, following the basic framework in Zohdi [3–42], let us consider the finite difference nodes (i):

$$\boldsymbol{v}_{fi}^{L+1,K} = \boldsymbol{v}_{fi}^{L} + \Delta t\left(\phi \boldsymbol{L}_i^{L+1,K-1} + (1-\phi)\boldsymbol{L}_i^{L}\right), \tag{E.50}$$

where i is the node counter, which is of the form

$$\boldsymbol{v}_{fi}^{L+1,K} = \mathcal{G}(\boldsymbol{v}_{fi}^{L+1,K-1}) + R_i, \tag{E.51}$$

where $K = 1, 2, 3, \ldots$ is the index of iteration within time-step $L+1$ and

- $\mathcal{G}(\boldsymbol{v}_{fi}^{L+1,K-1}) = \phi\Delta t \boldsymbol{L}_i^{L+1,K-1}$ and
- $R_i = \boldsymbol{v}_{fi}^{L} + \Delta t(1-\phi)\boldsymbol{L}_i^{L}$.

The term R_i is a remainder term that does not depend on the solution. The convergence of such a scheme is dependent on the behavior of $\mathcal{G}$. Namely, a sufficient condition for convergence is that $\mathcal{G}$ is a contraction mapping for all $\boldsymbol{v}_{fi}^{L+1,K}$, $K = 1, 2, 3, \ldots$ In order to investigate this further, we define the iteration error as

$$\varpi_i^{L+1,K} \stackrel{\text{def}}{=} \boldsymbol{v}_{fi}^{L+1,K} - \boldsymbol{v}_{fi}^{L+1}. \tag{E.52}$$

A necessary restriction for convergence is iterative self-consistency, i.e., the "exact" (discretized) solution must be represented by the scheme, $\boldsymbol{v}_i^{L+1} = \mathcal{G}(\boldsymbol{v}_i^{L+1}) + R_i$. Enforcing this restriction, a sufficient condition for convergence is the existence of a contraction mapping

$$\begin{aligned}\underbrace{||\,\boldsymbol{v}_{fi}^{L+1,K} - \boldsymbol{v}_{fi}^{L+1}\,||}_{\varpi_i^{L+1,K}} &= ||\mathcal{G}(\boldsymbol{v}_{fi}^{L+1,K-1}) - \mathcal{G}(\boldsymbol{v}_{fi}^{L+1})||\\ &\le \eta^{L+1,K}||\boldsymbol{v}_i^{L+1,K-1} - \boldsymbol{v}_{fi}^{L+1}||,\end{aligned} \tag{E.53}$$

where if $0 \leq \eta^{L+1,K} < 1$ for each iteration K, then $\varpi_i^{L+1,K} \rightarrow \mathbf{0}$ for any arbitrary starting value $\boldsymbol{r}_i^{L+1,K=0}$, as $K \rightarrow \infty$, which is a contraction condition that is sufficient, but not necessary, for convergence. The convergence of Eq. E.50 is scaled by $\eta \propto \frac{(\phi \Delta t)^2}{m_i}$. Therefore, we see that the contraction constant of $\mathcal{G}$ is:

- directly dependent on the magnitude of the interaction forces ($||\boldsymbol{L}||$),
- directly proportional to $(\Delta t)^2$.

Thus, decreasing the time-step size improves the convergence. *In order to maximize the time-step sizes (to decrease overall computing time) and still meet an error tolerance on the numerical solution's accuracy*, we build on an approach originally developed for continuum thermochemical multifield problems (Zohdi [3–42]), where one assumes: (1) $\eta^{L+1,K} \approx S(\Delta t)^p$, ($S$ is a constant) and (2) the error within an iteration behaves according to $(S(\Delta t)^p)^K \varpi^{L+1,0} = \varpi^{L+1,K}$, $K = 1, 2, \ldots$, where $\varpi^{L+1,0} = \boldsymbol{v}_{fi}^{L+1,K=1} - \boldsymbol{v}_{fi}^{L}$ is the initial norm of the iterative (relative) error and S is intrinsic to the system. For example, for second-order problems, due to the quadratic dependency on Δt, $p \approx 2$. The objective is to meet an error tolerance in exactly a preset (the analyst sets this) number of iterations. To this end, one writes $(S(\Delta t_{\mathrm{tol}})^p)^{K_d} \varpi^{L+1,0} = TOL$, where TOL is a tolerance and K_d is the number of desired iterations. If the error tolerance is not met in the desired number of iterations, the contraction constant $\eta^{L+1,K}$ is too large. Accordingly, one can solve for a new smaller STEP size, under the assumption that S is constant,

$$\Delta t_{\mathrm{tol}} = \Delta t \underbrace{\left(\frac{\left(\frac{TOL}{\varpi^{L+1,0}}\right)^{\frac{1}{pK_d}}}{\left(\frac{\varpi^{L+1,K}}{\varpi^{L+1,0}}\right)^{\frac{1}{pK}}} \right)}_{\stackrel{\mathrm{def}}{=} \Lambda_K}. \tag{E.54}$$

The assumption that S is constant is not critical, since the time-steps are to be recursively refined and unrefined throughout the simulation. Clearly, the expression in Eq. E.54 can also be used for time-step enlargement, if convergence is met in less than K_d iterations (typically chosen to be between five to ten iterations). Specifically, the solution STEPS are within a time-step:

- (1): Start a global fixed iteration (set $i = 1, \ldots, N_n$ (node counter) and $K = 0$ (iteration counter))
- (2): If $i > N_n$, then go to (4)
- (3): If $i \leq N_n$, then:

  (a) Compute the velocity $\boldsymbol{v}_{fi}^{L+1,K}$
  (b) Go to (2) for the next node ($i = i + 1$)

- (4): Repeat STEPS 1-3 for the particles, $i = 1, \ldots, N_p$, and compute the drag forces.
- (5): Measure error (normalized) quantities (where w_f is a weight on the fluid contribution and w_p is a weight on the particle contribution)

(a) $\varpi_K \stackrel{\text{def}}{=} w_f \frac{\sum_{i=1}^{N_n} ||\boldsymbol{v}_{fi}^{L+1,K} - \boldsymbol{v}_{fi}^{L+1,K-1}||}{\sum_{i=1}^{N_n} ||\boldsymbol{v}_{fi}^{L+1,K}||} + w_p \frac{\sum_{i=1}^{N_p} ||\boldsymbol{v}_{pi}^{L+1,K} - \boldsymbol{v}_{pi}^{L+1,K-1}||}{\sum_{i=1}^{N_p} ||\boldsymbol{v}_{pi}^{L+1,K}||}$

(b) $Z_K \stackrel{\text{def}}{=} \frac{\varpi_K}{TOL_r}$

(c) $\Lambda_K \stackrel{\text{def}}{=} \left(\frac{(\frac{TOL}{\varpi_0})^{\frac{1}{pK_d}}}{(\frac{\varpi_K}{\varpi_0})^{\frac{1}{pK}}} \right)$.

Note: As an option, one could select to only use the error metric from the finite difference grid, since the particle velocities have already been projected onto the grid.

- (6): If the tolerance is met: $Z_K \leq 1$ and $K < K_d$, then
  (a) Increment time: $t = t + \Delta t$,
  (b) Construct the next time-step: $\Delta t^{new} = \Lambda_K \Delta t^{old}$, and
  (c) Select the minimum size: $\Delta t = MIN((\Delta t)^{lim}, \Delta t^{new})$, and go to (1)
- (7): If the tolerance is not met: $Z_K > 1$ and $K < K_d$, then
  (a) Update the iteration counter: $K = K + 1$,
  (b) Reset the node counter: $i = 1$, and
  (c) Go to (2)
- (8): If the tolerance is not met ($Z_K > 1$) and $K = K_d$, then
  (a) Construct a new time-step: $\Delta t^{new} = \Lambda_K \Delta t^{old}$ and
  (b) Restart at time t, and go to (1)

Time-step size adaptivity is critical, since the system's dynamics and configuration can dramatically change over the course of time, possibly requiring quite different time-step sizes to control the iterative error. However, to maintain the accuracy of the time-stepping scheme, one must respect an upper bound dictated by the discretization error, i.e., $\Delta t \leq \Delta t^{lim}$. Note that in STEP (5), Λ_K may enlarge the time-step if the error is lower than the preset tolerance. At a given time, once the process is complete, then the time is incremented forward and the process is repeated. The overall goal is to deliver solutions, where the iterative error is controlled and the temporal discretization accuracy dictates the upper limit on the time-step size (Δt^{lim}). Clearly, there are various combinations of solution methods that one can choose from. For example, for the overall field coupling, one may choose implicit or explicit staggering and within the staggering process, either implicit ($0 < \phi \leq 1$) or explicit time-stepping ($\phi = 0$), and, in the case of implicit time-stepping, iterative or direct solvers for the Navier–Stokes equations. Furthermore, one could employ internal iterations for each field equation, then update more sophisticated metrics for certain components of the error, etc. For example, we utilized an error measure that used the velocities at the nodes of the finite difference grid, which were partially induced by the particles. We could have isolated the error in velocities of the particles, used a combined metric, etc. For details, see Zohdi [3–42].

Remark 1 To accelerate the multiparticle calculations, one can utilize nearest neighbor interaction lists, which indicate with which neighboring particles a specific particle interacts with, for a given number of time-steps. All other particle-to-particle interaction is truncated. The interaction lists are updated periodically during the computations (see Pöschel and Schwager [51]). Such lists were used in the upcoming calculations and follow this procedure: (1) An initial full-blown contact search is performed at the beginning of the simulation and a list of nearest neighbors for each particle is generated, (2) the nearest neighbor lists are used for all inter-particle calculations for a preset subinterval of the total simulation time, $0 \le \Delta T \le T$, and (3) after a subinterval of the total simulation time, $0 \le \Delta T \le T$, elapses, the nearest neighbor list for each particle is updated. This type of link list strategy was used throughout the simulations. The results were repeatedly tested against simulations utilizing direct calculations (no list list truncation) to determine the proper settings for the link lists to ensure insensitivity to the truncations.

Remark 2 As the physics changes, the field that is most sensitive (exhibits the largest amount of relative nondimensional change during the iterations) dictates the time-step size. Because the internal system solvers within the staggering scheme are also iterative and use the previously converged solution as their starting value to solve the system of equations, a field that is relatively insensitive at a given stage of the simulation will converge in very few internal iterations (perhaps even one).

Remark 3 A series of detailed numerical examples can be found in Zohdi [34]. Because of the extreme computational difficulty and expense of a conforming spatial discretization needed for large numbers of embedded particles, simplifying assumptions were made allowing a reduced-order model for the coupling, based on semianalytical computation of drag coefficients for the coupling of the particles and the fluid with coarse meshes. The strongly coupled system was solved iteratively within each time-step using a recursive staggering scheme, which employed temporal adaptivity to control the error. The developed approach is straightforward to implement and can be easily incorporated within any standard computational fluid mechanics code based on finite difference, finite element, finite volume, or discrete/particle element discretization (see Labra and Onate [52], Onate et al. [53, 54], Rojek et al. [55]). However, while the computational model can provide qualitative a priori information for further computationally intensive large-scale simulations, extensions are invariably going to require complex spatial discretization, which entails resolving particle–fluid interaction in detail. This is quite important if one is interested in particle clustering and agglomeration (see Zohdi [3–42]). Accordingly, it is important to revisit one of the primary simplifying assumptions, in the context of extensions and improvements. One of the primary assumptions was to assume that each particle is small enough that their rotation with respect to their mass centers was deemed insignificant. More detailed analyses of fluid-particle interaction can be achieved in a direct, brute-force, numerical schemes, treating the particles as part of the fluid continuum (as another fluid or solid phase), and thus meshing them in a detailed manner. In such an approach (e.g., see Avci and Wriggers [56]):

- A fluid-only problem is solved, with (instantaneous) boundary conditions of $\boldsymbol{v}_f(\boldsymbol{x}) = \boldsymbol{v}_{pi}(\boldsymbol{x})$ at each point on the fluid-particle boundaries, where the velocity of the points on the boundary is given by

$$\boldsymbol{v}_{pi}(\boldsymbol{x}) = \boldsymbol{v}_{pi}^{cm} + \boldsymbol{\omega}_i \times \boldsymbol{R}_{cm\rightarrow surf.}(\boldsymbol{x}), \tag{E.55}$$

  where $\boldsymbol{v}_{pi}^{cm}$ is the center of mass, for each of the individual particles, and $\boldsymbol{R}_{cm\rightarrow surf.}$ is a vector from the mass center to the surface.
- For each particle, one would solve:

$$m_i \dot{\boldsymbol{v}}_{pi} = \boldsymbol{\Psi}_i^{drag} + other\ forces \tag{E.56}$$

  and

$$I_i \dot{\boldsymbol{\omega}}_i = \boldsymbol{M}_i^{drag} + other\ moments, \tag{E.57}$$

  where the forces and moments would have a contribution from the fluid drag (with particle occupying domain Ω_i and outward surface normal $\boldsymbol{n}$) is defined as

$$\boldsymbol{\Psi}_i^{drag} = \int_{\partial\Omega_i} \boldsymbol{\sigma} \cdot \boldsymbol{n}\, dA, \tag{E.58}$$

  and

$$\boldsymbol{M}_i^{drag} = \int_{\partial\Omega_i} \boldsymbol{R}_{cm\rightarrow surf.} \times \boldsymbol{\sigma} \cdot \boldsymbol{n}\, dA. \tag{E.59}$$

- At a time-step, the process is iteratively driven by solving the fluid-only problem first, then the particles-only problem, and repeated until convergence in an appropriate norm.

References

1. Winslow, W.M.: Method and means for translating electrical impulses into mechanical force. U.S. Patent 2,417,850 (1947)
2. Winslow, W.M.: Induced fibration of suspensions. J. Appl. Phys. **20**(12), 1137–1140 (1949)
3. Zohdi, T.I.: Genetic design of solids possessing a random-particulate microstructure. Philos. Trans. R. Soc. Math. Phys. Eng. Sci. **361**(1806), 1021–1043 (2003)
4. Zohdi, T.I.: On the compaction of cohesive hyperelastic granules at finite strains. Proc. R. Soc. **454**(2034), 1395–1401 (2003)
5. Zohdi, T.I.: Computational design of swarms. Int. J. Numer. Methods Eng. **57**, 2205–2219 (2003)
6. Zohdi, T.I.: Constrained inverse formulations in random material design. Comput. Methods Appl. Mech. Eng. **192**(28–30), 3179–3194 (2003)

7. Zohdi, T.I.: Staggering error control for a class of inelastic processes in random microheterogeneous solids. Int. J. Nonlinear Mech. **39**, 281–297 (2004)
8. Zohdi, T.I.: Modeling and simulation of a class of coupled thermo-chemo-mechanical processes in multiphase solids. Comput. Methods Appl. Mech. Eng. **193**(6–8), 679–699 (2004)
9. Zohdi, T.I.: Modeling and direct simulation of near-field granular flows. Int. J. Solids Struct. **42**(2), 539–564 (2004)
10. Zohdi, T.I.: A computational framework for agglomeration in thermo-chemically reacting granular flows. Proc. R. Soc. **460**(2052), 3421–3445 (2004)
11. Zohdi, T.I.: Statistical ensemble error bounds for homogenized microheterogeneous solids. J. Appl. Math. Phys. (Z. Angew. Math. Phys.) **56**(3), 497–515 (2005)
12. Zohdi, T.I.: Charge-induced clustering in multifield particulate flow Int. J. Numer. Methods Eng. **62**(7), 870–898 (2005)
13. Zohdi, T.I.: On the optical thickness of disordered particulate media. Mech. Mater. **38**, 969–981 (2006)
14. Zohdi, T.I., Kuypers, F.A.: Modeling and rapid simulation of multiple red blood cell light scattering. Proc. R. Soc. Interface **3**(11), 823–831 (2006)
15. Zohdi, T.I.: Computation of the coupled thermo-optical scattering properties of random particulate systems. Comput. Methods Appl. Mech. Eng. **195**, 5813–5830 (2006)
16. Zohdi, T.I.: Computation of strongly coupled multifield interaction in particle-fluid systems. Comput. Methods Appl. Mech. Eng. **196**, 3927–3950 (2007)
17. Zohdi, T.I.: Particle collision and adhesion under the influence of near-fields. J. Mech. Mater. Struct. **2**(6), 1011–1018 (2007)
18. Zohdi, T.I.: On the computation of the coupled thermo-electromagnetic response of continua with particulate microstructure. Int. J. Numer. Methods Eng. **76**, 1250–1279 (2008)
19. Zohdi, T.I.: Mechanistic modeling of swarms. Comput. Methods Appl. Mech. Eng. **198**(21–26), 2039–2051 (2009)
20. Zohdi, T.I.: Charged wall-growth in channel-flow. Int. J. Eng. Sci. **48**, 1520 (2010)
21. Zohdi, T.I.: On the dynamics of charged electromagnetic particulate jets. Arch. Comput. Methods Eng. **17**(2), 109–135 (2010)
22. Zohdi, T.I., Kuypers, F.A., Lee, W.C.: Estimation of red blood cell volume fraction from overall permittivity measurement. Int. J. Eng. Sci. **48**, 1681–1691 (2010)
23. Zohdi, T.I.: Simulation of coupled microscale multiphysical-fields in particulate-doped dielectrics with staggered adaptive FDTD. Comput. Methods Appl. Mech. Eng. **199**, 79–101 (2010)
24. Zohdi, T.I.: Dynamics of clusters of charged particulates in electromagnetic fields. Int. J. Numer. Methods Eng. **85**, 1140–1159 (2011)
25. Zohdi, T.I.: Joule-heating field phase-amplification in particulate-doped dielectrics. Int. J. Eng. Sci. **49**, 30–40 (2011)

26. Zohdi, T.I.: Estimation of electrical-heating load-shares for sintering of powder mixtures. Proc. R. Soc. **468**, 2174–2190 (2012)
27. Zohdi, T.I.: Modeling and simulation of the optical response rod-functionalized reflective surfaces. Comput. Mech. **50**(2), 257–268 (2012)
28. Zohdi, T.I.: On the reduction of heat generation in lubricants using microscale additives. Int. J. Eng. Sci. **62**, 84–89 (2013)
29. Zohdi, T.I.: Numerical simulation of charged particulate cluster-droplet impact on electrified surfaces. J. Comput. Phys. **233**, 509–526 (2013)
30. Zohdi, T.I.: Rapid simulation of laser processing of discrete particulate materials. Arch. Comput. Methods Eng. **20**, 309–325 (2013)
31. Zohdi, T.I.: A direct particle-based computational framework for electrically-enhanced thermo-mechanical sintering of powdered materials. Math. Mech. Solids **19**(1), 93–113 (2014)
32. Zohdi, T.I.: On cross-correlation between thermal gradients and electric fields. Int. J. Eng. Sci. **74**, 143–150 (2014)
33. Zohdi, T.I.: Additive particle deposition and selective laser processing-a computational manufacturing framework. Comput. Mech. **54**, 171–191 (2014)
34. Zohdi, T.I.: Embedded electromagnetically sensitive particle motion in functionalized fluids. Comput. Part. Mech. **1**, 27–45 (2014)
35. Zohdi, T.I.: Rapid computation of statistically-stable particle/feature ratios for consistent substrate stresses in printed flexible electronics. J. Manuf. Sci. Eng. ASME. MANU-14-1476 (2015). https://doi.org/10.1115/1.4029327
36. Zohdi, T.I.: A computational modelling framework for high-frequency particulate obscurant cloud performance. Int. J. Eng. Sci. **89**, 75–85 (2015)
37. Zohdi, T.I.: On necessary pumping pressures for industrial process-driven particle-laden fluid flows. J. Manuf. Sci. Eng. ASME (2015). https://doi.org/10.1115/1.4030620
38. Zohdi, T.I.: On the thermal response of a laser-irradiated powder particle in additive manufacturing. CIRP J. Manuf. Sci. Technol. **10**, 7783 (2015)
39. Zohdi, T.I.: Modeling and simulation of the post-impact trajectories of particles in oblique precision shot-peening. Comput. Part. Mech. (2015). https://doi.org/10.1007/s40571-015-0048-5
40. Zohdi, T.I.: Modeling and simulation of cooling-induced residual stresses in heated particulate mixture depositions. Comput. Mech. **56**, 613–630 (2015)
41. Zohdi, T.I.: Modeling and efficient simulation of the deposition of particulate flows onto compliant substrates. Int. J. Eng. Sci. **99**, 74–91 (2015). https://doi.org/10.1016/j.ijengsci.2015.10.012
42. Zohdi, T.I.: Modeling and simulation of laser processing of particulate-functionalized materials. Arch. Comput. Methods Eng. 1–25 (2015). https://doi.org/10.1007/s11831-015-9160-1
43. Albrecht, T., Bhrer, C., Fhnle, M., Maier, K., Platzek, D., Reske, J.: First observation of ferromagnetism and ferromagnetic domains in a liquid metal. Appl. Phys. A: Mater. Sci. Process. **65**(2), 215 (1997)
44. Andelman, D., Rosensweig, R.E.: The phenomenology of modulated phases: from magnetic solids and fluids to organic films and polymers. In: Tsori, Y.,

Steiner, U. (eds.) Polymers, Liquids and Colloids in Electric Fields: Interfacial Instabilities, Orientation and Phase Transitions. World Scientific. pp. 156 (2009). ISBN 978-981-4271-68-4

45. Berger, P., Adelman, N.B., Beckman, K.J., Campbell, D.J., Ellis, A.B., Lisensky, G.C.: Preparation and properties of an aqueous ferrofluid. J. Chem. Educ. **76**(7), 943–948 (1999). ISSN 0021-9584. https://doi.org/10.1021/ed076p943
46. Feynman, R.P., Leighton, R.B., Sands, M.: The Feynman Lectures on Physics 2 (2006). ISBN 0-8053-9045-6
47. Cullity, B.D., Graham, C.D.: Introduction to Magnetic Materials, 2 edn., p. 103. Wiley-IEEE Press (2008). ISBN 0-471-47741-9
48. Boyer, T.H.: The force on a magnetic dipole. Am. J. Phys. **56**(8), 688–692 (1988). Bibcode: 1988AmJPh.56.688B. http://doi.org/10.1119/1.15501
49. Jackson, J.D.: Classical Electrodynamics, 3rd edn. Wiley (1998)
50. Chow, C.Y.: An Introduction to Computational Fluid Dynamics. Wiley, New York (1980)
51. Pöschel, T., Schwager, T.: Computational Granular Dynamics. Springer (2004)
52. Labra, C., Onate, E.: High-density sphere packing for discrete element method simulations. Commun. Numer. Methods Eng. **25**(7), 837–849 (2009)
53. Onate, E., Idelsohn, S.R., Celigueta, M.A., Rossi, R.: Advances in the particle finite element method for the analysis of fluid-multibody interaction and bed erosion in free surface flows. Comput. Methods Appl. Mech. Eng. **197**(19–20), 1777–1800 (2008)
54. Onate, E., Celigueta, M.A., Idelsohn, S.R., Salazar, F., Surez, B.: Possibilities of the particle finite element method for fluid-soil-structure interaction problems. Comput. Mech. **48**, 307–318 (2011)
55. Rojek, J., Labra, C., Su, O., Onate, E.: Comparative study of different discrete element models and evaluation of equivalent micromechanical parameters. Int. J. Solids Struct. **49**, 1497–1517 (2012). https://doi.org/10.1016/j.ijsolstr.2012.02.032
56. Avci, B., Wriggers, P.: A DEM-FEM coupling approach for the direct numerical simulation of 3D particulate flows. J. Appl. Mech. **79**, 010901-1–7 (2012)

MIX
Papier aus verantwortungsvollen Quellen
Paper from responsible sources
FSC® C105338

If you have any concerns about our products,
you can contact us on
ProductSafety@springernature.com

In case Publisher is established outside the EU,
the EU authorized representative is:
Springer Nature Customer Service Center GmbH
Europaplatz 3, 69115 Heidelberg, Germany

Printed by Libri Plureos GmbH
in Hamburg, Germany